SCOPE 48

Sulphur Cycling on the Continents

SCOPE 27: Climate Impact Assessment: Studies of the Interaction of Climate and Society, 1985, 650 pp

SCOPE 28: Environmental Consequences of Nuclear War Volume I Physical and Atmospheric Effects, 1986, 400 pp, Volume II Ecological and Agricultural Effects, 1985, 563 pp

SCOPE 29: The Greenhouse Effect, Climatic Change, and Ecosystems, 1986, 574 pp

SCOPE 30: Methods for Assessing the Effects of Mixtures of Chemicals, 1987, 928 pp

SCOPE 31: Lead, Mercury, Cadmium and Arsenic in the Environment, 1987, 384 pp

SCOPE 32: Land Transformation in Agriculture, 1987, 552 pp

SCOPE 33: Nitrogen Cycling in Coastal Marine Environment, 1988, 478 pp

SCOPE 34: Practitioner's Handbook on the Modelling of Dynamic Change in Ecosystems, 1988, 196 pp

SCOPE 35: Scales and Global Change: Spatial and Temporal Variability in Biospheric and Geospheric Processes, 1988, 376 pp

SCOPE 36: Acidification in Tropical Countries, 1988, 424 pp

SCOPE 37: Biological Invasions: A Global Perspective, 1989, 528 pp

SCOPE 38: Ecotoxicology and Climate with Special Reference to Hot and Cold Climates, 1989, 432 pp

SCOPE 39: Evolution of the Global Biogeochemical Sulphur Cycle, 1989, 224 pp

SCOPE 40: Methods for Assessing and Reducing Injury from Chemical Accidents, 1989, 320 pp

SCOPE 41: Short-Term Toxicity Tests for Non-genotoxic Effects, 1990, 353 pp

SCOPE 42: Biogeochemistry of Major World Rivers, 1991, 356 pp

SCOPE 43: Stable Isotopes: Natural and Anthropogenic Sulphur in the Environment, 1991, 472 pp

SCOPE 44: Introduction of Genetically Modified Organisms into the Environment, 1990, 224 pp

SCOPE 45: Ecosystem Experiments, 1991, 296 pp

SCOPE 46: Methods for Assessing Exposure of Human and Non-human Biota, 1991, 448 pp

SCOPE 47: Long-term Ecological Research. An International Perspective, 1991, 312 pp

SCOPE 48: Sulphur Cycling on the Continents: Wetlands, Terrestrial Ecosystems and Associated Water Bodies, 1992, 345 pp

SCOPE 49: Methods to Assess Adverse Effects of Pesticides on Non-target Organisms, 1992, 264 pp

Funds to meet SCOPE expenses are provided by contributions from SCOPE Committees, an annual subvention from ICSU (and through ICSU, from UNESCO), an annual subvention from the French Ministère de l'Environnement, contracts with UN Bodies, particularly UNEP, and grants from Foundations and industrial enterprises.

SCOPE 48

Sulphur Cycling on the Continents:

Wetlands, Terrestrial Ecosystems and Associated Water Bodies

Edited by

R. W. HOWARTH

Section of Ecology and Systematics, Corson Hall, Cornell University, Ithaca, NY 14853 USA

J. W. B. STEWART

Faculty of Agriculture, University of Saskatchewan, Saskatoon, Saskatchewan S7N 0W0, Canada

and

M. V. IVANOV

Institute of Microbiology, USSR Academy of Sciences, Prospect 60-letiya Oktjabrja 7A, Moscow 117811, USSR

Published on behalf of the Scientific Committee on Problems of the Environment (SCOPE) of the International Council of Scientific Unions (ICSU), and in collaboration with the United Nations Environment Programme (UNEP)

by

JOHN WILEY & SONS

Chichester · New York · Brisbane · Toronto · Singapore

Published by John Wiley & Sons Ltd,
Baffins Lane,
Chichester,
West Sussex PO19 1UD, England

Other Wiley Editorial Offices

John Wiley & Sons, Inc., 605 Third Avenue,
New York, NY 10158-0012, USA

Jacaranda Wiley Ltd, G.P.O. Box 859, Brisbane,
Queensland 4001, Australia

John Wiley & Sons (Canada) Ltd, 22 Worcester Road, Rexdale,
Ontario M9W 1L1, Canada

John Wiley & Sons (SEA) Pte Ltd, 37 Jalan Pemimpin 05-04,
Block B, Union Industrial Building, Singapore 2057

Library of Congress Cataloging-in-Publication Data

Sulphur cycling on the continents : wetlands, terrestrial ecosystems, and associated water bodies / edited by R.W. Howarth, J.W.B. Stewart, and M.V. Ivanov.
p. cm.—(SCOPE report : 48)
Includes bibliographical references and index.
ISBN 0 471 93153 5
1. Sulphur cycle. 2. Sulphur—Environmental aspects.
I. Howarth, R. W. (Robert W.) II. Stewart, J. W. B. III. Ivanov, M. V. (Mikhail Vladimirovich) IV. Series.
QH344.S85 1992
574.5′222—dc20

91–26109
CIP

British Library Cataloguing in Publication Data

A catalogue record for this book is available
from the British Library

ISBN 0 471 93153 5

Typeset by Acorn Bookwork, Salisbury, Wiltshire
Printed and bound in Great Britain by Courier International Ltd,
East Kilbride, Scotland.

International Council of Scientific Unions (ICSU)
Scientific Committee on Problems of the Environment (SCOPE)

SCOPE is one of a number of committees established by the non-governmental group of scientific organizations, the International Council of Scientific Unions (ICSU). The membership of ICSU includes representatives from 75 National Academies of Science, 20 International Unions and 29 other bodies called Associates. To cover multidisciplinary activities which include the interests of several unions, ICSU has established 13 Scientific Committees, of which SCOPE is one. Currently representatives of 35 member countries and 21 Unions, Scientific Committees and Associates participate in the work of SCOPE, which directs particular attention to the needs of developing countries. SCOPE was established in 1969 in response to the environmental concerns emerging at the time: ICSU recognized that many of these concerns required scientific inputs spanning several disciplines and ICSU Unions. SCOPE's first task was to prepare a report on Global Environmental Monitoring (SCOPE 1, 1971) for the UN Stockholm Conference on the Human Environment.

The mandate of SCOPE is to assemble, review, and assess the information available on man-made environmental changes and the effects of these changes on man; to assess and evaluate the methodologies of measurement of environmental parameters; to provide an intelligence service on current research; and by the recruitment of the best available scientific information and constructive thinking to establish itself as a corpus of informed advice for the benefit of centres of fundamental research and of organizations and agencies operationally engaged in studies of the environment.

SCOPE is governed by a General Assembly, which meets every three years. Between such meetings its activities are directed by the Executive Committee.

R. E. Munn
Editor-in-Chief
SCOPE Publications

Executive Secretary: V. Plocq-Fichelet

Secretariat: 51 boulevard de Montmorency
75016 Paris, France

Contents

Members of the Scientific Advisory Committee for the SCOPE/UNEP International Sulphur Project

V. M. Gorlenko
Institute for Microbiology, USSR Academy of Sciences, Prospect 60-letiya Oktjabrja 7A, Moscow 117811, USSR

R. W. Howarth
Section of Ecology and Systematics, Corson Hall, Cornell University, Ithaca, NY 14853, USA

M. V. Ivanov
Institute for Microbiology, USSR Academy of Sciences, Prospect 60-letiya Oktjabrja 7A, Moscow 117811, USSR

W. Jaeschke
Center of Environmental Protection, University of Frankfurt am Main, Frankfurt am Main, Germany

R. H. Krouse
Physics Department, University of Calgary, Calgary, Alberta T2N 1N4, Canada

A. G. Ryaboshapko
Institute of Applied Geophysics, Glebovskaja Street, 20b, Moscow, USSR

J. W. B. Stewart
Faculty of Agriculture, University of Saskatchewan, Saskatoon, Saskatchewan S7N 0W0, Canada

P. H. Trudinger
Baas-Becking Geobiological Laboratory, PO Box 378, Canberra, ACT 2601, Australia

List of Workshop Participants and Contributors to this Volume

S. L. Amarasiri
PO Box 5, 1 Sarasavi Mawatha, Peradeniya, Sri Lanka

M. O. Andreae
Max-Planck-Institut fur Chemie, Biogeochemistry Department, Saarstrasse 23, Postfach 3060, Mainz D-6500, Germany

A. Autry
University of Georgia, Department of Microbiology, Athens, GA 30602, USA

Pieter Bos
Delft University of Technology, P.O. Box 5057, Julianalaan 67, Delft 2600 GB, The Netherlands

T. Church
University of Delaware, College of Marine Studies, Lewes Complex, Lewes, DE 19958, USA

R. B. Cook
Oakridge National Laboratory, Environmental Sciences Division, Oak Ridge, TN 37831, USA

M. B. David
Department of Forestry, 1301 W. Gregory Drive, University of Illinois, Urbana, IL 61801, USA

Ben H. Ellert
Soil Science Department, University of Saskatchewan, Saskatoon, Saskatchewan S7N 0W0, Canada

John Fitzgerald
University of Georgia, Department of Microbiology, Athens, GA 30602, USA

J. R. Freney
CSIRO, P.O. Box 1600, Canberra City, ACT 2601, Australia

D. K. Friesen
International Fertilizer Development Center, P.O. Box 2040, Muscle Shoals, AL 35662, USA

J. J. Germida
University of Saskatchewan, Soil Science Department, Saskatoon, Saskatchewan S7N 0W0, Canada

Anne Giblin
Ecosystems Center, Marine Biological Laboratory, Woods Hole, MA 02543, USA

Vladimir Grinenko
Institute of Geochemistry, USSR Academy of Sciences, Kosygin Street 19, 117975 Moscow, USSR

V. V. S. R. Gupta
Saskatchewan Institute of Pedology, University of Saskatchewan, Saskatoon, Saskatchewan S7N 0W0, Canada

Bruce Haines
University of Georgia, Botany Department, Athens, GA 30602, USA

R. B. Harrison
College of Forest Resources, AR-10 University of Washington, Seattle, WA 98195, USA

Jamie Hern
Trent University, Chemistry Department, Peterborough, Ontario K9J 7B8, Canada

Robert Howarth
Section of Ecology and Systematics, Cornell University, Ithaca, NY 14853, USA

W. Jaeschke
Johann Wolfgang Goethe-Universitat, Robert-Mayer-Strasse 7–9, Postfach 11 19 32, DK-6000 Frankfurt am Main 11, Germany

Carol Kelly
University of Manitoba, Department of Microbiology, Winnipeg, Manitoba R3T 2N2, Canada

D. P. Kelly
University of Warwick, Department of Environmental Sciences, Coventry, CV4 7AL, UK

H. Roy Krouse
University of Calgary, Physics Department, Calgary, Alberta T2N 1N4, Canada

G. Lang
Department of Biology, West Virginia University, Morgantown, WV 26506, USA

Rod Lefroy
University of New England, Department of Agronomy, Armidale, NSW 2350, Australia

Zhong-qun Liu
Institute of Soil Science, Academia Sinica, P.O. Box 821, Nanjing, P. R. China

L. E. Lowe
University of British Columbia, Soil Science Department, Suite 248-2357 Main Mall, Vancouver, BC V6T 2A2, Canada

George Luther
University of Delaware, College of Marine Studies, Lewes Complex, Lewes, DE 19958, USA

C. P. Mamaril
International Rice Research Institute, P.O. Box 933, Manila, Philippines

Juanita Mamaril
University of Philippines, Biochemistry Department, Los Baños, Philippines

D. G. Maynard
Canadian Forestry Research Service, 5320-122 Street, Edmonton, Alberta T6H 3S5, Canada

Myron Mitchell
State University of New York, Environmental and Forest Biology, Syracuse, NY 13210, USA

Jerome Nriagu
Environment Canada, National Water Research Institute, P.O. Box 5050, Burlington, Ontario L7R 4A6, Canada

Terry L. Roberts
Agriculture Canada Research Station, P.O. Box 3000 Main, Lethbridge, Alberta T1J 4B1, Canada

Surrinder Saggar
Massey University, Soil Science Department, Palmerston North, New Zealand

J. J. Schoenau
University of Saskatchewan, Saskatchewan Institute of Pedology, Saskatoon, Saskatchewan S7N 0W0, Canada

John W. B. Stewart
University of Saskatchewan, Saskatchewan Institute of Pedology, Saskatoon, Saskatchewan S7N 0W0, Canada

T. C. Strickland
Oregon State University, Department of Soil Science, Corvallis, OR 97331, USA

M. A. Tabatabai
Iowa State University, Agronomy Department, Ames, IO 50011, USA

P. N. Takkar
Indian Institute of Soil Science, c/o CIAE-GTB Complex, TT Nagar, Bhopal 462003, India

Philip Trudinger
Baas-Becking Geobiological Laboratory, P.O. Box 378, Canberra, ACT 2601, Australia

Milton Wainwright
University of Sheffield, Department of Microbiology, Western Bank, Sheffield, S10 2TN, UK

D. Whelpdale
University of Toronto, Institute for Environmental Studies, Toronto, Ontario M5S 1A1, Canada

Kelman Wieder
Villanova University, Biology Department, Villanova, PA 19085, USA

Joseph Yavitt
Cornell University, Ecosystems Research Center, Wing Hall, Ithaca, NY 14853, USA

Many workshop participants provided reviews of chapters or portions of chapters. In addition, the following individuals provided constructive reviews:

Suzanne Bayley
Robert Berner
Brian Fry
Dale Johnson
Carl Jorden
Bo Barker Jorgensen
Steven Lindberg
Gary Lovett
Derek Lovely
L. E. Lowe
Elder Paul
Henning Rodhe
John Rudd
David Rudnick
S. Saggar
Alan Stam
Frank Stevenson
Tim Strictland
Noel Urban

Working Groups

I. Anaerobic Processes in Sulphur Cycle
   Chair P. Trudinger
   Rapporteur K. Wieder

II. Aerobic Processes in Sulphur Cycle
   Chair M. Wainwright
   Rapporteur J. Germida

III. Atmospheric Chemistry
   Chair J. Nriagu
   Rapporteur D. Whelpdale

IV. Sulphur Input (Fertilizer and Acid Input), Fate and Budgets
   Chair C. Mamaril
   Rapporteur D. Friesen

V. Measurement of Sulphur Forms and Sulphur Fluxes in Soil, Sediments, Water, Plant and Atmosphere
   Chair J. Freney
   Rapporteur G. Luther

VI. Elemental Interactions
   Chair J. Stewart
   Rapporteur R. Howarth

Working groups reports were incorporated in the chapters. All workshop participants participated in two working groups, as well as reviewing chapters or parts of chapters.

1 Introduction

J. W. B. STEWART
Faculty of Agriculture, University of Saskatchewan, Saskatoon, Canada

and

R. W. HOWARTH
Section of Ecology and Systematics, Cornell University, Ithaca, New York, USA

The Scientific Committee on Problems of the Environment (SCOPE) of the International Council of Scientific Unions (ICSU) launched a major series of projects on biogeochemical cycles in 1974. Initially these projects focused on the study of the cycles of carbon, nitrogen, sulphur, and phosphorus separately. It was however understood from the beginning that none of these cycles operates independently from the others and that attempts would be made to include studies of elemental interactions in each unit's activities. Much of the work on individual cycles has been coordinated by the SCOPE/UNEP international units working separately: carbon, under the direction of the late Professor Egon Degens in Hamburg, Germany; nitrogen, under the direction of Professor Thomas Rosswall in Uppsala, Sweden; sulphur, under the direction of Academician Mikhail Ivanov in Moscow, USSR; and phosphorus under the direction of Dr C. Vernon Cole in Fort Collins, USA. SCOPE also has sponsored efforts to have groups of scientists unravel many of the mechanisms of mutual interactions (SCOPE 17, edited by G.E. Likens, 1981, and SCOPE 21, edited by B. Bolin and R.B. Cook, 1983). Also various international units have met jointly to explore specific interactions between two element cycles, as between carbon and sulphur (SCOPE 42, edited by Degens *et al.* 1991, *Biogeochemistry of Major World Rivers*).

The SCOPE/UNEP International Sulphur Unit's first investigation, *The Global Biogeochemical Sulphur Cycle* (SCOPE 19, edited by Ivanov and Freney, 1983), brought together information on sulphur in all aspects of the globe and attempted to construct a simulation model of the modern sulphur cycle. Since many of the processes of global sulphur fluxes were not adequately related to their controls, the focus of the book was on areas where knowledge of processes and effects (actual or potential) of anthropogenic inputs of sulphur on the environment were poorly understood. The second

Sulphur Cycling on the Continents
Edited by R.W. Howarth, J.W.B. Stewart and M.V. Ivanov

phase of the SCOPE/UNEP International Sulphur Unit, started in 1982 and continuing until 1986, had two important tasks. The first was to examine the information on the global cycle and its perturbations that could be obtained from the sulphur isotopic composition of various reservoirs, fluxes and ecosystems. This material has been published as SCOPE 43, 1991, edited by H. R. Krouse and V. A. Grinenko, entitled *Stable Isotopes: Natural and Anthropogenic Sulphur in the Environment*. The second activity was a meeting in Tallinn, Estonia, which drew together information on the evolution of the sulphur cycle, allowing the current anthropogenic perturbations to be placed in a context related to past changes over geological time. This material was published in 1989 as SCOPE 39, edited by P. Brimblecombe and A. Yu. Lein, entitled, *Evolution of the Global Biogeochemical Sulphur Cycle*.

The third and final phase of the activities of the SCOPE/UNEP International Sulphur Unit was started in 1986 and focused on the cycling of sulphur in continental ecosystems. Continental was interpreted as including soils, wetlands, inland lakes, rivers, and estuaries. Sulphur cycling in oceans was dealt with in part in SCOPE 39.

Three meetings were held as part of this third phase. Two of these were in conjunction with the SCOPE/UNEP International Carbon Unit. The first, a joint workshop to deal with interactions in estuaries, was held at Texel in the Netherlands in September, 1987. The second, also a joint meeting between the International Carbon and Sulphur Units, entitled, The Biogeochemical Carbon and Sulphur Cycles in Lakes and Water Bodies, was held at Irktusk, Lake Baikal, in September 1988. The main objective of this workshop was a discussion and synthesis of the existing information on biogeochemical processes and fluxes of elements in lakes and water bodies of different types. Attempts were made to identify knowledge gaps in data on biogeochemical processes, and to discuss the evaluation of existing methods for investigation and modelling of freshwater aquatic ecosystems. The information accumulated from the first two workshops was synthesized and published as *Interactions of Biogeochemical Cycles and Ecosystems*, edited by E. T. Degens, S. Dempe, A. Lein, and Y. Sorokin, University of Hamburg Press.

The third UNEP/SCOPE workshop in this series was held at Trent University, Peterborough, Canada in May 1989. This workshop was designed so that in addition to an overview of atmospheric sulphur and biogenic fluxes of sulphur to the atmosphere, the major focus would be on sulphur cycling processes and fluxes in natural and cultivated wetlands, forest and agroecosystems. The outcome of this third meeting is the current book. It includes some material from the earlier meetings at Texel and Lake Baikal.

In organizing this workshop, it was realized that there was a need for scientists working in different disciplines and in different types of ecosystems to share information with one another. For instance, the community of scientists working in agricultural systems and that working in natural ecosystems do not always have the opportunity to interact, and therefore processes that are important in one system are not seen to have the same relevance in another. Most of the earlier work on the sulphur biogeochemistry of cultivated soils was carried out in regions where sulphur was a limiting nutrient, and attempts focused on understanding the natural cycling of the element before adding extra sulphur as a fertilizer. In areas of Australia and western Canada, an initial emphasis was on understanding the requirements of important agricultural crops such as legumes and/or oilseed crops, with some study of microbial processes in cycling sulphur in plant residues through the soil making it available as a nutrient. Similarly, studies on the application of elemental sulphur in soils concentrated on the rate at which it would be oxidized to the form more available to plants, and the microbial populations that aided this process. This was in complete contrast to another body of knowledge which had focused on the effect of anthropogenic additions of sulphur to natural forests and lakes from acid rain. Here the emphasis was not on the lack of sulphur in the ecosystem, but with excess sulphur. In some cases, the amounts of sulphur cycling through decomposer organisms was insignificant in comparison to the capacity of the soil to maintain its viability under heavy sulphur excess loading. Unless brought together to discuss the complete sulphur cycle, these two groups of scientists would rarely meet.

Similarly, the questions that were asked of scientists working in wetlands were completely different. In cultivated wetlands which are mostly used in rice production, the emphasis has again been on the efficient supply of sulphate and the efficiency of use of applied sulphur fertilizer. On the other hand, research on marshes and other wetlands has tended to be much more comprehensive and much more ambitious. A driving force behind research in natural wetlands has been an understanding of the overwhelming importance of sulphate reduction to the metabolism of these systems. In general, the emphasis of research in natural wetlands has been on microbial and biogeochemical processes and their controls rather than on the net effect of a sulphur input over the 90-day period of growth of a rice crop. Also, much more attention has been paid to diurnal changes in wetland conditions and properties, and the resulting effect this has on both emissions of sulphur to the atmosphere and in change of sulphur forms in the soil. For freshwater wetlands, it has been extremely important to understand sulphur cycling and fluxes in the context of how they have been affected by inputs of acid rain. These two groups of scientists, both working in wetlands, have rarely been in close contact. This workshop brought them together for such interaction.

A common problem faced by all scientists studying the sulphur cycle is the difficulty of the analytical chemistry of sulphur. This element is difficult to analyse accurately in pure solutions, and extremely difficult to understand in a heterogeneous system with solids and/or a changing oxidation–reduction potential. Our understanding of sulphur biogeochemistry has often been limited by these difficulties, and the literature is full of methodological disagreements. Scientists working in different types of ecosystems have frequently used different analytical methods for sulphur, often with different and contradictory underlying assumptions. This workshop allowed a cross-fertilization among these disciplines and furthered communication towards better approaches. There is always a need to use new developments in analytical chemistry, and/or isotope technology, to comprehend fully the processes associated with sulphur transformations under oxic or anoxic conditions, and to develop and use methods that give unequivocal results.

The workshop consisted of a mixture of lectures, discussions and small meetings of working groups. The insights developed by the working groups have been included in chapters of this book, and the chapters have benefited from this collective wisdom. The working groups were at least partly successful in breaking down some of our traditional disciplinary boundaries. We hope that this volume will further break down these boundaries and lead to progress in understanding the interaction of sulphur with other element cycles in a variety of ecosystems. Different disciplines use different units for presenting data on the sulphur cycle. To further communication, in this volume information in each chapter is provided in both mass and molar units.

The continuous support provided over the years by the United Nations Environment Programme has ensured the successful development of this study. We are indebted to the participants in the meeting for their interest and free discussion. Each of the chapters received critical comments from outside reviewers, and these reviewers were also major contributors to the overall book. This book would not have been possible without the capable editorial assistance of Ms Diane Sherman.

2 An Overview of the Atmospheric Sulphur Cycle

D. M. WHELPDALE
Institute for Environmental Studies, University of Toronto, Toronto, Canada

2.1 INTRODUCTION

This chapter has two objectives. The first is to describe briefly the atmospheric sulphur cycle—its processes, pathways and fluxes—to provide a context for the following chapters on sulphur cycling in terrestrial and wetland ecosystems. The second is to identify areas of difficulty in the study of the sulphur cycle which may arise, usually at the interface between different reservoirs in the cycle, and thus, at the interface between the different disciplines involved. These may include a lack of understanding of an exchange process or flux between two reservoirs, or a mismatch between the needs and capabilities of two disciplines, as happens, for example, in the choice of spatial scales of study. Identification of such issues may lead to new opportunities for interdisciplinary cooperation, aiding in their resolution.

The chapter is selective in its emphasis, rather than being a comprehensive, in-depth treatise on the atmospheric sulphur cycle. A number of these already exist (e.g. Galloway *et al.*, 1985; Ivanov and Freney, 1983). The intended audience is primarily non-atmospheric—those studying sulphur cycling in various terrestrial and wetland ecosystems, those concerned with nutrient or harmful acidic atmospheric inputs to a surface ecosystem, or those engaged in integrated ecosystem modelling.

The contents are arranged in the traditional sequence, proceeding through the parts of the atmospheric sulphur cycle—emissions, transport, transformations, and deposition—with emphasis on the first and last. Brief discussions of atmospheric sulphur budgets and atmospheric inputs to ecosystems are included to serve as linkages with following chapters.

2.2 EMISSIONS

On a global basis, natural emissions of sulphur into the atmosphere from biogenic sources and from volcanoes are perhaps some 60 Tg (S) a^{-1}

Sulphur Cycling on the Continents
Edited by R.W. Howarth, J.W.B. Stewart and M.V. Ivanov

(2.0 Tmol a^{-1}), with an estimated range of 46 to 124 Tg (S) a^{-1} (1.4 to 3.9 Tmol a^{-1}) (Andreae, 1989; Andreae and Jaeschke, this volume). The inclusion of sulphate produced by seaspray increases this estimate by a highly uncertain 35 to 315 Tg (S) a^{-1} (1.1 to 9.8 Tmol a^{-1}), but some 90% of this is cycled directly back into the ocean. The approximate percentage contributions of the main sources of natural sulphur emissions are: for volcanoes, 18%; open-ocean biogenic production, 46%; coastal zone and wetland biogenic sources, 3%; terrestrial plants and soils, 13%; biomass burning, at least 4%; and wind-raised dust, 16%. The magnitudes of all natural sources are highly uncertain, but aside from seaspray, the terrestrial plant and soil term is least well known. In recent years the estimated magnitude of this source has been decreasing as more representative measurements have become available. In large part uncertainties arise from the extrapolation of few point measurements to the global scale.

Natural emissions exhibit high spatial and temporal variability. Approximately one-half occur in the marine environment. Terrestrial biogenic emissions are correlated with temperature, and thus emission rates tend to be higher towards the tropics. As well, emissions from plants and soils are highly dependent on specific plant and soil type, and thus are highly variable geographically, often on a small scale. Volcanic emissions are of two kinds: those during non-eruptive periods contribute about 90% of the total, while those during eruptions contribute significantly less, on average.

Estimates of global anthropogenic sulphur emissions range from 60 to 110 Tg (S) a^{-1} (2 to 3 Tmol a^{-1}) for the years 1976 to 1985 (Moller, 1984; Hameed and Dignon, 1988), comparable to but probably somewhat larger than natural emissions (excluding seasalt spray). The combustion of fossil fuels accounts for 80 to 85% of the total, with the remainder coming from the smelting of ores and other industrial processes and burning. The approximate 1980 regional apportionment of fossil-fuel sulphur emissions is: China and Japan, 16%; USA and Canada, 28%; Europe, 34%; the remainder of the Northern Hemisphere, 15%; and the Southern Hemisphere, 7% (Hameed and Dignon, 1988). Annual emissions of sulphur in North America and Europe east to the Ural Mountains are known with an uncertainty of about 10–15% as a result of detailed emission inventories developed in recent acid rain programmes (Dovland and Saltbones, 1986; NAPAP, 1990). In these regions, emissions are resolved into a grid of approximately 100 km on a side. For other regions of the world, the spatial distributions of anthropogenic sulphur emissions are not as well known.

Historically, sulphur emissions in North America and Europe have shown a general upward trend from the early part of this century into the decade of the 1970s (Figure 2.1). As a result of slowing economies and a growing awareness of sulphur's role in acid deposition, sulphur emissions have been decreasing—from an estimated high of 17 Tg (S) a^{-1} (0.53 Tmol a^{-1}) in North America in

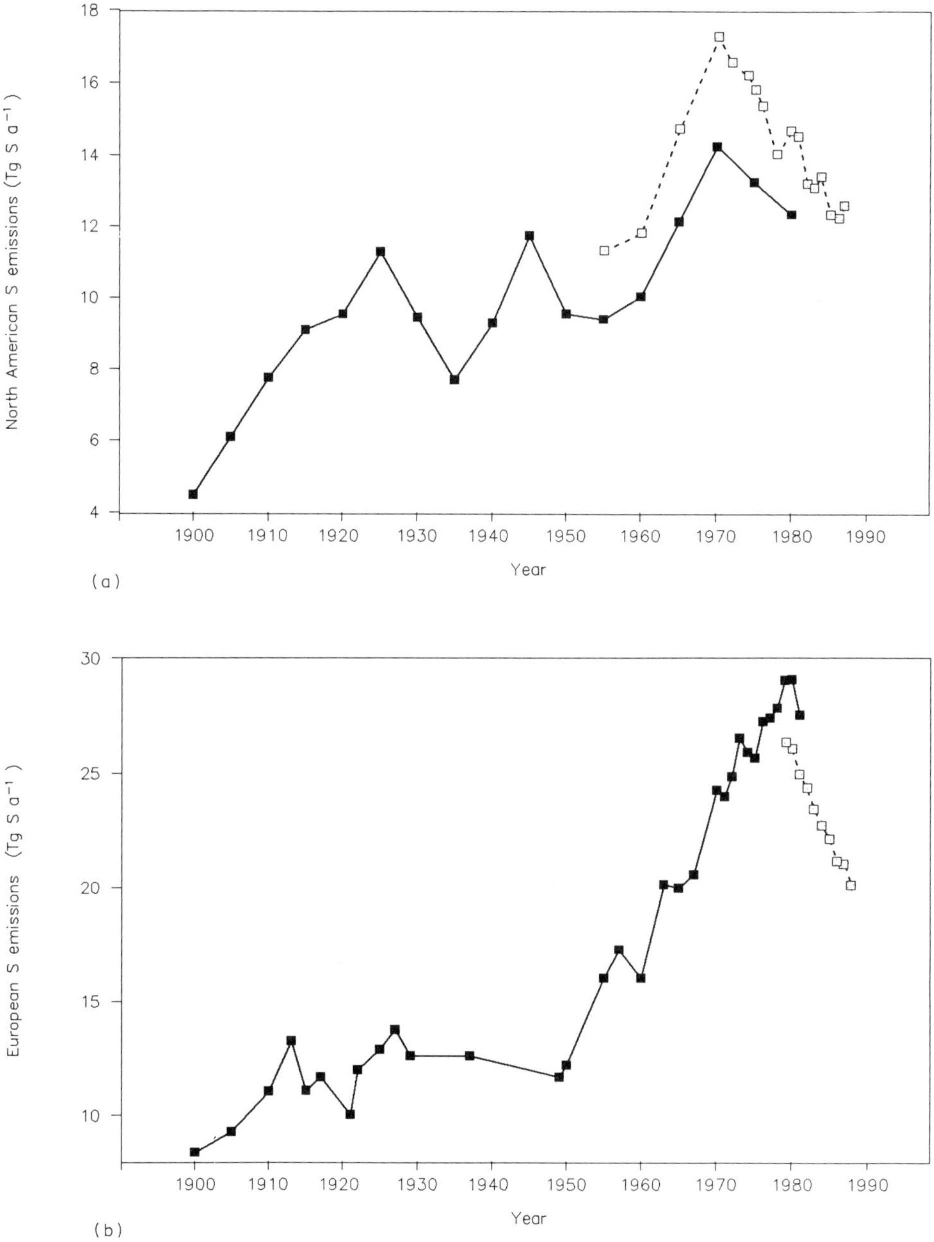

Figure 2.1. (a) Sulphur dioxide emissions (i) for the United States to 1980 (solid line) and (ii) for the United States plus Canada (broken line). (b) Sulphur dioxide emissions for (i) Europe from Fjeld (1975) updated by Semb (pers. comm., 1986) (solid line) and (ii) for EMEP Europe for Iversen *et al.* (1989) (broken line). Units are Tg (S) a^{-1}

1970 down to 12 Tg (S) a^{-1} (0.34 Tmol a^{-1}) in 1985, and in western Europe (those countries participating in the European Monitoring and Evaluation Program, EMEP), from a high of about 26 Tg (S) a^{-1} (0.81 Tmol a^{-1}) in 1979 down to 21 Tg (S) a^{-1} (0.66 Tmol a^{-1}) in 1988 (Iversen *et al.*, 1989).

Of great interest now is how much sulphur emissions in different regions of the world will increase in the coming decades. Most members of the Economic Commission for Europe signed a Convention on Long-Range Transboundary Air Pollution, one of whose protocols stipulated a decrease of at

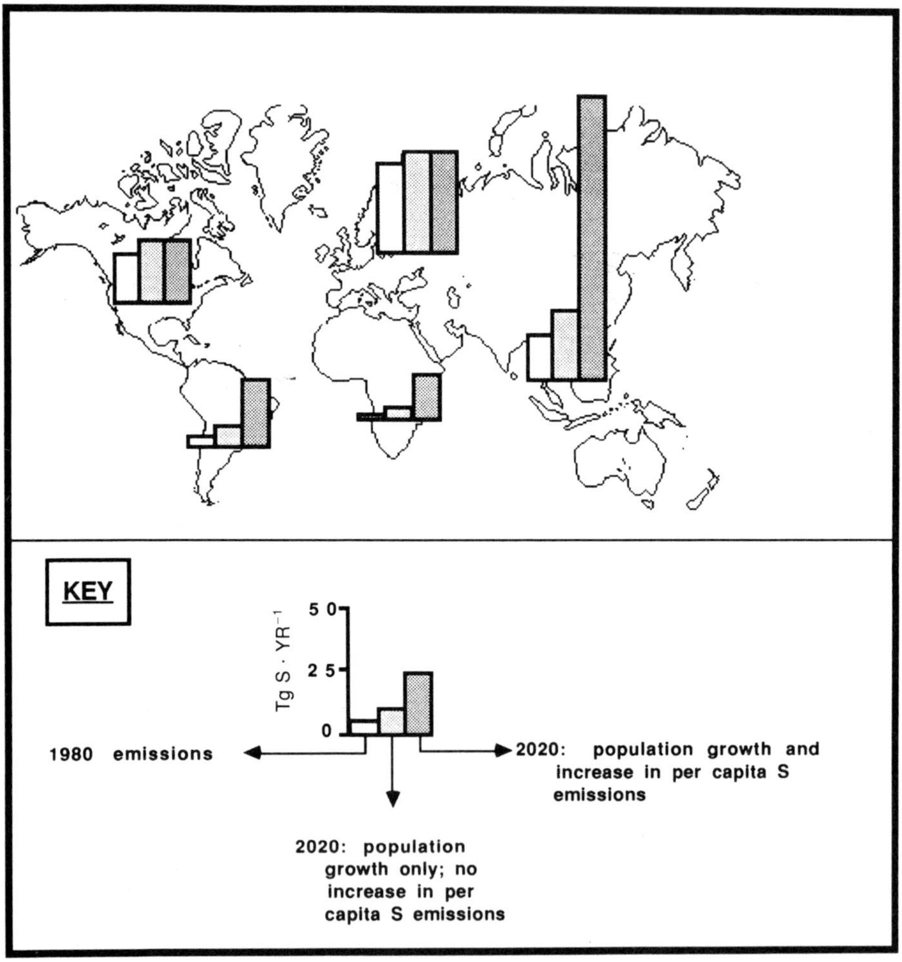

Figure 2.2. The emission of sulphur to the atmosphere in 1980 compared to 2020 with no change and with a change in per capita energy consumption (Galloway, 1989). Units are Tg (S) a^{-1}

least 30% in sulphur dioxide emissions from 1980 levels by 1993. Even though the full reduction has not yet been achieved, evidence is accumulating that ambient concentrations and sulphur deposition are decreasing in Western Europe (Leck and Rodhe, 1989; Mylona, 1989) and in North America (Dillon *et al.*, 1988).

Emissions from Asia, South America, and Africa may well increase dramatically. Galloway (1989) used United Nations estimates of population growth for 1980 to 2020 to estimate future sulphur emissions. In one scenario, the per capita emissions were held constant at 1980 levels while population alone increased. In a second, per capita emissions were allowed to increase by a factor of four relative to 1980 values, along with population growth. Figure 2.2 shows increases from 1980 for the two scenarios of approximately 2 and 6 times for South America, 2 and 7 times for Africa, and 1.5 and 6.5 times for Asia. The conclusion is that dramatic increases in sulphur emissions are likely from these continents in the next few decades, and, as a result, globally. As Galloway points out, the consequences are likely to be severely increased acidification in tropical regions, and significant disruption of the biogeochemical sulphur cycle, particularly in the tropics and sub-tropics. With reference to biogeochemical cycling, it is informative to compare emission densities and the relative importance of natural and anthropogenic emissions on different spatial scales. On a global basis, the average emission density for sulphur is about 0.4 g (S) m^{-2} a^{-1} (0.01 mol m^{-2} a^{-1}), and the ratio of anthropogenic to natural emissions is of order unity or slightly greater. By contrast, the average emission densities in the highly industrialized areas of Europe and North America may be higher by an order of magnitude, and the ratio of anthropogenic to natural emissions is approximately ten. In the high-emission grid squares of Europe or North America, the emission density may reach more than 40 g (S) m^{-2} a^{-1} (1.3 mol m^{-2} a^{-1}), and natural emissions are likely to be less by a factor of 100 to 1000. Although these comparisons are illustrative only, it is important to note that much of the current sulphur cycling work is being done on continents, often at sites in close proximity to areas of high emission density where anthropogenic emissions greatly outweigh natural ones.

2.3 TRANSPORT AND TRANSFORMATION

From the perspective of the study of chemical cycling and surface–atmosphere exchange in terrestrial and wetland ecosystems, the atmosphere serves as a delivery system to bring material to and take material away from the immediate vicinity of an ecosystem. The transport of chemical constituents through the atmosphere is effected by the several different scales of motion in the atmosphere, each dominated by different physical processes.

On the global scale, transport is controlled by the general circulation of the atmosphere, the westerlies of the mid-latitudes and the easterlies of the subtropics, for example, and the quasi-stationary high and low pressure systems such as the Bermuda–Azores high and the Greenland or Aleutian low. The time and distance scales for such transport are several days and several thousand kilometres, respectively. On the next smaller scale, transport is governed primarily by synoptic disturbances such as the travelling low- and high-pressure systems that migrate across the mid-latitudes of the Northern Hemisphere. Characteristic time and distance scales are of the order of a few days and a few thousand kilometres, respectively. At increasingly smaller scales—regional, local, and micro, which are characterized, respectively, by time and space scales of several hours and hundreds of kilometres, an hour and a kilometre, and minutes and metres—transport is increasingly influenced by orography and the nature of the underlying surface. These also have a strong influence on the vertical temperature structure of the lower atmosphere and thus on the amount of convective activity and turbulent mixing, which in turn affect transport on these smaller scales. Extensive reference material is available on meteorological regimes and factors affecting the transport of chemical constituents through the atmosphere (e.g. Hasse, 1983; Merrill, 1986; Whelpdale and Moody, 1989).

Several factors can influence the effectiveness of atmospheric transport of material to a receiving ecosystem. Most important are the magnitude and location of upwind sources of emissions. On larger scales these may be large areas of biomass burning in the trade-wind belt in either hemisphere, or emissions from the large industrialized regions of the North American and European continents. Both prevailing wind direction and distance to sources are important. In addition, seasonal changes in meteorological regime can result in the transport of material from completely different source regions. On the smaller scales, in addition to wind speed, direction, and distance to upwind sources, local influences on transport may become dominant: examples are areas of high convective activity, regions of extensive turbulent mixing, land–sea breeze circulations, or orographic channelling. Even the diurnal changes in atmospheric stability, with efficient day-time vertical mixing and strong night-time stratification may result in two distinct transport regimes for a particular location.

The same meteorological factors which influence the delivery of chemical constituents to a terrestrial or wetland system also control the dispersal and transport of natural emissions from the system away through the atmosphere. Of initial importance is the efficiency of vertical mixing which takes substances emitted at the surface to regions higher up in the atmosphere where more efficient transport occurs.

As noted earlier, the main anthropogenic sulphur species released into the atmosphere are gaseous sulphur dioxide and particulate sulphate. Species emitted from natural sources are more numerous, and include primarily the reduced sulphur species hydrogen sulphide, dimethyl sulphide, dimethyl disulphide, carbonyl sulphide, carbon disulphide and methyl mercaptain, in addition to sulphur dioxide and particulate sulphate.

The reduced sulphur compounds undergo odixation in the atmosphere to sulphur dioxide, primarily through reaction with the hydroxyl radical. The expected atmospheric residence time of most is a day or less. Sulphur dioxide undergoes oxidation in the gas phase, primarily with the hydroxyl radical, on the surface of particles, or in precipitation elements, through oxidation with the peroxyl radical or ozone, or in the presence of metal catalysts. The product in all cases is the sulphate form either in particles or in cloud and precipitation droplets.

Sulphur dioxide has an atmospheric residence time of about one day; for sulphate it is three to four days; and for total sulphur it is about two days (Rodhe, 1978). Residence times vary depending on the chemical and meteorological regime in the region of the world in question. The chemical form in which sulphur exists in the atmosphere has a significant influence on the efficiency with which it is removed to the surface. Consideration of meteorological regime and residence time for sulphur in the troposphere provides some insight into the 'zone of influence' that sulphur emissions may have. For example, in the mid-latitudes where the average lower troposphere wind speed may be about 10 m s^{-1}, sulphur will be deposited over a range of 1000 to 2000 km. Such distances are, in fact, comparable to the scale of current North American and European acidification problems, to which sulphur is the major contributor.

2.4 DEPOSITION

Chemical constituents in the atmosphere can be brought to the surface by a variety of processes (Figure 2.3). Those for which precipitation is the delivery mechanism are termed wet deposition processes. Gases, such as sulphur dioxide, can dissolve in cloud and rain drops or adsorb on to frozen precipitation elements. Sulphate particles are efficient condensation nuclei and are incorporated into precipitation by nucleation or as a result of scavenging by cloud droplets and falling drops. The efficiency of wet removal of sulphur species depends on the form (sulphate is more efficiently removed than sulphur dioxide), and on the characteristics of the precipitation in a given location: type, intensity, duration, frequency, etc. In principle, wet deposition is easily measured. With proper attention to sampling location, cleanliness, sample handling and chemical analysis procedures, reliable measurements of wet sulphur deposition can be routinely obtained.

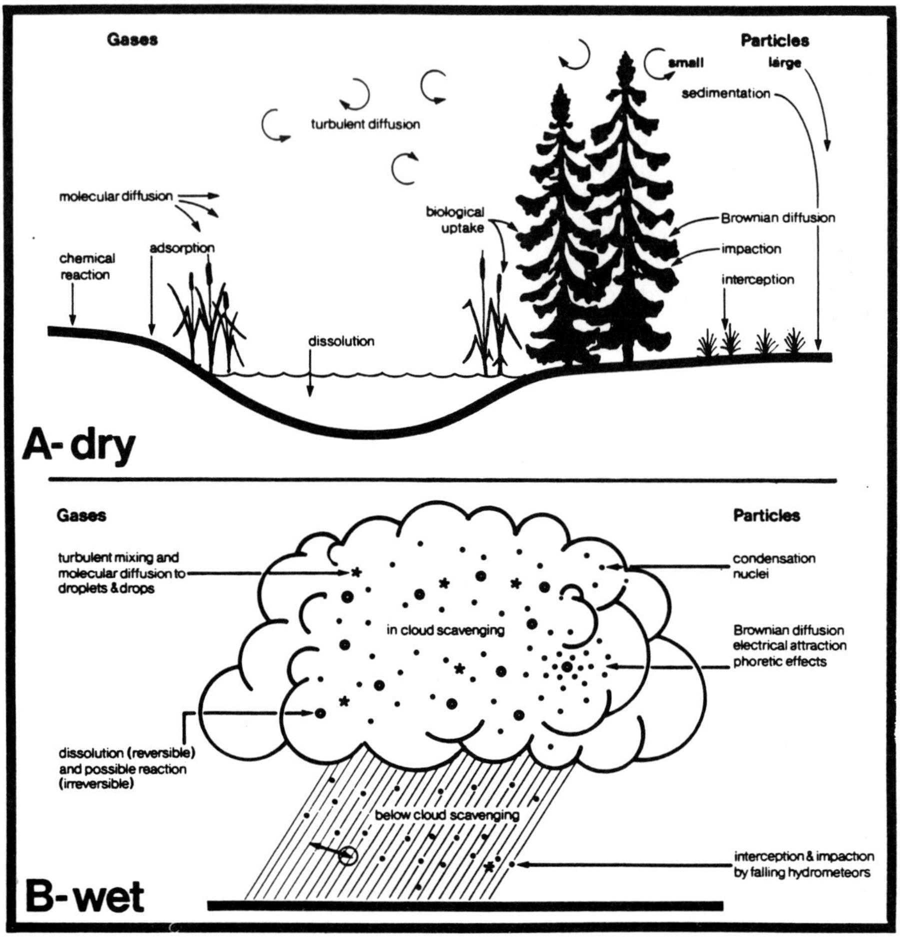

Figure 2.3. Schematic representation of the deposition processes

Those deposition processes which do not involve precipitation are collectively termed dry deposition. Particles larger than about 10 μm in diameter may be removed by gravitational sedimentation. Smaller particles and gases, however, are more efficiently brought to the near-surface region by turbulent atmospheric motions, where they may subsequently be brought into contact with surface elements by molecular-scale processes. Actual uptake is accomplished by chemical reaction, dissolution, adsorption, etc. The efficiency of transfer to the near-surface region depends primarily on turbulence intensity in the several metres above the surface. Transfer to the surface depends on the characteristics of the surface (e.g. roughness, wetness, chemical properties) and on the properties of the constituent (e.g. state, reactivity, solubility). Sulphur dioxide is dry deposited more efficiently than particulate sulphate because it is more readily taken up at the surface. In contrast to wet, dry deposition is not easily measured directly. It can be measured in a research mode by the flux-gradient or eddy-correlation approach; but these techniques are not used for routine network measurements. The more usual approach is to parameterize the dry deposition process using a 'deposition velocity', and calculate deposition as a product with a measured or modelled ambient concentration.

A third category of deposition process includes mass transfer to the surface by impaction of fog or cloud droplets, and by riming. These are not usually included as either wet or dry deposition. However, in high-elevation forested ecosystems or in areas with frequent fog, they can be very efficient. For example, Lovett, Reiners and Olson (1982) estimated that chemical inputs to a balsam fir forest in New Hampshire by this mechanism exceeded those by ordinary precipitation by 50 to 300%. Deposition by these processes and by dry deposition may be extremely efficient locally, for example on the windward edge of a forested area. Techniques to quantify such deposition are limited, although some progress is being made by examining throughfall in forests.

In addition to the measurement of deposition, progress has been made in calculating deposition using atmospheric models. The important processes of deposition are reasonably well understood, but due to their complexity, they are parameterized in rather simple terms in most regional and larger scale models. In the case of wet deposition, the limitations to more sophisticated simulation lie in the fine temporal and spatial detail required in describing the cloud microphysics, and in the limited resolution available in the input fields (such as precipitation amount). Commonly, scavenging (or washout) ratio or scavenging (or washout) coefficient parameters are used in connection with rainfall amount and atmospheric concentrations to model wet deposition. Modelling limitations in the case of dry deposition are the detail required in the various process descriptions and the characteristics of the underlying surfaces (e.g. roughness, moisture, temperature, etc.). The deposition velocity parameter is derived on the basis of measurements, and is a function of atmospheric stability, underlying surface characteristics, and chemical species. A typical range of sulphur dioxide deposition velocity values for summer season day-time at mid-latitudes is 0.5–0.8 cm s^{-1}. For particulate sulphate, values under comparable conditions would be 0.1–0.4 cm s^{-1}. Values could well be an order of magnitude lower at night and in winter. Only limited modelling of fog, cloud, and rime deposition inputs has been carried out.

Estimates of wet and dry deposition of sulphur are available for most parts of the world. In much of the marine environment, in the Southern Hemisphere, and in other remote areas, such estimates are often based on very limited precipitation and ambient concentration measurements. In many developed countries of the Northern Hemisphere the phenomenon of acid deposition has prompted intensive routine measurement programmes which produce reliable wet deposition information and rather more limited estimates of dry deposition of sulphur. In eastern North America and west-central Europe annual wet deposition may range up to more than 4 g (S) m^{-2} (0.13 mol m^{-2}), and annual dry deposition may exceed 5 g (S) m^{-2} (0.16 mol m^{-2}). In contrast, Figure 2.4 (Galloway, 1985) shows that annual non-sea-salt sulphur deposition in remote continental areas of the world is approximately 0.2 and 0.1 g (S) m^{-2} (6 and 3 mmol m^{-2}) for wet and dry deposition, respectively, and in remote marine areas, 0.1 and 0.02 g (S) m^{-2} (3 and 0.6 mmol m^{-2}) for wet and dry deposition, respectively.

For the heavily industrialized regions of the world like eastern North America and west-central Europe, wet and dry deposition are approximately equal on an annual basis as an area average. Dry deposition is greater close to source regions where sulphur dioxide is the prevalent species, usually in relatively high concentrations (Figure 2.5). Wet deposition tends to dominate further from source regions as sulphate becomes dominant. Wet deposition is the main pathway to the surface for sulphur in remote regions of the world. As noted above, in special circumstances, deposition by the direct impaction of droplets on to vegetation may far outweigh the two major pathways.

Sea-salt sulphur is a major constituent of the global, and particularly the marine, sulphur cycle. However, most sea-salt sulphur (an estimated 90%) is

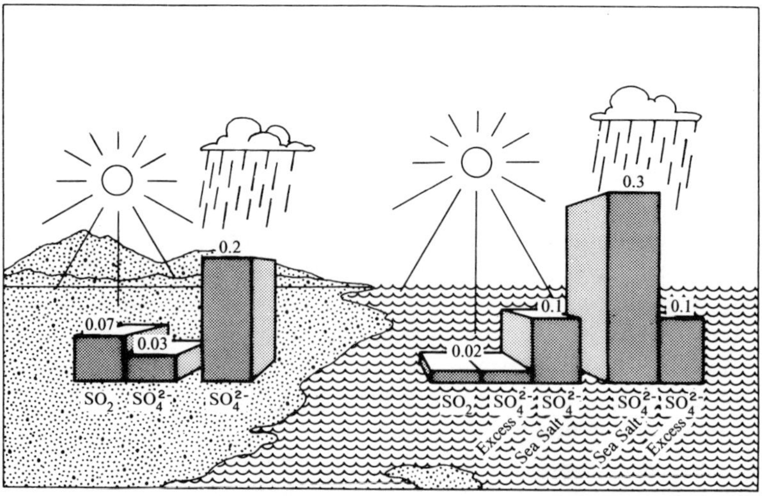

Figure 2.4. Wet and dry deposition of sulphur species in remote areas (Galloway, 1985). Units are g (S) m^{-2} a^{-1}

more or less directly redeposited to the ocean (Andreae and Jaeschke, this volume). Of that which does reach the continents most is deposited within about 100 km of the coast. In many uses of atmospheric-sulphur data, sulphate concentrations in precipitation and air are corrected for the sea-salt component by subtracting off an amount determined by ratio to the sodium ion which is assumed to have an exclusively seawater origin.

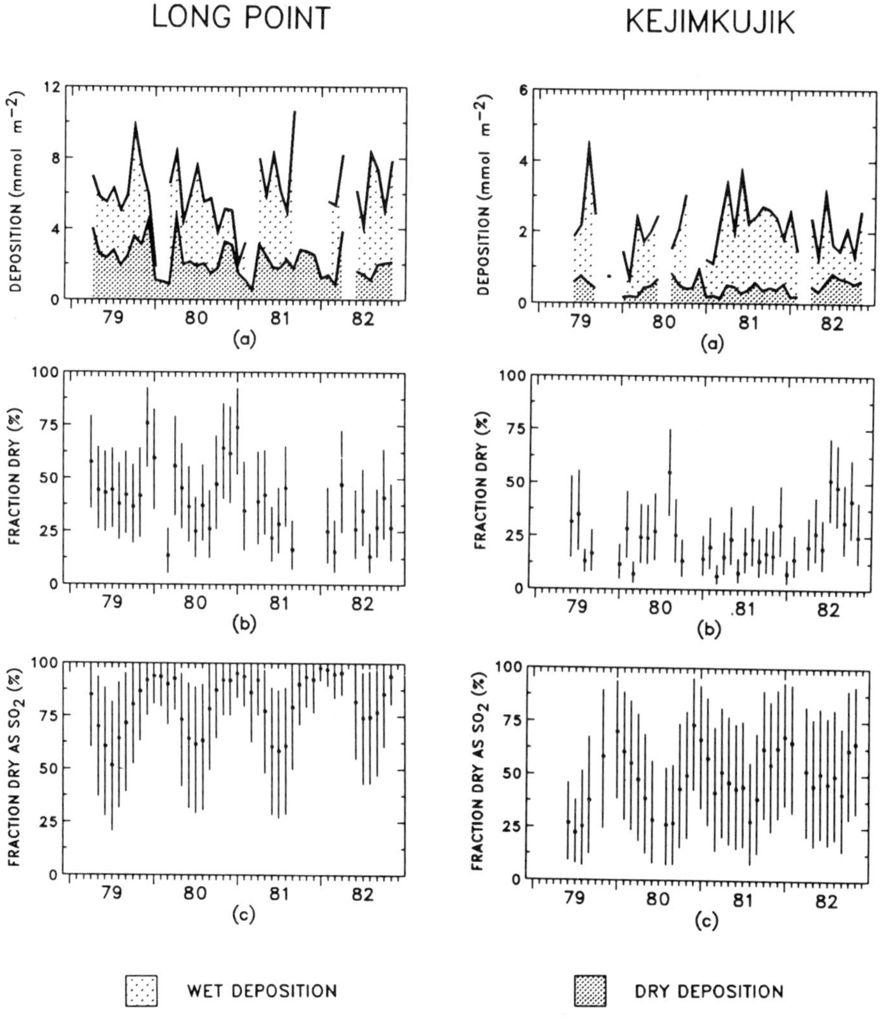

Figure 2.5. Monthly variation of sulphur deposition: (a) its wet and dry components; (b) the dry fraction; and (c) the fraction of dry as sulphur dioxide. The left panel is for Long Point, Ontario, a site close to strong emissions, and the right panel is for Kejimkujik, Nova Scotia, a site distant from sources (Sirois and Barrie, 1988)

In most well established networks which measure wet sulphur deposition on a routine basis, station density is in the range of one station per 10 000 to 100 000 km^2. Better resolution is occasionally achieved in special studies. Networks for ambient concentration measurements, from which dry deposition estimates are obtained, generally have a lower density of stations (an exception is the EMEP network in Europe). Temporal resolution in wet deposition measurements is usually daily or weekly; in ambient concentration measurements it is usually daily. Most regional-scale models which have been developed to address the problem of long-range transport of sulphur and acid deposition have a spatial resolution of 80 to 150 km which is dictated by the resolution of input data, and a temporal resolution of not much less than a day. However, with the exception of the complex grid models under development and testing at the present time, regional models are most suitable for use on a seasonal or annual basis, thus providing an effective temporal resolution of this degree.

A characteristic of both wet and dry deposition is their high variability. Data from several years of observation at Canadian stations show that both wet and dry deposition are highly episodic. It is not unusual for 50% of the annual deposition (wet or dry) to come from 20% of the events (or daily measurements) (Figure 2.6; Sirois and Barrie, 1988). Depending on possible ecosystem response, this observation might lead to some concern about temporal and spatial scales of deposition that are smaller than those resolved by current measurements. Differences of an order of magnitude are not uncommon between consecutive daily measurements, although this may be less in remote regions. The variability arises primarily from differences in meteorological and scavenging processes, and the location of the measurement site with respect to sources. One implication of the high temporal and spatial variability in deposition measurements is that care must be taken to ensure that single sites often used in ecosystem studies are representative of the area for which the deposition data are being used. For example, dry deposition determined from ambient concentration measurements made near the edge of a forest may not be representative of a complete watershed; wet deposition measurements made on the crest of a hill with frequent upslope flow may not be representative of an entire catchment. A comprehensive short-term testing programme is often required to establish the representativeness of a proposed site.

In summary, for many parts of the world where the sulphur cycle has been studied in the context of the acid rain problem, extensive deposition data bases are available as are numerous model results. These provide useful large-scale definition of the sulphur deposition fields in these regions. However, in most other regions of the world, sulphur deposition estimates are based on limited measurements at relatively few sites. In the context of sulphur cycling in terrestrial and wetland ecosystems, one would want to have

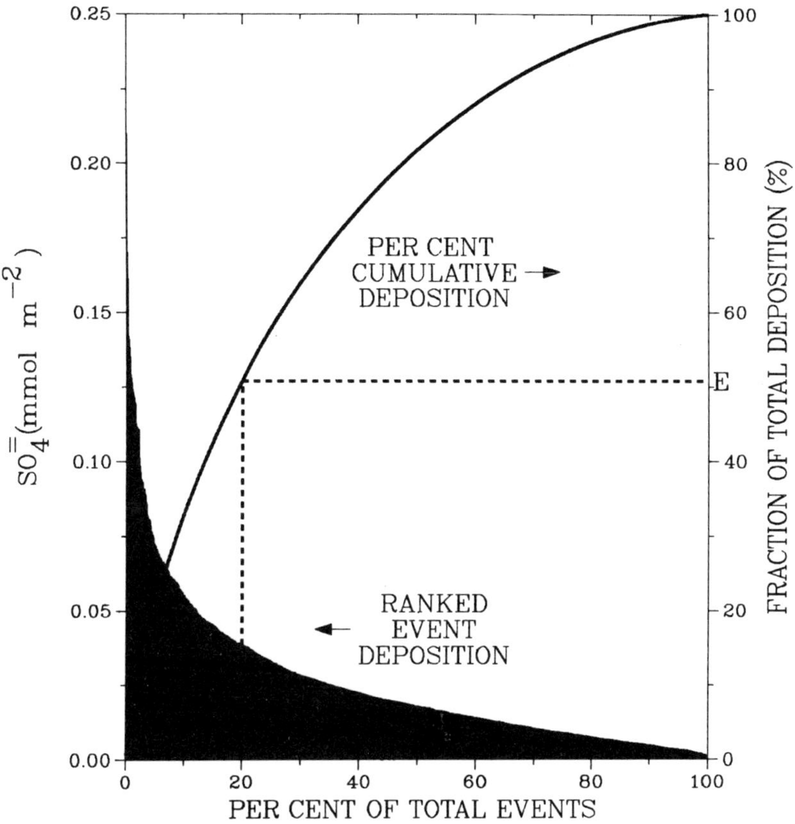

Figure 2.6. Total dry sulphate deposition at Chalk River, Ontario, ranked by percentage of total events (dark histogram), and percentage cumulative deposition (solid line). The horizontal dashed line shows the fraction of total deposition contributed by 20% of events (Sirois and Barrie, 1988)

reliable, representative measurements of wet deposition, and ambient concentration at least, in the particular ecosystem of interest. Often this requires that such studies establish their own atmospheric measurement stations. In these cases, comparison studies should be undertaken with neighbouring network stations and use made of available modelled deposition fields.

2.5 BUDGETS

Freney, Ivanov and Rodhe (1983) summarized previous work on the global sulphur cycle. They showed that the atmosphere is one of the smaller reservoirs for sulphur, containing less than 5 Tg (S) (0.16 Tmol) at any one time. Figure 2.7, from Freney, Ivanov and Rodhe (1983), is a schematic representation of the major fluxes of the global sulphur cycle. As more recent information has become available, individual flux terms have been updated. For example, the review by Andreae and Jaeschke (this volume), which summarizes most recent information on natural sources within the global sulphur cycle, suggests that volcanic and terrestrial biogenic emissions are lower by a factor of two, while marine biogenic emissions are twice as large, compared to values given in Figure 2.7.

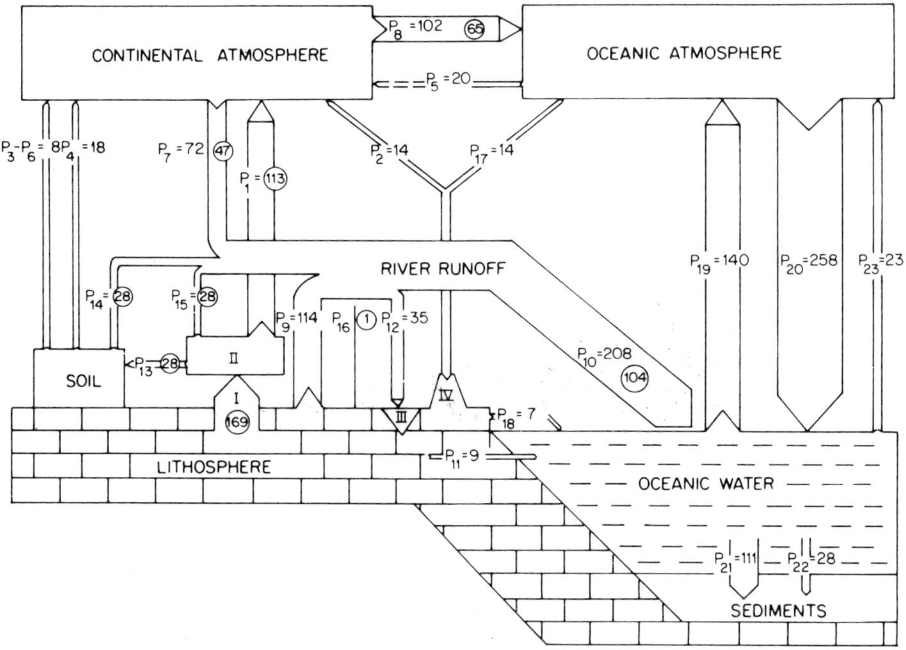

Figure 2.7. The major natural and anthropogenic (circled) fluxes of the global biogeochemical sulphur cycle (Tg (S) a^{-1}) (Ivanov and Freney, 1983). The Roman numerals denote: I, output of sulphur-containing minerals; II, industrial treatment of the sulphur-containing raw materials; III, inland water bodies; and IV, volcanoes. The fluxes are denoted by: P1, fuel combustion and metal smelting; P2 and P17, volcanic emissions; P3, aeolian dust; P4, biogenic emission from land; P5, sea–air–land transport; P6, deposition of large dust particles; P7, wash-out and dry deposition; P8, land–air–sea air transport; P9, weathering; P10, river runoff to oceans; P11, transport to ocean in underground water; P12, runoff to inland water bodies; P13, input in fertilizers; P14, leaching of fertilizers; P15, efflux from chemical industries; P16, efflux of acid mine water; P18, abrasion of shores; P19, sea spray; P20, wash-out and dry deposition; P21, sedimentation of reduced sulphur; P22, sedimentation of sulphate; P23, biogenic emission from oceans

A budget of the inflows and outflows for a specified region is an instructive means of organizing knowledge of the cycling of sulphur through the atmosphere and other compartments of the environment. Ideally, all the important pathways into and out of the region of study would be known; all the individual processes that comprise each pathway would be understood; and some quantitative measure of the importance of each pathway, i.e. the flux of sulphur, would be available. All this information is usually not available, but budgets help to account for the various inflows and outflows, and help to identify and determine the importance of missing pathways.

The major input terms in an atmospheric sulphur budget are usually natural and anthropogenic inputs from the earth's surface, and a horizontal advection flux from upwind sources. Outflow terms are usually wet and dry deposition, and a horizontal advection outflow. On occasion, exchange at the upper surface of the atmospheric 'box' is taken into account. Uniform conditions are assumed within the atmospheric region chosen; i.e. smaller scale processes and variability within the region are not considered. For horizontal advection, the transit time for the region of the atmosphere chosen must be large in comparison to the atmospheric residence time of sulphur, in order not to have a unacceptably large error in the difference between inflow and outflow. This requirement means that global and regional budgets are the most suitable for this approach (Rodhe, 1978).

Figure 2.8 is an example of a regional atmospheric sulphur budget for eastern North America (Galloway and Whelpdale, 1980). The individual fluxes are shown, and it is of interest to note that anthropogenic emissions exceed natural ones by a factor of about twenty. Dry and wet deposition are approximately equal, and, as well, are similar to the horizontal outflow. That is, one-third of eastern North American emissions are deposited wet, one-third are deposited dry, and one-third flow out of the region to the east with the prevailing winds. The atmospheric sulphur budget for northern Europe gives similar results (Rodhe, 1978), as does a recent budget for southeastern Europe (Katsoulis and Whelpdale, 1990). A budget for the North Atlantic Ocean demonstrates another use of this approach (Figure 2.9; Galloway and Whelpdale, 1987). In this case the wet deposition to the ocean is known reasonably well, but dry deposition is quite uncertain (0.9 ± 0.9 Tg (S) a^{-1}; 0.03 ± 0.03 Tmol a^{-1}), pointing to the need for additional work to determine this term more accurately. In this case, the eastward flux from the region is determined by difference using all other budget terms. Regional budgets provide a useful approach to organizing and summarizing knowledge on this scale. This has been done for several regions of the world now, but additional work is needed for the Mediterranean, the eastern Atlantic, southeast Asia and the southwest Pacific.

On the spatial scale of many countries, resolution of the inflow and outflow advection fluxes becomes difficult because the transit time across the country

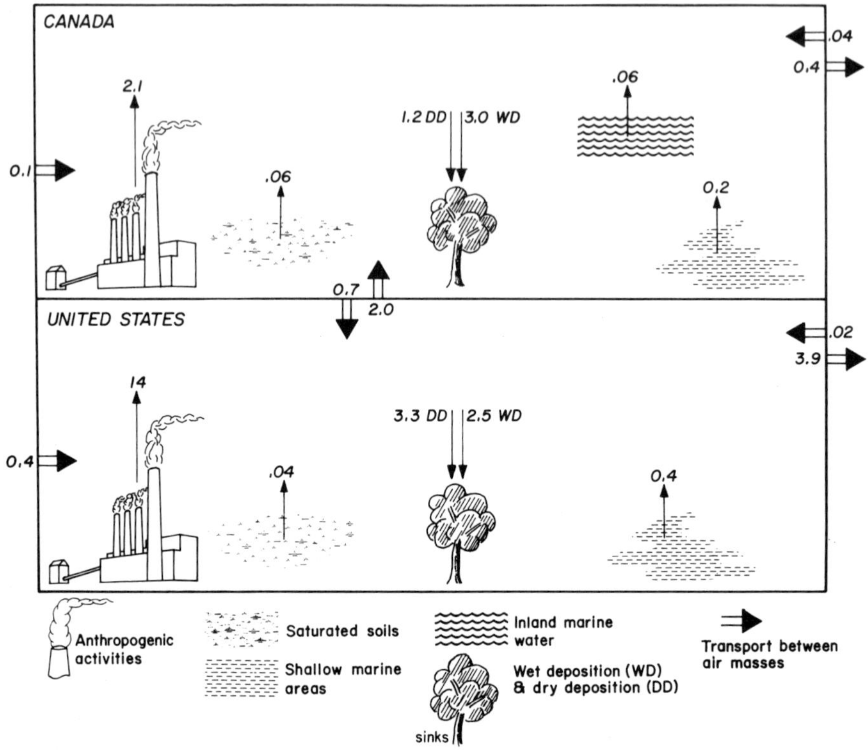

Figure 2.8. Schematic representation of the atmospheric sulphur budget for eastern North American *c.* 1980 (Tg (S) a^{-1}) (Galloway and Whelpdale, 1980)

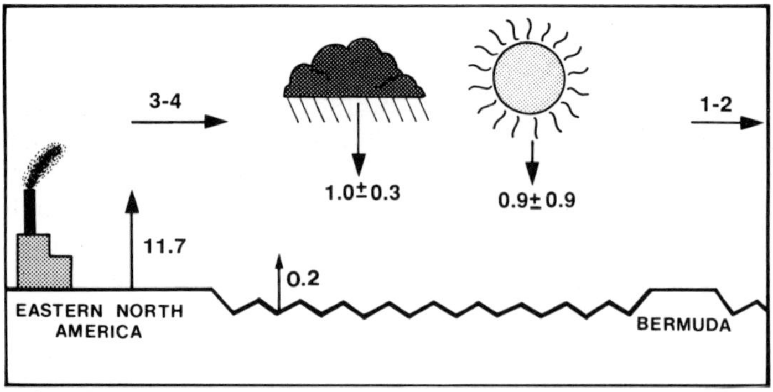

Figure 2.9. Schematic representation of the amospheric sulphur budget for the western North Atlantic Ocean atmosphere (Tg (S) a^{-1}) (Galloway and Whelpdale, 1987)

is less than the atmospheric residence time of sulphur. The recent use of long-range transport models has provided a remedy to this limitation. Country budgets have been constructed for Canada, the United States, the Soviet Union, Hungary, Czechoslovakia, Poland and the United Kingdom. Within the EMEP programme, budgets have been determined for all European countries using a modelling approach (Eliassen *et al.*, 1988). Country budgets are useful in an accounting sense, identifying those terms in which there is most confidence and those in which there is least.

Budgets for terrestrial ecosystems and wetlands usually include reservoirs in addition to the atmosphere: for example, biota, surface waters, soil. The main atmospheric fluxes of interest are usually natural emissions (and anthropogenic ones if present), wet deposition, and dry deposition. Unless the ecosystem is extensive the advection flux terms are not considered. Identification of the specific processes of natural emission, and quantification of these three flux terms is the subject of much of the remainder of the volume, and this topic will be touched on only briefly here.

2.6 ATMOSPHERIC INPUTS TO ECOSYSTEMS

In this section the magnitude and possible significance of atmospheric sulphur deposition to a variety of natural and managed ecosystems are considered. Of primary interest are the several terrestrial systems and wetlands discussed in subsequent chapters. These include natural systems such as temperate and tropical forestland, savanna, freshwater lake ecosystems, and freshwater and saline wetlands. Managed systems include rice paddies, managed forestlands and many other agricultural systems. Possible inputs of sulphur to these systems, in addition to the atmosphere, include weathering, groundwater, runoff, flooding, river inflow, fertilizer application, irrigation water, and mine drainage.

It is difficult to associate a spatial scale with many of these systems. Some are extensive, such as the temperate and tropical forest systems, covering large portions of continents. Some are numerous, such as temperate lake and freshwater wetland systems, but they not necessarily contiguous, being more 'patchy' in nature, and thus of a smaller characteristic size. Systems in regions remote from industrial emissions are likely to experience sulphur deposition of uniform magnitude, while those close to highly populated and industrialized areas, such as much of the inland lakes and freshwater wetlands in the temperate Northern Hemisphere, may expeience a large range of atmospheric inputs. In the latter case, where these systems are studied as 'calibrated watersheds' or 'selected forest ecosystems', the magnitude of sulphur deposition and its role in biogeochemical cycling may differ significantly from location to location.

Table 2.1. Sulphur inputs to lakes, expressed per unit area of lake surface. Atmospheric inputs are those directly to the lake surface. Runoff is via overland (not channel) flow. Seepage is the estimated inflow by groundwater; only a small fraction (< 3%) of the total input of water to Little Rock Lake is derived from seepage

Site	Atmosphere	Runoff	Stream inflow	Acid addition	Seepage	Reference
	$(g\ (S)\ m^{-2}\ a^{-1})$					
Lake 223	0.26	1.47	0.48	7.90	—	Cook *et al.* (1986)
Lake 227	0.35	2.27	—	—	—	Cook (1981)
Little Rock Lake	0.61	—	—	—	0.03	Baker *et al.* (1989)
Lake Anna	0.61	0.35	205	—	—	Herlihy *et al.* (1987)

Andreae and Jaeschke (this volume) have reviewed wet deposition fluxes in tropical regions, primarily areas of rainforest and moist savanna. Wet sulphur fluxes range from 0.07 to 0.68 g (S) $m^{-2}\ a^{-1}$ (2 to 21 mmol $m^{-2}\ a^{-1}$), with a median value of 0.17 g (S) $m^{-2}\ a^{-1}$ (5.3 mmol $m^{-2}\ a^{-1}$). These values are in agreement with the estimate of Galloway (1985), 0.2 g (S) $m^{-2}\ a^{-1}$ (6 mmol $m^{-2}\ a^{-1}$), for wet deposition in remote continental areas. For such systems, atmospheric deposition is usually the only sulphur input to the system, aside from internal cycling.

A second example of atmospheric sulphur inputs is for northern temperate lakes, where once again, the atmosphere is usually the only source of sulphur. Although such lake systems cover a small portion of the earth's surface, there are hundreds of thousands of them in the northern temperate zone, typically of area up to some 100 ha. Atmospheric inputs, direct and from watershed runoff, are the major sources of sulphur to freshwater lakes (Table 2.1); these amount to between 0.6 and 2.6 g (S) $m^{-2}\ a^{-1}$ (19 to 81 mmol $m^{-2}\ a^{-1}$) for the examples given. The table also contains two unusual cases: one where sulphuric acid was added in a whole-lake acidification experiment, and another which is an acid-mine drainage reservoir. In such non-natural situations, these other inputs far outweigh the natural ones.

An interesting comparison (R.B. Cook, pers. comm.) is the contribution of atmospheric and land sulphur to rivers, based on global river data from Holland (1978, pp. 67, 96, and 140). Holland calculates that of the global average river flow to the oceans of 2 g (S) $m^{-2}\ a^{-1}$ (63 mmol $m^{-2}\ a^{-1}$), about 40% was derived from the atmosphere, and about 60% was derived from terrestrial sources, primarily weathering of sulphate and sulphide minerals.

Some measurements of atmospheric sulphur deposition and comparisons with other inputs have been made for freshwater and saline wetlands (Giblin and Wieder, this volume). Freshwater wetlands cover about 5.3×10^8 ha of the earth, and saline wetland about 0.5×10^8 ha. Atmospheric inputs to freshwater bogs, compiled by Giblin and Wieder (this volume), range from 0.3 to 3.2 g (S) m^{-2} a^{-1} (9.4 to 100 mmol m^{-2} a^{-1}). Some receive other inputs, such as groundwater or overland flow, but their magnitudes are not known. In saline wetlands, the atmospheric inputs are insignificant, because the salt-water source of sulphur dominates.

The importance of atmospheric inputs of sulphur to rice paddies and agricultural ecosystems is dependent on the degree of management of the system. In the case of systems that are neither intentionally flooded nor fertilized, the atmosphere may be the only source of sulphur (and other nutrients). Rice paddies in southeast Asia, for example, receive between 0.5 and 2.3 g (S) m^{-2} a^{-1} (16 and 72 mmol m^{-2} a^{-1}) (Lefroy *et al.*, this volume). In some systems the fertilizer application may be calculated to supplement that delivered by the atmosphere. In others, where large amounts of fertilizer sulphur are applied, atmospheric inputs may not be important.

This section has dealt only with sulphur inputs from the atmosphere to various types of ecosystems. Emissions of reduced sulphur compounds from these various ecosystems to the atmosphere have been measured much less frequently. Estimates for sulphur gas fluxes from wetlands and terrestrial ecosystems are discussed elsewhere in this volume (Andreae and Jaeschke, this volume; Giblin and Wieder, this volume). Emissions to the atmosphere are generally small compared to rates of cycling within the ecosystem. With the possible exception of marine wetlands (saltmarshes and mangrove swamps), emissions to the atmosphere are also small compared to atmospheric inputs to terrestrial and wetland ecosystems.

2.7 CONCLUDING REMARKS

This summary of the atmospheric sulphur cycle has attempted to present what is known about the cycle that is of most relevance to the study of sulphur cycling in terrestrial systems and wetlands.

Emission rates of sulphur to the atmosphere from anthropogenic sources are known with reasonable certainty for most industrialized regions of the Northern Hemisphere. Most problematic at present are those from southeast Asia, Africa and South America. Natural emissions are less well known because estimates are derived from scattered point measurements of sulphur fluxes; these demonstrate high temporal and spatial variability in both marine and terrestrial systems.

Extensive networks for the routine measurement of wet deposition are in place in many areas of the world. Data quality ranges from very good in some of the 'research-grade' and intensively managed networks, to less than acceptable in others. One must exercise caution about the quality of much of the available data. Dry deposition is determined from measurements of ambient concentration combined with a parameter, the deposition velocity. Such measurements are fewer than are those for wet deposition, and in general, the derived dry fluxes are less certain. Deposition by other processes, such as impaction of cloud droplets and riming, is determined only on a limited basis, even though the inputs may be very important in certain ecosystems. Models of atmospheric transport and deposition have improved greatly over the past decade. They are able to provide reliable calculated estimates of wet and dry deposition for large portions of Eurasia and North America. These models are perhaps under-used by the ecosystem community at present. Results from both networks and model applications are limited in spatial and temporal resolution. The characteristic variability of wet and dry deposition in time and space requires that care be taken in interpolating to small space and time scales for small-scale ecosystem studies.

In the context of using information from the atmospheric portion of the sulphur cycle for ecosystem studies, the more important unknowns are the magnitude of other than wet deposition fluxes, the magnitude of natural emissions in a specific location, the lack of comprehensive modelled deposition values for small scales, and—more a problem than an unknown—the mismatch between the scales of interest for much of the atmospheric and ecosystem work. Improved dry and other deposition flux values will require that additional effort be devoted to the development of measurement techniques for reliable field operation, and measurement strategies to elucidate deposition in special situations where such inputs may dominate. More reliable natural emission fluxes will also require improved techniques and many more measurements over a variety of representative surfaces. The two remaining issues, both pertaining to the scales of interest, require the development of common interest and understanding in the concerned disciplines.

The following are among the important future challenges for interaction between the atmospheric scientist and those working on other portions of the biogeochemical cycle of sulphur:

(1) Obtaining reliable estimates of likely changes in anthropogenic sulphur emissions—decreases in North America and Europe, and probable large increases in southeast Asia, Africa and South America.
(2) Evaluating the implications of climate warming for the sulphur cycle—changes in variability of atmospheric inputs, increased stresses on marginal systems, and increases in natural emissions as a result of temperature rise.
(3) Making more extensive use of 'integrated models', i.e. models which couple simulations of processes in all reservoirs of the sulphur cycle—atmosphere, biosphere, soil, waters—to enhance understanding of link-

ages within the cycle and the overall functioning of the biogeochemical sulphur cycle.

(4) Assessing the impact of current and planned reductions in anthropogenic sulphur emissions in regions of the world affected by acid deposition, to determine whether these reductions are adequate to protect the biogeochemical sulphur cycle from unacceptable perturbation.

REFERENCES

Andreae, M. O. (in press). The global biogeochemical sulfur cycle: a review. In: Moore, B., Schimel, D. and Eddy, J. *Trace Gases and the Biosphere*. Office for Interdisciplinary Earth Studies, UCAR, Boulder, CO.

Andreae, M. O. and Jaeschke, W. A. This volume.

Baker, L. A., Urban, N. R., Brezonik, P. L. and Sherman, L. A. (1989). Sulfur cycling in a seepage lake. In: Saltzman, E. and Cooper, W. (Eds) *Biogenic Sulfur in the Environment*, American Chemical Society, Washington, pp. 79–100.

Cook, R. B., Kelly, C. A., Schindler, D. W. and Turner, M. A. (1986). Mechanisms of hydrogen ion neutralization in an experimentally acidified lake. *Limnology and Oceanography*, **31**, 134–148.

Cook, R. B. (1981). The biogeochemistry of sulfur in two small lakes. PhD Thesis, Columbia University, New York, 248 pp.

Dillon, P. J., Lusis, M., Reid, R. and Yap, D. (1988). Ten-year trends in sulphate, nitrate and hydrogen deposition in central Ontario. *Atmos. Environ.*, **22**, 901–5.

Dovland, H. and Saltbones, J. (1986). Emissions of sulphur dioxide in Europe in 1980 and 1983. *EMEP/CCC-Report: 1/86*, Norwegian Institute for Air Research, Lillestrøm, 34 pp.

Eliassen, A., Hov, Ø., Iversen, T. Saltbones, J. and Simpson, D. (1988). Estimates of airborne transboundary transport of sulphur and nitrogen over Europe. *EMEP/MSC-W Report 1/88*, Norwegian Meteorological Institute, Oslo, 80 pp.

Fjeld, B. (1975). Forbruk av fossilt brensel i Europa og utslipp av SO_2 i perioden 1900–1972. *NILU Teknisk Notat Nr 1/76*, 6pp.

Freney, J. R., Ivanov, M. V. and Rodhe, H. (1983). The sulphur cycle. Section 2.5. In: Bolin, B. and Cook, R. B. (Eds). *The Major Biogeochemical Cycles and their Interactions*. Wiley, New York, pp. 56–61.

Galloway, J. N. (1985). The deposition of sulfur and nitrogen from the remote atmosphere: background paper. Chapter 8. In: Galloway, *et al.* (Eds). *The Biogeochemical Cycling of Sulfur and Nitrogen in the Remote Atmosphere*, pp. 143–72. Reidel, Boston.

Galloway, J. N. (1989). Atmospheric acidification: projections for the future. *Ambio*, **18**, 161–6.

Galloway, J. N. and Whelpdale, D. M. (1980). An atmospheric sulfur budget for eastern North America. *Atmos. Environ.*, **14**, 409–18.

Galloway, J. N. and Whelpdale, D. M. (1987). WATOX-86 overview and western North Atlantic Ocean S and N atmospheric budgets. *Global Biogeochem. Cycles*, **1**, 261–81.

Galloway, J. N., Charlson, R. J., Andreae, M. O. and Rodhe, H. (1985). *The Biogeochemical Cycling of Sulfur and Nitrogen in the Remote Atmosphere*. Reidel, Boston, 249pp.

Giblin, A. and Wieder, D. (1990). This volume.

Hameed, S. and Dignon, J. (1988). Changes in the geographical distributions of global emissions of NO_x and SO_x from fossil-fuel combustion between 1966 and 1980. *Atmospheric Environment*, **22**, 441–449.

Hasse, L. (1983). Introductory meteorology and fluid mechanics. Chapter 1. In: Liss, P. S. and Slinn, W. G. N. (Eds), *Air–Sea Exchange of Gases and Particles*. Reidel, Boston, pp. 1–51.

Herlihy, A. T., Mills, A. L. Hornberger, G. M. and Bruckner, A. E. (1987). The importance of sediment sulfate reduction to the sulfate budget of an impoundment receiving acid mine drainage. *Water Resources Research,* **23**, 287–292.

Holland, H. D. (1978). *The Chemistry of the Atmosphere and Oceans*. Wiley, New York, 351pp.

Ivanov, M. V. and Freney, J. R. (1983). *The Global Biogeochemical Sulphur Cycle*. Wiley, New York, 470pp.

Iversen, T., Saltbones, J., Sandnes, H., Eliassen, A. and Hov, Ø. (1989). Airborne transboundary transport of sulphur and nitrogen over Europe—model descriptions and calculations. *EMEP MSC-W Report 2/89*, Norwegian Meteorological Institute, Oslo, 92pp.

Katsoulis, B. D. and Whelpdale, D. M. (1990). Atmospheric sulfur and nitrogen budgets for southeast Europe. *Atmos. Environ.*, **24A**, 2959–70.

Leck, C. and Rodhe, H. (1989). On the relation between anthropogenic SO_2 emissions and concentration of sulfate in air and precipitation. *Atmospheric Environment*, **23**, 959–66.

Lefroy, R. D. B., Mamaril, C. P., Blair, G. J. and Gonzales, P. B. (1990). This volume.

Lovett, G. M., Reiners, W. A. and Olson, R. K. (1982). Cloud droplet deposition in subalpine balsam fir forests: hydrological and chemical inputs. *Science*, **218**, 1303–4.

Merrill, J. T. (1986). Atmospheric pathways to the oceans. Chapter 2. In: Buat-Ménard, (Ed.). *The Role of Air–Sea Exchange in Geochemical Cycling*. Reidel, Boston, pp. 35–63.

Moller, D. (1984). Estimation of the global man-made sulphur emission. *Atmos. Environ.*, **18**, 19–27.

Mylona, S. N. (1989). Detection of sulphur emission reductions in Europe during the period 1979–1986. *EMEP MSC-W Report 1/89*, Norwegian Meteorological Institute, Oslo, 149pp.

NAPAP (1990). Emissions involved in acidic deposition processes. *State-of-Science/Technology Report 1*, January 1990 draft, National Acid Precipitation Assessment Program, Washington DC.

Rodhe, H. (1978). Budgets and turn-over times of atmospheric sulfur compounds. *Atmos. Environ.*, **12**, 671–80.

Sirois, A. and Barrie, L. A. (1988). An estimate of the importance of dry deposition as a pathway of acidic substances from the atmosphere to the biosphere in eastern Canada. *Tellus*, **40B**, 50–80.

Whelpdale, D. M. and Moody, J. L. (1989). Large-scale meteorological regimes and transport processes. Chapter 2. In: Knap, A. H. (Ed.). *The Long Range Atmospheric Transport of Substances from Continents to Continents and Continents to Oceans*, Reidel, Boston, pp. 3–36.

3 Exchange of Sulphur between Biosphere and Atmosphere over Temperate and Tropical Regions

M. O. ANDREAE
Max-Planck-Institut für Chemie, Mainz, Germany

and

W. A. JAESCHKE
Johann Wolfgang Goethe-Universität, Frankfurt am Main, Germany

3.1 INTRODUCTION

Estimates of man-made sulphur emissions to the atmosphere fall into a relatively narrow range of about 93 ± 15 Tg (S) a^{-1} (2.9 ± 0.47 Tmol a^{-1}) (Cullis and Hirschler, 1980). In contrast, the characteristics of the natural biogeochemical sulphur cycle in the atmosphere–biosphere–ocean system are much less well known. The existence of natural sources of atmospheric sulphur compounds has long been known, particularly for such source processes as volcanic emissions of SO_2 and emission of H_2S from anaerobic marshlands and estuaries (Jaeschke *et al.*, 1978). In the last 15 years, a number of additional sulphides have been discovered in the atmosphere: dimethyl sulphide (DMS; Hill, Aneja and Felder, 1978; Andreae *et al.*, 1985), carbonyl sulphide (COS; Hanst *et al.*, 1975), and carbon disulphide (CS_2; Sandalls and Penkett, 1977). Other mercaptans have been identified over cattle feed lots, where they arise from rotting manure (Burnett, 1969; Stephens, 1971). These discoveries have shifted the emphasis away from H_2S and have led towards a much better understanding of biogenic sulphur emissions.

3.2 BIOLOGICAL PROCESSES LEADING TO BIOGENIC SULPHUR FLUXES TO THE ATMOSPHERE

Reduced volatile sulphur compounds, which are released to the oxygen-rich atmosphere, are chemically oxidized during their atmospheric lifetime and

Sulphur Cycling on the Continents
Edited by R.W. Howarth, J.W.B. Stewart and M.V. Ivanov

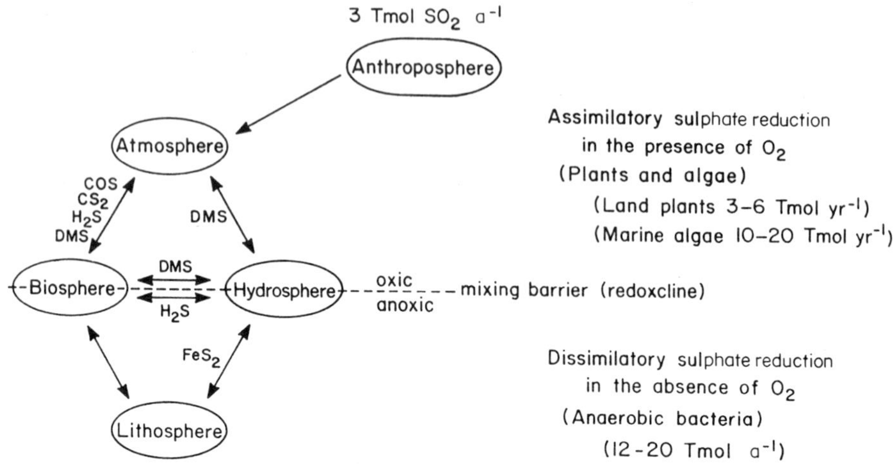

Figure 3.1. Interactions in the global biogeochemical sulphur cycle

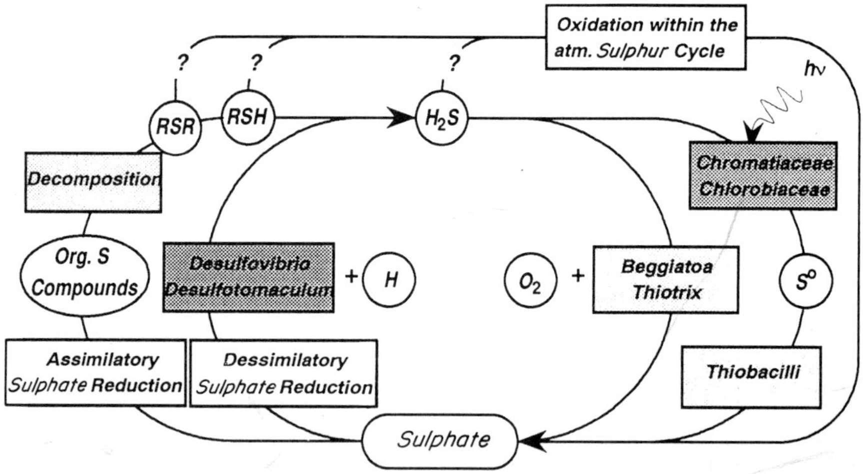

$$8\,[H] + SO_4^{2-} \rightarrow H_2S + 2\,H_2O + 2\,OH^-$$ anaerobic

$$2\,CO_2 + H_2S + 2\,OH^- \xrightarrow{h\nu} 2 < CH_2O > + SO_4^{2-}$$ anaerobic, photothropic

$$2\,H_2S + O_2 \rightarrow 2\,H_2O + 2\,S$$ aerobic

$$S + H_2O + 1\tfrac{1}{2}\,O_2 \rightarrow SO_4^{2-} + 2\,H^+$$ aerobic

Figure 3.2. Scheme of the microbiological cycle of sulphur and its possible influence on the atmosphere

end up finally as sulphur dioxide (oxidation state +4) and sulphuric acid and particulate sulphate (oxidation state +6) and methane sulphonate (oxidation state +6; Andreae *et al.*, 1988; Saltzman *et al.*, 1986). It is mainly these compounds that are removed from the atmosphere and brought back to the earth by dry and wet deposition.

Since the oxidation state +6 is the most stable under oxic conditions, sulphate is the predominant form of sulphur in oxic waters and soils. Thus, the reduction of sulphate to a more reduced sulphur species is a necessary prerequisite for the formation of volatile sulphur compounds and their emission to the atmosphere. Biochemical processes which lead to this reduction can be considered as the driving force of the atmospheric sulphur cycle. A simplified, conceptual overview of the global biogeochemical cycle of sulphur is shown in Figure 3.1. The global environment is subdivided into four compartments: atmosphere, biosphere, hydrosphere, and lithosphere, the latter including sediments and rocks of the earth's crust. Two types of biochemical pathways of sulphate reduction are important in the global cycles: dissimilatory and assimilatory sulphate reduction. The influence of these two pathways on sulphur cycling between biosphere and atmosphere is shown in Figure 3.2.

3.2.1 DISSIMILATORY SULPHATE REDUCTION

Dissimilatory reduction of sulphate is a strictly anaerobic process which takes place only in anoxic environments. Sulphate-reducing bacteria reduce sulphate and other sulphur oxides to support respiratory metabolism, using sulphate as a terminal electron acceptor instead of molecular oxygen (see chapters by Howarth and Stewart, Cooke and Kelly, and Giblin and Wieder, this volume). Since the process is strictly anaerobic, dissimilatory sulphate reduction occurs largely in stratified, anoxic water basins and in sediments of wetlands, lakes, and coastal marine ecosystems. The process is particularly important in marine ecosystems, including salt marshes, because sulphate is easily available due to its high concentration in seawater (28 mM; 900 g (S) l^{-1}). The interface between oxic and anoxic regimes is called the redoxcline and is indicated in Figure 3.1 by a dashed line through the biosphere and hydrosphere compartments. The redoxcline is maintained by physical barriers to oxygen movement and by rapid rates of oxygen consumption. As can be seen from Figure 3.2, dissimilatory sulphate reduction is the major pathway for the production of H_2S globally.

Under favourable conditions, the rate of dissimilatory sulphate reduction to H_2S in anoxic environments can be quite high, ranging up to 100 mmol m^{-2} day^{-1} (3 mg (S) m^{-2} day^{-1}) and more in shallow near-shore marine sediments and in salt marshes (Jørgensen, 1982; Howarth, 1984; Ivanov *et al.*, 1989). However, since the occurrence of this process is dependent on the existence of a mixing barrier which prevents oxygen from entering the system, the escape of H_2S from the system will be limited by the same barrier. Most of the sulphate reduced is re-oxidized before ever leaving the sediment (Howarth, 1984; Giblin and Wieder, this volume; Luther and Church, this volume). In the presence of oxygen, H_2S and other reduced sulphur compounds provide excellent substrates for microbial oxidation from which certain bacteria can obtain a substantial amount of energy (Howarth, 1984, and references therein). Such mircoorganisms tend therefore to be present in high numbers at the oxic-anoxic interface. They are very efficient in removing H_2S and can completely oxidize this compound in a layer only a fraction of a millimetre thick (Jørgensen and Revsbech, 1983). Consequently, the very large amounts of H_2S which are produced in coastal marine ecosystems cannot usually be transferred to the atmosphere (Andreae, 1986, and references therein; Giblin and Wieder, this volume), but are either re-oxidized at the oxic/anoxic interface, or precipitated in the form of iron sulphides. In spite of this limitation, near-shore environments like estuaries and salt marshes are the major source of H_2S emissions from the oceans.

3.2.2 ASSIMILATORY SULPHATE REDUCTION

In contrast to animals, which are dependent on organosulphur compounds in their food to supply their sulphur requirement, other biota (bacteria, cyanobacteria, fungi, eukaryotic algae, and vascular plants) can obtain sulphur from assimilatory sulphate reduction for synthesis of organosulphur compounds (see review by Anderson, 1980). Sulphate is assimilated from the environment, reduced inside the cell, and fixed into sulphur-containing amino acids and other organic compounds. The process is ubiquitous in both oxic and anoxic environments. Most of the reduced sulphur is fixed by the intracellular assimilation process and only a minor fraction of the reduced sulphur is released as volatile gaseous compounds as long as the organisms are alive. However, after death of the organisms, microbial degradation liberates reduced sulphur compounds (mainly in the form of H_2S, but also as organic sulphides) to the environment. During this stage, volatile sulphur compounds may escape to the atmosphere. However, as with sulphides formed from dissimilatory sulphate reduction, the sulphides released during decomposition are chemically unstable in an oxic environment and are re-oxidized to sulphate by a variety of microorganisms (Figure 3.2; Schoenau and Germida, this volume).

The biochemistry of assimilatory sulphate reduction has been studied mostly in the green alga *Chlorella*. Therefore, most of the following discussion refers specifically to this organism. It is not altogether clear at this time how far these conclusions can be generalized to other organisms.

The assimilation of sulphate to cysteine, the first organosulphur metabolite produced, is a complex, multi-step process (Figure 3.3). Sulphate is taken up

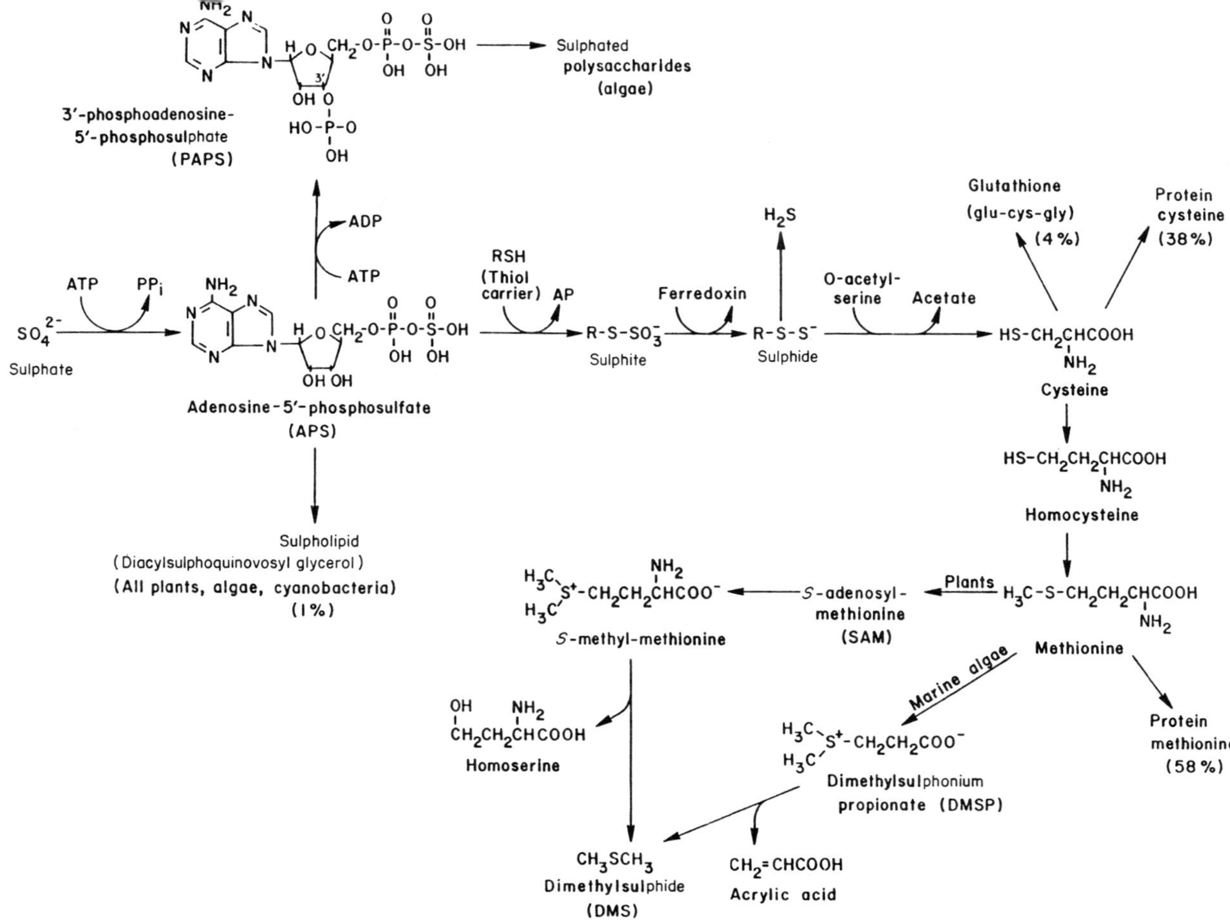

Figure 3.3. Major metabolic pathways of sulphur in algae and plants. The percentages represent the approximate distribution of the major organosulphur compounds in *Chlorella*

into the cell by an active transport mechanism, and inserted into an energetically activated molecule, APS (adenosine-5′-phosphosulphate), which can be further activated at the expense of one more ATP molecule to PAPS (3′-phosphoadenosine-5′-phosphosulphate). It is then transferred to a thiol carrier (RSH) and reduced to the −2 oxidation state. In contrast to nitrate assimilation, where the various intermediates are present free in the cytoplasm, sulphur remains attached to a carrier during the reduction sequence. In a final step, the carrier-bound sulphide reacts with *O*-acetyl-serine to form cysteine. Wilson, Bressan and Filner (1978) have suggested that under conditions when availability of this or other endogenous sulphide acceptors is limiting the rate of cysteine synthesis, the volatilization of H_2S could serve as a mechanism for removing excess reduced sulphur. Such volatilization has been observed from vascular plants (Winner *et al.*, 1981), but its possible occurrence in marine algae has yet to be investigated.

Cysteine serves as the starting compound for the biosynthesis of all other sulphur metabolites, especially the sulphur-containing amino acids homocysteine and methionine (Figure 3.3). Cysteine and methionine are the major sulphur amino acids in plants and usually represent a very large fraction of the sulphur content of biological materials (Giovanelli, Mudd and Datko, 1980). Glutathione (L-glutamyl-L-cysteyl-L-glycine) plays a variety of biochemical roles, including redox transfer reactions and the removal of H_2O_2 in chloroplasts. Methionine reacts with ATP to form *S*-adenosyl-methionine (SAM), the most important methyl group donor in methyl group transfer reactions in plants and algae. Transfer of a methyl group from SAM to methionine yields *S*-methyl-methionine, the precursor of dimethylsulphide in plants. In some marine algae, dimethylsulphonium propionate (DMSP) is formed in a multistep process from methionine.

Table 3.1. Biochemical origin of volatile sulphides produced in soils by the microbial degradation of organic matter under aerobic and anaerobic conditions (Warneck, 1988)

Volatile	Biochemical precursors
H_2S	Proteins, polypeptides, cystine, cysteine, gluthathione
CH_3SH	Methionine, methionine sulphoxide, methionine sulphone, *S*-methyl cysteine
CH_3SCH_3	Methionine, methionine sulphoxide, methionine sulphone, *S*-methyl cysteine, homocysteine
CH_3SSCH_3	Methionine, methionine sulphoxide, methionine sulphone, *S*-methyl cysteine
CS_2	Cysteine, cystine, homocysteine, lanthionine, djenkolic acid
COS	Lanthionine, djenkolic acid

Information about the biochemical precursors of volatile sulphur compounds has been obtained from artificial culture studies (reviewed by Kadota and Ishida, 1972), and from incubation studies of natural and amended soils (reviewed by Bremner and Steele, 1978). A compilation of the most prominent biochemical precursors of volatile sulphides produced in soils by microbial degradation of organic matter under aerobic and anaerobic conditions is given in Table 3.1.

3.3 BIOGENIC SULPHUR EMISSIONS FROM THE OCEANS

In recent years, the knowledge about biogenic sulphur emissions from the oceans has improved considerably and now there is a reasonably consistent picture of the types and amounts of sulphur gases emitted from the sea and their fate in the atmosphere. The main gases which are released from the ocean surface are dimethylsulphide (DMS), H_2S, carbon disulphide (CS_2), and carbonyl sulphide (COS). In this section, we concentrate on emissions from open, oceanic waters. Biogenic sulphur fluxes from salt marshes are discussed by Giblin and Wieder (this volume).

3.3.1 DIMETHYLSULPHIDE

In open ocean waters, DMS is the predominant volatile sulphur compound, a result of its association with phytoplankton production (Barnard *et al.*, 1982; Nguyen, Bergeret and Lambert, 1984; Bates *et al.*, 1987; Turner *et al.*, 1988; Turner, Malin and Liss, 1989). The main precursor of DMS in algae is DMSP, a ternary sulphonium compound that is thought to be generated from methionine (Challenger, 1951). Although DMSP is known to undergo enzymatic cleavage to DMS in the living algal cell, not all DMS appears to be released directly. Many observations suggest that DMS in aquatic environments derives to a significant extent from the bacterial decomposition of DMSP leaked from aged cells, or from zooplankton grazing on these cells. For example, as discussed by Bremner and Steele (1978), very high concentrations of DMS have been observed during the bacterial putrefraction of algae following algal blooms.

Figure 3.4 shows a typical vertical distribution of particulate (intracellular) DMSP, dissolved DMSP and DMS, and chlorophyll (an indicator of phytoplankton biomass) in the marine water column for the example of data from the northwestern Atlantic. The vertical distribution of DMS and DMSP in seawater as shown in Figure 3.4 is typical for these compounds as well as for a number of other phytoplankton metabolites, e.g. dimethylsulphoxide (DMSO) and the methylarsenates (Andreae, 1979, 1980). The characteristic

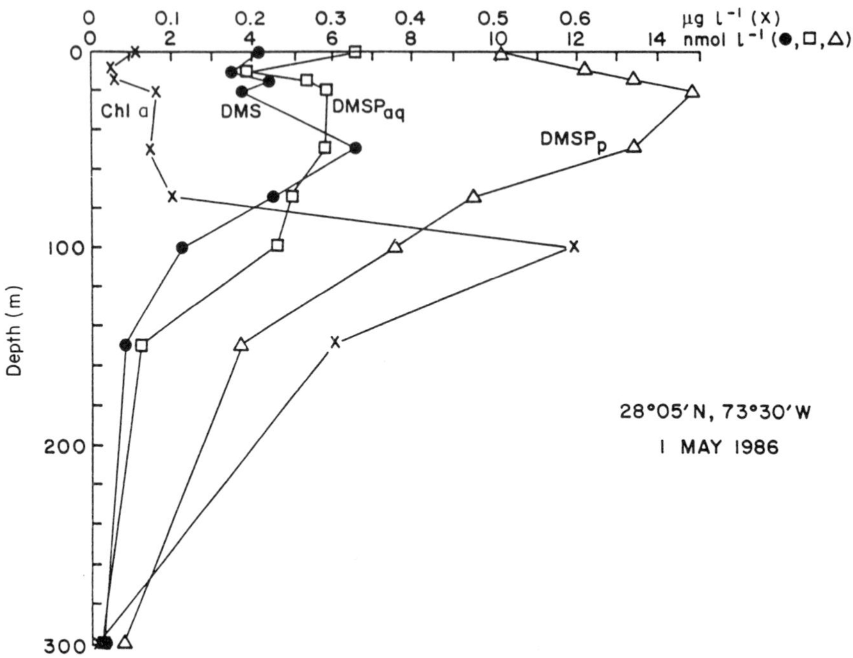

Figure 3.4. Typical vertical distribution of particulate DMSP, dissolved DMSP and DMS during the April/May 1986 cruise of R/V *Columbus Iselin* in the northwestern Atlantic Ocean. The vertical distribution of chlorophyll a is also included because it indicates the stratification of phytoplankton biomass

features of this distribution are a maximum at or a few metres below the sea surface, and a sharp decrease in DMS concentration near the level of 1% light transmission. This depth represents the lower limit of growth for most phytoplankton species.

Volatile substances like DMS are transferred across the air/sea interface by a combination of molecular and turbulent diffusion processes, which are still poorly understood and for which no entirely satisfactory physical and mathematical models are available. For a discussion of the state of the art in this field, see the review by Liss and Merlivat (1986). The sea-to-air flux is proportional to the air/sea concentration difference. The atmospheric concentration of DMS is several orders of magnitude below the value in equilibrium with seawater predicted by Henry's law (Andreae *et al.*, 1985). It can therefore be ignored for the purpose of estimating the sea-to-air concentration gradient and only the concentration of DMS in seawater is required to estimate the emission flux of DMS.

Table 3.2 summarizes the distribution of DMS in the world's oceans as measured on numerous cruises conducted by several groups (Barnard *et al.*,

Table 3.2. Concentrations and fluxes of DMS, CS_2 and COS for the world oceans

Biogeographic region	Area (10^6 km^2)	Mean concentration ($pmol\ l^{-1}$)			Flux/area ($nmol\ m^{-2}\ day^{-1}$)			Total flux ($Gmol\ a^{-1}$)		
		DMS	CS_2	COS	DMS	CS_2	COS	DMS	CS_2	COS
Oligotrophic (tropical/ low productivity)	148	2400		11.3	4 000–11 000		14.0	200–600		0.8
Temperate	83	2100	16	20.3	3 300– 9 900	45	45.0	100–300	5.1	1.4
Upwelling (coastal and equatorial)	86	4900		24.1	6 300–22 200		64.0	200–700		2.0
Coastal/shelf	49	2800	33	95.0	5 500–10 100	90	373.0	100–200	1.6	6.7
Mean		3000	18	27.6		Total		600–1800	6.7	10.9
In terms of Tg: (S) a^{-1}						Total		19–58	0.21	0.35

1982; Andreae, Barnard and Ammons, 1983; Cline and Bates, 1983; Andreae and Barnard, 1984; Nguyen, Bergeret and Lambert, 1984; Bingemer, 1984; Bates *et al.*, 1987). Table 3.2 is based on the compilation of Andreae (1986). The data are organized by biogeographical regions as defined by Koblentz-Mishke, Volkevinsky and Kabanova (1970); averages for each of these regions are used together with an estimate of their areal extent for the prediction of the flux of DMS from each region. The data base used for Table 3.2 contains no measurements from the Southern Ocean; recent work by Berresheim (1987) has shown, however, that oceanic emissions of DMS in this region are similar to those found in temperate regions.

To obtain the DMS transfer velocities used in the flux calculations in Table 3.2, the radon transfer velocities of Peng *et al.* (1979) and of Smethie *et al.* (1985) were adjusted by assuming that the transfer velocity is proportional to the square root of the diffusivity. If we use the global average $^{14}CO_2$ transfer velocity (*c.* 21 cm h^{-1}; Liss and Merlivat, 1986), we obtain a significantly higher flux: a mean estimate of 51 instead of 38 Tg (S) a^{-1} (1.6 instead of 1.2 Tmol a^{-1}). This is probably due to the fact that the radon transfer velocity is based almost entirely on summer data, when wind speeds are lower. In view of the extensive data on DMS concentrations in the surface ocean (Table 3.2), it appears that the major error in estimating the sea-to-air flux of DMS rests in the uncertainties associated with the use of the 'stagnant-film' model, and in particular with the assumption of the transfer velocities. This uncertainty may be as large as a factor of two.

There are only relatively small differences in total DMS fluxes from different oceanic regions (Table 3.2). One reason is that the tropical regions have relatively high DMS concentrations year-round, whereas in temperate regions, especially the temperate coastal areas, there is a pronounced seasonality with quite low values during the cold season. Secondly, the large areas of low and moderate biological productivity contribute amounts of DMS to the atmosphere comparable to those from the relatively small regions of high productivity in the upwelling regions and the coastal areas. This is in contrast to earlier views (Graedel, 1979), which had assumed that the biogenic sulphur flux from the oceans would be dominated by localized 'hot spots' of biological productivity.

We find that the estimate for the global ocean DMS flux has by now become very stable, in the sense that the addition of new data even from a number of different groups does not change it significantly. Whereas the number of data points in Table 3.2 is about 2.5 times greater than in a comparable table in Andreae and Raemdonck (1983), the estimate for the global mean DMS concentration has only changed from 3.2 to 3.1 nmol l^{-1} (102 to 99 ng l^{-1}), and that for the global flux remains unchanged at *c.* 38 Tg a^{-1} (1.2 Tmol a^{-1}). Using a data set from the Pacific Ocean only, Bates *et al.* (1987) obtain a lower mean DMS concentration (*c.* 1.8 nmol l^{-1}; 58 ng l^{-1}), and a correspondingly lower flux of 16 Tg a^{-1} (0.5 Tmol a^{-1}) with an estimated uncertainty of a factor of two. In view of the large uncertainties associated with the 'stagnant-film' model, it seems very important that independent methods should be developed to test the predictions based on this model. However, alternative methods to determine the flux, e.g. the eddy-correlation or gradient techniques, still face large experimental difficulties. No rapid-response sensor which would make the eddy-correlation technique possible is available for DMS, or in fact any of the reduced sulphur gases. The gradient method has been used on board ship by Bingemer (1984) and by Nguyen, Bergeret and Lambert (1984) by sampling at different levels above the waterline. While the results compare well to predictions from gas transfer calculations, they may contain substantial error due to the influence of the ship on the air flow characteristics. Due to the difficulty of simulating a realistic wave climate inside a flux chamber, direct measurements of sulphur gas fluxes across the air–sea interface by the chamber technique have not been attempted.

3.3.2 HYDROGEN SULPHIDE

There are few data on the concentration of dissolved H_2S in surface seawater, and a small number of reliable measurements of H_2S in the marine atmosphere. Therefore, the air–sea exchange flux of this compound is difficult to estimate. H_2S is oxidized rapidly in oxygenated seawater; chemical half-lives on the order of a few hours are reported (Almgren and Hagstrom, 1974); other workers have found values as high as 50 hours (Chen and Morris, 1972). The most reliable measurements appear to be those of Millero *et al.* (1987), who found a half-life of 26 hours at 25° C. Dissolved sulphide concentrations of 0.1 to 1.0 nmol l^{-1} (3 to 32 ng l^{-1}) have been found in surface seawater from the Atlantic Ocean (T.W. Andreae *et al.*, 1991).

The production mechanism of this H_2S remains unclear, but its vertical distribution in the ocean suggests that bacterial reduction in microbial microenvironments may play an important role (Jørgensen, Hansen and Ingvorsen, 1978). It must, however, be remembered that biological processes in plants can result in the production and release of substantial amounts of H_2S even in the presence of oxygen. This is especially true in the presence of high ambient sulphate concentrations, as is the case in seawater. The hydrolysis of COS may also result in the production of significant amounts of H_2S in seawater (T.W. Andreae *et al.*, 1991).

H_2S has been observed in the marine atmosphere at levels of a few pptv to a few tens of pptv (Slatt *et al.*, 1978; Delmas and Servant, 1982; Herrmann and Jaeschke, 1984; Cooper and Saltzman, 1987; T. W. Andreae *et al.*, 1991). All authors used the same method of trapping the H_2S on $AgNO_3$-impregnated filters and subsequent determination of the enriched Ag_2S by the quenching of the fluorescence of fluorescein mercuric acetate (Jaeschke and Herrmann, 1981). However, Cooper and Saltzman (1987) found a positive interference in the determination of H_2S by this method and suggested that the mean concentration of H_2S in the marine boundary layer does not exceed 10 pptv (Saltzman and Cooper, 1988; T.W. Andreae *et al.*, 1990). At these levels, H_2S in the atmosphere is near thermodynamical equilibrium with the concentrations in surface seawater (T.W. Andreae *et al.*, 1991).

To obtain an estimate of the rate of H_2S oxidation in the marine atmosphere, we can simply use an average concentration of 6 pptv (Saltzman and Cooper, 1988; T.W. Andreae *et al.*, 1990, 1991) with a scale height of 2 km, a diurnal average OH concentration of 8×10^5 mol cm^{-3}, and the measured reaction rate for the oxidation of H_2S by OH (5×10^{-12} cm^{-3} mol^{-1} s^{-1}; Cox and Sheppard, 1980). The resulting estimate, 0.5 Tg (S) a^{-1} (0.015 Tmol a^{-1}), is an upper limit for the sea-to-air flux of H_2S, and is much smaller than the DMS flux of *c.* 38 Tg (S) a^{-1} (1.2 Tmol a^{-1}). Based on their measurements in the Caribbean and the Gulf of Mexico, Saltzman and Cooper (1988) suggest that the oxidation of H_2S accounts for only 11% of the production of biogenic non-seasalt sulphate in the remote marine boundary layer, the rest being due to the oxidation of DMS. It is not clear, however, if the source of the H_2S found in the marine troposphere is necessarily the ocean surface or if other processes could be responsible for its presence. For example, advection from coastal regions, where H_2S is emitted from salt marshes (Giblin and Wieder, this volume), may supply some of this H_2S. This is supported by a correlation between H_2S and radon observed over the northern Atlantic (T.W. Andreae *et al.*, 1991). Based on simultaneous measurements of H_2S in seawater and air, T.W. Andreae *et al.* (1991) estimate that the oceanic emission of H_2S is less than 0.3 Tg (S) a^{-1} (0.01 Tmol a^{-1}). On the other hand, McElroy, Wofsy and Sze (1980) have speculated that atmospheric reactions of COS and CS_2 with OH radical could produce the necessary amounts of H_2S. However, this suggestion has not yet been verified experimentally.

3.3.3 CARBON DISULPHIDE

The presence of CS_2 in seawater was first observed by Lovelock (1974), who measured an average concentration of 14 pmol CS_2 l^{-1} (0.45 ng (S) l^{-1}) in 35 samples taken in the open Atlantic Ocean. Inshore values were about an order of magnitude higher. Turner and Liss (1985) also report the presence of high levels of CS_2 in coastal waters of England, but they give quantitative information for only a few samples with values near 300 pmol CS_2 l^{-1} (9.6 ng (S) l^{-1}). They found substantially higher concentrations in the low salinity region of an estuary (up to *c.* 2 nmol CS_2 l^{-1}; 64 ng (S) l^{-1}). It is possible that much of the CS_2 found in coastal waters is the result of the diffusion of this substance from the porewater of the underlying sediments. This would be consistent with the relatively high concentrations and fluxes of CS_2 observed in coastal marsh environments (Adams *et al.*, 1981a,b; Steudler and Peterson, 1984). CS_2 could be formed there either by fermentation reactions of organosulphur compounds or by 'pulp-mill' type reactions of terrestrial plant matter, including dead salt-marsh grasses with dissolved polysulphides originating from bacterial dissimilatory sulphate reduction (see Luther and Church, this volume). Kim and Andreae (1987) have recently determined CS_2 in open ocean and coastal seawater from the North Atlantic, and have observed mean concentrations of 16 ± 8 and 33 ± 19 pmol CS_2 l^{-1} respectively (0.51 ± 0.25 and 1.1 ± 0.6 ng (S) l^{-1}) (Table 3.2), somewhat higher than Lovelock's results.

From these data, we estimate a flux of *c.* 0.23 Tg (S) a^{-1} (0.007 Tmol a^{-1}) in the form of CS_2 from the world ocean surface, about 0.6% of the DMS flux. The photochemical oxidation of CS_2 produces one molecule each of SO_2 and COS per molecule of CS_2 oxidized. Thus, the marine emission of CS_2 provides a significant indirect source of COS, whereas it is clearly inconsequential as a source of tropospheric SO_2 or sulphate.

3.3.4 CARBONYL SULPHIDE

Carbonyl sulphide (COS) is the most abundant atmospheric sulphur species in the remote troposphere, with an average concentration near 500 pptv. Due to its low reactivity in the troposphere, and its correspondingly long residence time (of the order of one year), it is the only sulphur compound which can enter the stratosphere (with the exception of SO_2 injections during violent volcanic eruptions). The input of COS is considered responsible for the maintenance of the sulphate aerosol layer in the stratosphere during volcanically quiescent periods (Crutzen, 1976; Jaeschke, Schmidt and Georgii, 1976). Therefore, even a relatively small COS source flux can be of considerable importance in atmospheric chemistry.

COS is present in surface seawater at concentrations of about 0.03 to 1.0 nmol l^{-1} (1 to 32 ng l^{-1}) (Rasmussen, Khalil and Hoyt, 1982; Ferek and Andreae, 1983, 1984; Turner and Liss, 1985; Andreae, 1991b; T.W. Andreae *et al.*, 1991). The observed concentrations are almost always higher than the equilibrium concentration relative to the overlying atmosphere, so that a net sea-to-air flux exists essentially from the entire ocean surface. Johnson (1981) has speculated that the ocean should be a sink for COS due to its hydrolysis at the slightly alkaline pH of seawater. This suggestion is clearly not supported by the measured COS supersaturation ratios across the air–sea interface. Pronounced diurnal variations of the COS concentration in surface seawater (Figure 3.5) suggest that COS is produced there by photochemical reactions (Ferek and Andreae, 1984). Laboratory experiments with seawater and with

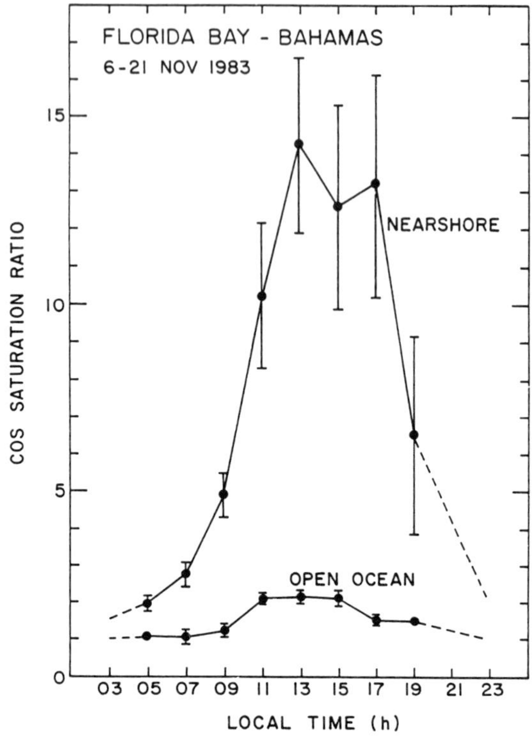

Figure 3.5. Mean diurnal variation of COS in surface seawater during a cruise of R/V *Bellows* in November 1983. The concentration of COS is indicated as a saturation ratio, i.e. the ratio between the measured concentration and the concentration in equilibrium with ambient air with about 500 pptv COS

solutions of organosulphur compounds in distilled water showed that seawater sulphate does not participate in the reaction, and that only the presence of dissolved organic sulphur compounds, dissolved oxygen, and light are necessary to produce COS (Ferek and Andreae, 1984; Zepp and Andreae, 1989, 1990). Carbonyl sulphide is formed by irradiation of a variety of organic sulphur compounds commonly found in biological materials, e.g. cysteine, methionine, glutathione, and dimethylsulphonium propionate. The mechanism to this reaction is not known at this time, but it is likely that short-lived, photochemically produced radicals (e.g. OH) are involved. The presence or absence of living microorganisms like planktonic algae or bacteria is of no influence on the rate of formation of COS in seawater. It appears that the role of organisms in the production of COS in seawater is limited to the synthesis of dissolved organic sulphur compounds which are then abiotically photolysed to COS. The dependence of the rate of COS formation on the concentration of dissolved organic sulphur in seawater is reflected by the difference between the COS supersaturation measured in coastal and open ocean waters (Figure 3.5).

An attempt to obtain a representative estimate of the sea-to-air flux of COS is also presented in Table 3.2, where the ocean surface is divided into the same biogeographic regions as used for DMS. Then, based on the diurnally averaged data on the supersaturation of COS in the surface seawater relative to the overlying atmosphere, and the average temperature of the surface ocean in these regions, the flux of COS across the air–sea interface for these regions was calculated (the piston velocities for COS are a factor of 1.3 higher than for DMS, due to the higher diffusivity of COS). We see that, in contrast to DMS, the flux of COS is dominated by the high productivity regions, especially the coastal and shelf areas. Due to the low levels of COS in oligotrophic areas, they contribute little to the global flux, which is estimated to be *c.* 0.35 Tg (S) a^{-1} (0.11 Tmol a^{-1}), similar to previous flux estimates (Rasmussen, Khalil and Hoyt, 1982; Ferek and Andreae, 1983).

3.4 BIOGENIC EMISSION FROM FRESHWATER WETLANDS AND INLAND SOILS

In contrast to the oceanic emissions, our knowledge about biogenic sources of volatile sulphur compounds from the continents is extremely limited. At this time, the terrestrial biogenic sources represent the largest uncertainty in the global atmospheric sulphur budget (Andreae, 1985a, 1986). This uncertainty is related to several problems: (1) the diversity of the sulphur species emitted, which include hydrogen sulphide (H_2S), dimethylsulphide (DMS), dimethyldisulphide (DMDS), methylmercaptan (MeSH), carbon disulphide (CS_2), carbonyl sulphide (COS), and others; (2) the diversity of terrestrial ecosystems, including freshwater wetlands; (3) the patchiness in time and space of determined sulphur gas fluxes from terrestrial ecosystems; (4) analytical problems with some of the most important species, especially H_2S, as well as problems with flux measurement methodology; and (5) the logistical problem of obtaining representative measurements in several important regions, especially with respect to variations caused by climatical and microbiological circumstances at the chosen measuring site.

As an example, in Figure 3.6 the diurnal variation of measured H_2S flux is shown, as determined by Jaeschke, Claude and Herrmann (1980) in a freshwater marshland region in northern Germany. After sunset the H_2S flux rose and reached a maximum shortly before sunrise. This variation was observed on several occasions during intensive studies of H_2S fluxes in German wetlands. In the example shown in Figure 3.6, the H_2S flux varied during the 24 h measuring period by a factor of 1000. If the flux had been determined only during one or two hours, it would have been highly misleading.

The concentrations of dissolved volatile sulphur compounds in freshwater bodies tend to be much lower than those in the surface layer of the oceans (Adams *et al.*, 1981a,b; Froelich *et al.*, 1985; Iverson, Nearhoof and Andreae, 1989). However, based on measurements of DMS concentrations in waters from freshwater wetlands in Ontario, Nriagu, Holdway and Coker (1987) recently proposed a DMS emission rate of 150 ng (S) m^{-2} min^{-1} (4.7 nmol m^{-2} min^{-1}) for such ecosystems, and suggested that H_2S emissions of a similar magnitude may be occurring. These fluxes may, however, have been overestimated since the exchange coefficient appropriate for the hydrodyna-

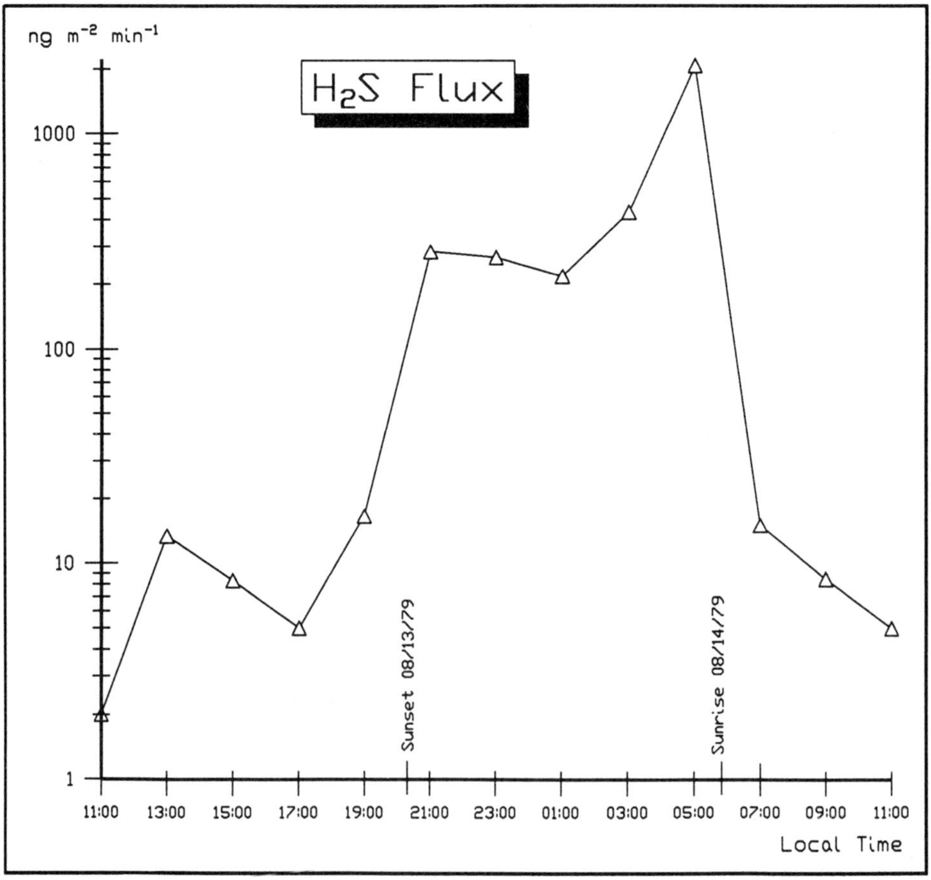

Figure 3.6. Diurnal variation of the H_2S flux measured at marshland soils in the Ems River region of northern Germany

mic regime of a bog is not known, and a rather high value was selected. It is also possible that the very high sulphate loading from anthropogenic pollution may have enhanced sulphur fluxes from these wetlands. Recently, Staubes (1986) has studied the temperature dependence of biogenic sulphur exhalation at several sites over Germany. These studies suggest extremely low emissions of sulphur volatiles from bog and wetlands in cold northerly regions.

Most studies of biogenic sulphur emissions from terrestrial ecosystems have concentrated on H_2S. Delmas *et al.* (1980) measured an average flux of 134 ng (S) m^{-2} min^{-1} (4.2 nmol m^{-2} min^{-1}) from oxic lawn soils in France. Jaeschke, Claude and Herrmann (1980) found fluxes between 2 and 2000 ng (S) m^{-2} min^{-1} (0.06 to 63 nmol m^{-2} min^{-1}) from marshland soils in the Ems River region of northern Germany (Figure 3.6). Somewhat higher average fluxes were determined from tropical soils in the Ivory Coast by Delmas *et al.* (1980). They estimated the mean annual flux from these soils to be in the range of 600 to 1700 ng (S) m^{-2} min^{-1} (19 to 53 nmol m^{-2} min^{-1}).

In the first comprehensive study of sulphur gas emissions from inland soils, Adams *et al.* (1981a,b) determined the fluxes of H_2S, COS, MeSH, DMS, CS_2, and dimethyldisulphide (DMDS) from soils at 35 sites in the eastern and southeastern United States. Their flux data showed large variability between sites and between sampling periods at the same site (several orders of magnitude). For non-saline soils, fluxes ranged from 3.8 to 640 ng (S) m^{-2} min^{-1} (0.12 to 20 nmol m^{-2} min^{-1}). The average flux from inland soils was *c.* 38 ng (S) m^{-2} min^{-1} (1.2 nmol m^{-2} min^{-1}) for the latitude zone from 25° N to 47° N. Most of the flux from inland soils was in the form of H_2S (66%); the rest was accounted for by COS (13%), CS_2 (13%), DMS (7%), and a trace of DMDS (2%). MeSH was not found to be emitted from non-saline soils.

Most recent studies have shown much lower flux estimates from soils. In a large study of soil emissions conducted partly at the same sites as had been used by Adams and co-workers, values about ten times lower than those reported previously were found by Goldan *et al.* (1987) and Lamb *et al.* (1987). The differences between these values and those reported previously are largely attributed to experimental problems in the older studies. Staubes, Ockelmann and Georgii (1986) measured DMS and CS_2 fluxes from soils in Germany. For soil temperatures above 21° C, they observed DMS fluxes in the range of 6.4 to 16 ng (S) m^{-2} min^{-1} (0.2 to 0.5 nmol m^{-2} min^{-1}) and CS_2 fluxes in the range of 3.2 to 6.4 ng (S) m^{-2} min^{-1} (0.1 to 0.2 nmol m^{-2} min^{-1}). In the rainforest at La Selva, Costa Rica, Haines *et al.* (1984, pers. comm. quoted by Delmas and Servant, 1988) measured a mean flux of 4 ng (S) m^{-2} min^{-1} (0.1 nmol m^{-2} min^{-1}) for DMS, with possibly some contribution by methylmercaptan.

Andreae and Andreae (1988) made some preliminary measurements of soil emissions in wet tropical forests using a chamber technique at the Ducke

Reserve near Manaus, Brazil during the Amazon Boundary Layer Experiment (ABLE-2). During the 1985 dry season (ABLE-2A), fluxes of DMS were 12.8 ± 1.3 ng (S) m^{-2} min^{-1} (0.4 ± 0.04 nmol m^{-2} min^{-1}), fluxes of MeSH were 1.3 ± 0.3 ng (S) m^{-2} min^{-1} (0.04 ± 0.009 nmol m^{-2} min^{-1}), and fluxes of H_2S were 2.2 ± 1.0 ng (S) m^{-2} min^{-1} (69 ± 0.03 nmol m^{-2} min^{-1}). These values combine to a total flux of short-lived sulphur species (ignoring CS_2, COS, etc.) of 16 ± 2.6 ng (S) m^{-2} min^{-1} (0.5 ± 0.08 nmol m^{-2} min^{-1}), in good agreement with the soil emission fluxes obtained by the authors cited above. During the 1987 wet season (ABLE-2B), even lower values were observed. However, due to the small number of measurements and sites involved, our ability to extrapolate from these data is very limited.

3.5 EMISSIONS BY LIVING PLANTS, AND COMBINED MEASUREMENTS OF FLUXES FROM PLANTS AND SOILS

The emission of sulphur gases from vegetation, including trees, is well established. The emission of H_2S from plant leaves is light-dependent: it increases strongly with the intensity of light flux, and drops to very low values in the absence of light (Wilson, Bressan and Filner, 1978; Filner *et al.*, 1984). H_2S emission also increases as a result of root injury, high levels of atmospheric SO_2, and increased levels of soil-water sulphate or bisulphite. Based on studies with various crop plants, Filner *et al.* (1984) have estimated a worldwide emission of volatile sulphur from plants of the order of 7.4 Tg (S) a^{-1} (0.23 Tmol a^{-1}), whereas Winner *et al.* (1981) gave an estimate of 54 Tg (S) a^{-1} (1.7 Tmol a^{-1}). Adjusting these values to the leaf areas and biomass values characteristic of tropical forests, fluxes of about 160 to 1600 ng (S) m^{-2} min^{-1} (5 to 50 nmol m^{-2} min^{-1}) can be predicted. Filner *et al.* (1984) also report the emission of MeSH by plants under some conditions. The emission of DMS from trees was observed by Lovelock, Maggs and Rasmussen (1972); from their data a flux of 32 to 640 ng (S) m^{-2} min^{-1} (1 to 20 nmol m^{-2} min^{-1}) can be estimated for a forest with a leaf mass of *c.* 2 kg (dry weight) m^{-2}. In a recent study, Lamb *et al.* (1987) found emission fluxes in the range of 6.4 to 64 ng (S) m^{-2} min^{-1} (0.2 to 2 nmol m^{-2} min^{-1}) from various crops and trees. The mean sulphur flux from deciduous trees was 29 ng (S) m^{-2} min^{-1} (0.9 nmol m^{-2} min^{-1}) during summer conditions (mean sample temperature: 29.5° C). Emissions from soybeans and corn were dominated by DMS, while deciduous trees emitted H_2S and DMS in similar amounts. These fluxes are comparable to or greater than sulphur gas fluxes which have been reported from vegetated inland soils.

Several authors have used micrometeorological or budget techniques which estimate a total flux from the soil-vegetation system. In temperate latitudes, Jaeschke, Claude and Herrmann (1980) measured an H_2S flux of 5.2 ng (S)

m^{-2} min^{-1} (0.17 nmol m^{-2} min^{-1}) from wetland areas near the German coast. In the Ivory Coast, Delmas *et al.* (1980) obtained H_2S fluxes of 208 and 5120 ng (S) m^{-2} min^{-1} (6.5 and 160 nmol m^{-2} min^{-1}) from forests on oxic and flooded anoxic soils, respectively. Delmas and Servant (1983) reported fluxes of 960 ng (S) m^{-2} min^{-1} (30 nmol m^{-2} min^{-1}) in the dry season and 4480 ng (S) m^{-2} min^{-1} (140 nmol m^{-2} min^{-1}) in the wet season from equatorial Africa, using aircraft data and a budget method.

Recent laboratory studies on corn, alfalfa, and wheat plants showed DMS to be the main sulphur species emitted, followed by variable amounts of MeSH, H_2S, and CS_2. The emission rates for all sulphur gases increased with temperature and illumination (Fall *et al.*, 1988). Comparison of the emission of COS from bare and vegetated soils showed that the vegetation canopy was a sink for this gas, reducing the flux of COS from soils to the atmosphere (Goldan *et al.*, 1987; Fall *et al.*, 1988). The uptake of COS by vegetation has been proposed as the major global sink for this compound (Brown and Bell, 1986).

The fluxes of DMS, H_2S, and MeSH from the soil/plant system of the Amazon forest were determined during the 1985 dry season by a gradient/flux technique (Andreae and Andreae, 1988). The mean fluxes were 26 and 13 ng (S) m^{-2} min^{-1} (0.8 and 0.4 nmol m^{-2} min^{-1}) for DMS and MeSH, respectively. For H_2S, an upper limit of 50 ng (S) m^{-2} min^{-1} (1.6 nmol m^{-2} min^{-1}) was found. During the wet season a similar approach resulted in estimates of 6.4 and 35 ng (S) m^{-2} min^{-1} (0.2 and 1.1 nmol m^{-2} min^{-1}) for the emission fluxes of DMS and H_2S respectively (Andreae *et al.*, 1990a). Since these fluxes are much larger than the observed soil fluxes, most of the sulphur emissions must come from the plant canopy. In the case of DMS, the vertical profiles through the forest canopy suggest that emission takes place only during the day-time, and that the forest canopy acts as a sink for this species during the night.

In Figure 3.7, the DMS fluxes obtained from the gradient measurements are compared with fluxes estimated by a model which incorporates transport and photochemical oxidation of DMS. Both model and measurements show the same diel variation of the DMS flux, and suggest that the fluxes at night are low. In view of the uncertainties of both the measurements and the model, the observed and predicted values agree reasonably well. Sulphur gas measurements in and over the Congo rainforest suggest that similar emission fluxes prevail in the African humid tropics (Bingemer *et al.*, 1991).

Assuming that a sulphur emission flux of about 30 to 100 ng (S) m^{-2} min^{-1} (0.9 to 3.1 nmol m^{-2} min^{-1}) is representative of wet tropical South America and that this flux can be extrapolated to the wet tropics world-wide, we can assess the importance of the wet tropics in the global atmospheric sulphur cycle. Using 14.5×10^{12} m^2 as an estimate for the area covered by tropical forests (Cicerone *et al.*, 1984), we predict a sulphur emission flux of 0.23 to

0.76 Tg (S) a^{-1} (0.007 to 0.024 Tmol a^{-1}) from this region. If the rest of the vegetated land surface (88×10^{12} m^2) has an emission rate as estimated by Guenther, Lamb and Westberg (1989) for the United States (including wetlands), the sulphur emissions from this area are about 3.5 Tg (S) a^{-1} (0.11 Tmol a^{-1}). For the total land area, this corresponds to 3.8 to 4.3 Tg (S) a^{-1} (0.12 to 0.13 Tmol a^{-1}), considerably less than the total oceanic flux of about 38 Tg (S) a^{-1} (1.2 Tmol a^{-1}). It is also small relative to the anthropogenic emissions from fossil fuel burning (about 96 Tg (S) a^{-1}), so that increasing development can be expected to significantly perturb the atmospheric sulphur cycle in tropical regions (McDowell, 1987).

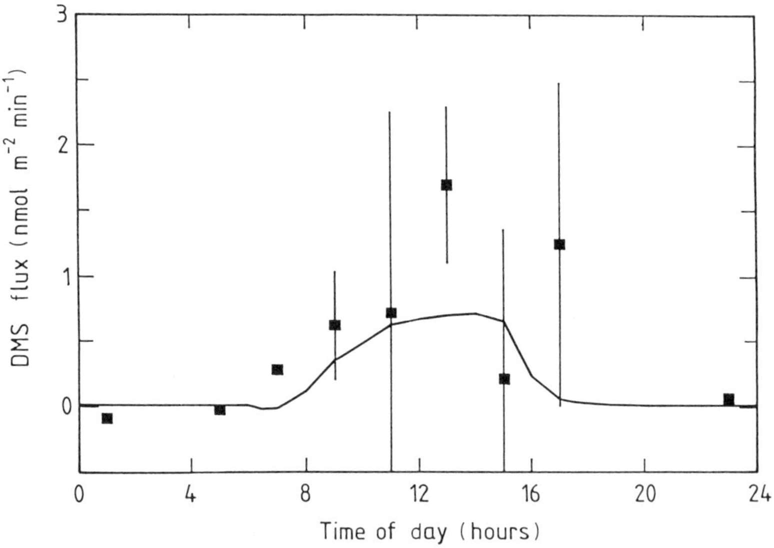

Figure 3.7. Observed DMS fluxes from the Amazon forest near Manaus, Brazil, during the ABLE-2B wet season experiment in July/August 1987, obtained by gradient/flux measurements compared with fluxes estimated by a model incorporating transport and photochemical oxidation of DMS (solid line)

If these observations are representative for the sulphur cycle over remote continents world-wide, an assumption which is supported by the existing data for sulphate deposition (Galloway, 1985; Andreae *et al.*, 1990b, and references therein), the biogenic sulphur emission from continents must be near the low end of the range given in Table 3.4, and emissions from land biota must play a subordinate role in the continental sulphur cycle over the continents world-wide, being overshadowed by marine and anthropogenic sources essentially everywhere.

3.6 SULPHUR EMISSIONS FROM BIOMASS BURNING

Biomass burning cannot be classified clearly as either a natural or an anthropogenic source. Most of the biomass burning takes place as a result of agricultural practices (including tropical deforestation) and of fuel wood burning (totalling *c.* 8700 Tg dry matter per year; Andreae, 1991a). Since most of the burning occurs in relatively undeveloped regions, especially in the tropics, we briefly discuss its importance as a sulphur source.

The average sulphur content of land plants is *c.* 0.2% (Bowen, 1979), which corresponds to a potential sulphur flux of *c.* 17 Tg (S) a^{-1} (0.53 Tmol a^{-1}) at a burning rate of 8700 Tg (dry matter) a^{-1}. Not all of this sulphur enters the atmosphere, since a significant fraction is retained in the ash (35–67% in the case of the savanna fires described by Delmas (1982)). The chemical forms in which sulphur is emitted from fires and their relative proportions are still quite uncertain. It is to be assumed that SO_2 is a major product, but H_2S, COS, and sulphate aerosol have also been observed. SO_2 was found to be more abundant than particulate sulphur in emissions from a wood-burning stove (Cooper, 1980). There are no studies at this time in which sulphur gases and aerosols have been measured together in biomass burning emissions except from highly polluted regions. The data from controlled burns in the western United States cannot be considered globally representative due to the high anthropogenic and marine sulphur deposition in this region, much of which appears to be remobilized in the burning process (Hegg *et al.*, 1987).

The global emission rate for COS from biomass burning has been estimated to be 0.09 Tg (S) a^{-1} (0.003 Tmol a^{-1}), only a minor fraction of the sulphur flux from fires, but a significant component in the atmospheric cycle of COS (Crutzen *et al.*, 1979; Andreae, 1991a). Preliminary studies from our laboratory show that only a very small fraction of the fuel sulphur is emitted in the form of H_2S during biomass burning. The global emission of SO_2 from biomass burning can be estimated based on the total amount of biomass burned and an emission factor, i.e. the amount of SO_2 emitted per unit of fuel burned. Based on our studies of biomass burning in Brazil, we have obtained an emission factor of 0.32×10^{-3} (mol SO_2/mol C). From Delmas' (1982) data on the sulphur yield in savanna fires, an emission factor of 0.16×10^{-3} can be calculated, somewhat lower than our estimate.

The rate of sulphur emission from biomass burning, derived from measurements over Amazonia, corresponds to a global source flux of 2.8 Tg (S) a^{-1} (0.088 Tmol a^{-1}). This estimate is lower than previous ones, e.g. that of Andreae (1985a), who had estimated a global flux of 7 Tg (S) a^{-1} (0.2 Tmol a^{-1}), based on a global burning rate of 6.8 Tg dry matter per year, a sulphur content of 0.2 wt.%, and a burning yield of 50%. The value of 2.8 Tg (S) a^{-1} appears to be more reliable, however, since it is based on actual field measurements.

In comparison to the global emissions of sulphur from the land and ocean biota, *c*. 4 and 38 Tg (S) a^{-1} (0.13 and 1.2 Tmol a^{-1}) respectively, and from fossil fuel burning, biomass burning may be considered a minor source of atmospheric sulphur on a global scale. However, the fact that biomass burning occurs predominantly in tropical regions during a defined burning season suggests that it may be the dominant sulphur source to the atmosphere in the remote tropics during a significant part of the year. Observations from Africa (Lacaux *et al.*, 1988; Bingemer *et al.*, 1991) and from Amazonia (Andreae and Andreae, 1988; Andreae *et al.*, 1990a) appear to substantiate this hypothesis.

3.7 EXHALATION OF SULPHUR GASES BY VOLCANOES

Although volcanism is a geological phenomenon, sulphur gases emitted by volcanoes are ultimately also a product of biogenic sulphur reduction. As pointed out earlier, large quantities of reduced sulphur gases, especially H_2S, are not able to escape to the atmosphere, because they are precipitated by metal ions forming sulphide minerals in the sediment. Due to tectonical movement, sulphide-containing sedimentary rocks are subducted into the upper mantle of the earth from where volatile sulphur compounds are re-emitted to the atmosphere by volcanic activity. The exhalations contain sulphur mainly in the form of SO_2 and H_2S in the gas phase and sulphate in the particle phase. Thermodynamic equilibrium calculations by Heald, Naughton and Barnes (1963) indicate that in the anoxic environment of the magma, SO_2 is dominant at high temperatures, whereas H_2S is preferred at low temperatures. Accordingly, most active volcanoes are expected to release sulphur mainly as SO_2. Upon entering the atmosphere, both compounds are oxidized as they react with oxygen; H_2S is partially converted to SO_2, and SO_2 to sulphuric acid and sulphate.

A first estimate for the global rate of sulphur emissions from volcanoes was made by Kellog *et al.* (1972). Their assessment was based on the total volume of lava and ashes ejected during the time period from 1500 to 1914 according to Sapper (1927), the estimate by MacDonald (1955) of the weight of the gases evolved from fresh lava on Hawaii, and the analysis of Shepherd (1938), who found 10% by weight of the gases to be SO_2. The average flux of volcanic SO_2 derived in this manner is 0.75 Tg (S) a^{-1} (0.023 Tmol a^{-1}). The result actually is a lower limit to the real value, because it considers only SO_2 released during eruptions, whereas the much longer quiescent periods of volcanic activity are not taken into account. The same deficiency is inherent in later attempts by Friend (1973) and Cadle (1975) to improve on the estimate of Kellog *et al.* (1972).

In the meantime, more direct information has been obtained on volcanic SO_2 emissions by means of aircraft sampling of volcanic plumes (Jaeschke, Berresheim and Georgii, 1982), and by remote measurement techniques. Table 3.3 presents a compilation of observational data prepared by Berresheim and Jaeschke (1983). To determine average emission rates from these data requires a correlation with the intensity of volcanic activity at the time of observation. Berresheim and Jaeschke (1983) have performed the analysis and thereby achieved a more realistic assessment of the situation than was hitherto available. Five categories of volcanic activities were distinguished, namely eruptive, pre-eruptive, intra-eruptive, post-eruptive, and extra-eruptive phases. The first category was further subdivided into nine groups of differing intensities. According to Tsuya (1955) the intensity can be expressed by the total volume of ejected material, often reported in connection with the eruptive events. The nine categories of the intensity scale represent the following ranges: I $\leqslant 10^{-5}$ km^3, II = 10^{-5} to 10^{-4} km^3, etc., up to VIII = 10^1 to 10^2 km^3 and IX $\geqslant 10^2$ km^3. After this categorization, the world-wide frequency of eruptions in the period between 1961 and 1979 was considered and individual eruptions were classified by this intensity scale. On average, about 55 volcanoes undergo eruptions each year. This is about one-tenth of all active volcanoes currently in existence. By summing over the entire amount of SO_2 released within each intensity class, Berresheim and Jaeschke (1982) found an average emission rate of 0.5 Tg (S) a^{-1} (0.015 Tmol a^{-1}) for the time period considered. The value differs little from that of Kellog *et al.* (1972).

A much greater total emission rate was obtained, however, for the steadier, non-eruptive activities. Among some 520 live volcanoes, there are about 365 that may be classified as post-eruptive, and an additional 100 in the extra-eruptive phase. Volcanoes in the latter category may be neglected to a first approximation because they emit only a small fraction of the sulphur released by the others. The pre-eruptive and intra-eruptive phases are relatively short and may also be disregarded. According to Table 3.3, the average SO_2 flux for a single volcano in the post-eruptive activity class is roughly 3×10^5 kg (SO_2) day^{-1}. The use of this latter value leads to a global sulphur emission rate of 20 Tg (S) a^{-1} (0.63 Tmol a^{-1}).

Berresheim and Jaeschke (1983) adopted a more conservative attitude. On the basis of MacDonald's (1972) description of 516 active volcanoes, they assigned a source strength of the magnitude suggested by the average of the data in Table 3.3 to not more than 90 volcanoes. The remaining 275 volcanoes with post-eruptive activity were considered to produce only 5×10^4 kg (SO_2) day^{-1} each. The global sulphur flux from non-erupting volcanoes is then reduced to 7 Tg (S) a^{-1} (0.22 Tmol a^{-1}). The rate is still an order of magnitude greater than that associated with erupting volcanoes. Based on their own measurements of H_2S and sulphate in the plume of Mt Etna,

Table 3.3. Measured or estimated rates of SO_2 emissions from volcanoes

Authors	Volcano	Emission rate (10^3 kg (SO_2) day^{-1})	Activity[a]
Italy			
Haulet, Zettwoog and Sabroux (1977)	Etna	3740	Eruptive (I)
Zettwoog and Haulet (1978)	Etna	3325	Pre-eruptive
Zettwoog and Haulet (1978)	Etna	1130	Post-eruptive
Malinconico (1979)	Etna	1000	Intra-eruptive
Jaeschke, Berresheim and Georgii (1982)	Etna	142	Post-eruptive
Jaeschke, Berresheim and Georgii (1982)	Etna	955/1600	Pre-eruptive
Nicaragua			
Stoiber and Jepson (1973)	Telica	20	Post-eruptive
Stoiber and Jepson (1973)	Momotombo	50	Post-eruptive
Stoiber and Jepson (1973)	San Christobal	360	Post-eruptive
Stoiber and Jepson (1973)	Masaya	180	Pre-eruptive
Taylor and Stoiber (1973)	Cerro Negro	120[b]	Eruptive (I)
Taylor and Stoiber (1973)	Cerro Negro	2000[b]	Eruptive (II–IV)
Guatemala			
Stoiber and Jepson (1973)	Pacaya	260	Eruptive (I)
Stoiber and Bratton (1978)	Pacaya	300	Post-eruptive
Stoiber and Jepson (1973)	Santiaguito	420	Eruptive
Stoiber and Jepson (1973)	Santiaguito	300/1500	Eruptive (I)
Rose *et al.* (1973)	Fuego	10 000[b]	Eruptive (II–IV)
Stoiber and Jepson (1973)	Fuego	40	Post-eruptive
Crafford (1975)	Fuego	55 000[c]	Eruptive (V)
Crafford (1975)	Fuego	423	Intra-eruptive
Stoiber and Bratton (1978)	Fuego	300/1500	Eruptive (I)

Hawaii			
Naughton *et al.* (1975)	Kilauea	1280[d]	Eruptive (I–V)
Naughton *et al.* (1975)	Kilauea	280	Undetermined
Stoiber and Malone (1975)	Mauna Ulu	30	Extra-eruptive
Stoiber and Malone (1975)	Sulfur Banks	7	Extra-eruptive
Alaska			
Stith, Hobbs and Radke (1978)	St Augustine	86 400[c]	Eruptive (V)
Stith, Hobbs and Radke (1978)	St Augustine	432[c]	Post-eruptive
Stith, Hobbs and Radke (1978)	St Augustine	8640	Eruptive (II–IV)
Stith, Hobbs and Radke (1978)	St Augustine	86	Post-eruptive
Stith, Hobbs and Radke (1978)	Mt Martin	3	Extraeruptive
Washington State			
Hobbs *et al.* (1981)	St Helens	10/50	Pre-eruptive
Casadeval *et al.* (1981)	St Helens	1000/1900	Pre-eruptive
Casadeval *et al.* (1981); Hobbs *et al.* (1981)	St Helens	1300/2400	Eruptive (V)
Casadeval *et al.* (1981)	St Helens	900	Intra-eruptive
Casadeval *et al.* (1981)	St Helens	600/3600	Post-eruptive
Japan and USSR			
Okita (1971)	Mihara	345	Post-eruptive
Okita and Shimozuru (1975)	Asama	142	Post-eruptive
Okita and Shimozuru (1975)	Asama	780	Post-eruptive
Crafford (1975)	Karimski	173	Undetermined

[a]Classification of activities: eruptive, further subdivided by the categories I–IX of the intensity scale after Tsuya (1955) in terms of total volume of ejected material; pre-eruptive, intensification of activity before an eruption; intra-eruptive, phase of repose between two paroxy small eruptions; post-eruptive, permanent fuming or fumarolic activity; extra-eruptive, exclusive fumarolic and solphataric activity (Berresheim and Jaeschke, 1983).
[b]Calculated from the SO_4^{2-} content of ash particles.
[c]Estimated value.
[d]Calculated from the SO_2 content of lava.

Berresheim and Jaeschke (1982) also estimated the rate of H_2S and sulphate emissions from all volcanoes with post- and extra-eruptive activities as in the order of 1 ± 1 Tg (S) a^{-1} (0.03 ± 0.03 Tmol a^{-1}) and 3 ± 1 Tg (S) a^{-1} (0.1 ± 0.03 Tmol a^{-1}), respectively. Measurements at the southern Italian volcanoes Etna, Stromboli, and Vulcano also showed that the contribution of COS and CS_2 is negligible compared to the other volcanic sulphur exhalations.

3.8 SUMMARY

A summary of the most important naturally emitted sulphur compounds from all sources is given in Table 3.4. This table presents our best estimates of these fluxes based on current information. However, it must be emphasized that most of these estimates are rather uncertain. This applies especially to the emissions of particulate sulphur in the form of dust and seaspray and to the emissions from soil and plants on the continents. The main reasons for the uncertainty regarding continental emissions of biogenic sulphur compounds are: (1) the difficulty of accurately determining the various biogenic sulphur species, particularly hydrogen sulphide (H_2S), at the low levels found in unpolluted environments; (2) the technical problems of measuring emission fluxes from forest and brush ecosystems; and (3) the inadequate geographical coverage of existing data.

As can be seen from Table 3.4, the total flux of sulphur compounds into the atmosphere, including anthropogenic sources, is of the order of 160 to 500 Tg (S) a^{-1} (5 to 16 Tmol a^{-1}). The flux which is most uncertain is the emission of

Table 3.4. Estimates of natural sulphur emissions (Tg (S) a^{-1})

Compound Source	H_2S	DMS	CS_2	COS	SO_2	Sulphate
Oceans	<0.5	19–51	0.3	0.3	—	38.4–320 (Seaspray)
Coastal wetlands	1.0	0.6	0.06	0.1	—	—
Soils and plants	3.2–10.0	0.2–3.8	0.6–0.8	0.3–0.4	—	2–4 (Dust)
Biomass burning	?	—	?	0.01	2.6	?
Volcanoes	0.5–1.5	—	—	—	7.4–9.3	2–4
Total	4.7–13	19.2–54.4	0.1–1.3	0.1–1.3	10–13	4–8
Total natural sulphur emissions (without seasalt sulphate)						65 ± 25
Anthropogenic sulphur emissions						93 ± 15
Total global sulphur emissions (without seasalt sulphate)						158 ± 30

particulate sulphate as seaspray. Without this uncertain source, the flux of sulphur compounds into the atmosphere would be in the range of 125 to 185 Tg (S) a^{-1} (3.9 to 5.8 Tmol a^{-1}).

In order to consider an atmospheric sulphur budget, the emissions of sulphur compounds can be compared with the global sulphur deposition fluxes estimated by several authors. Since the main part of the emitted reduced sulphur gases undergo atmospheric oxidation before they are deposited, only the finally deposited oxidation products SO_2 and sulphate are taken into account in the budget estimation. A summary of the most recent estimates of sulphur deposition fluxes is given in Table 3.5 (Jaeschke, 1988, and references therein). The fluxes are subdivided with respect to dry and wet deposition and to continental and oceanic regions. Moreover, the fluxes of seaspray sulphate and excess sulphate are considered separately. In this context, excess sulphate is defined as that part of sulphate which is generated by the oxidation of reduced sulphur gases in the atmosphere. Therefore, excess sulphate is mainly occurring in the particle size range below 1 μm radius.

A diagram of the total global atmospheric sulphur budget is presented in Figure 3.8. The thickness of the arrows represents the rate of the respective fluxes. It can be seen from this graph that the emission and deposition of seasalt-sulphate is a rather independent, closed cycle. Therefore, the seasalt-sulphate can be ignored when analysing the rest of the global biogeochemical cycle of sulphur. A comparison of the global deposition flux of sulphur compounds of 190 ± 40 Tg (S) a^{-1} (5.9 ± 1.3 Tmol a^{-1}) in Table 3.5 with the global emission rates of 155 ± 30 Tg (S) a^{-1} (4.8 ± 0.94 Tmol a^{-1}) in Table 3.4 shows that ignoring the emission and deposition of seasalt-sulphate, the

Table 3.5. Estimates of global sulphur depositions (Tg (S) a^{-1})

Deposition	Region	Compound	Flux
Dry	Continents	SO_2	28 ± 12
	Oceans	SO_2	5 ± 3
Dry	Continents	Sulphate	5 ± 4
	Oceans	Excess sulphate	13 ± 4
	Oceans	Seasalt sulphate	68 ± 28
Wet	Continents	Sulphate	70 ± 30
	Oceans	Excess sulphate	65 ± 25
	Oceans	Seasalt sulphate	110 ± 40
Dry and wet	Continents	Seasalt sulphate	15 ± 5
Global deposition (without seasalt sulphate)			186 ± 41
Global emission[a] (without seasalt sulphate)			158 ± 30

[a]Table 3.4.

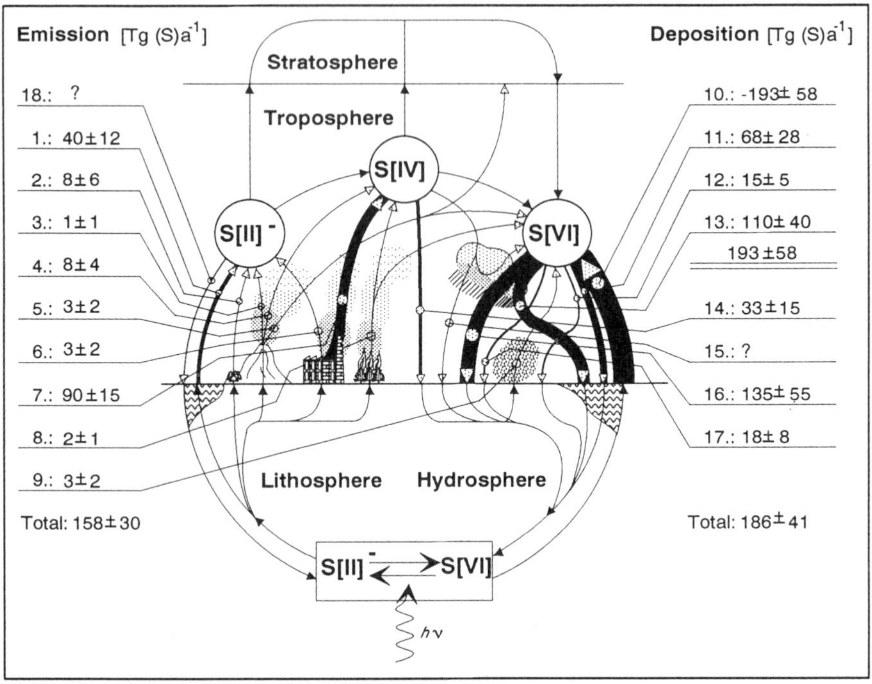

Figure 3.8. Scheme of the atmospheric sulphur cycle, together with numbers of the following annual fluxes: Emissions: 1. $S(II)^-$ oceans; 2. $S(II)^-$ soil and wetlands; 3. $S(II)^-$ volcanoes; 4. S(IV) volcanoes; 5. S(VI) volcanoes; 6. $S(II)^-$ anthropogenic; 7. S(IV) anthropogenic; 8. S(IV) biomass burning; 9. S(VI) dust; 10. S(VI) sea salt. Depositions: 11. S(VI) sea salt dry on the oceans; 12. sea salt dry and wet on the continents; 13. S(VI) wet on the oceans; 14. S(IV) dry; 15. S(IV) wet; 16. S(VI) excess sulphate wet; 17. S(VI) excess sulphate dry; 18. $S(II)^-$ dry and wet. The budget of the fluxes is nearly balanced when the cycle of the seasalt sulphate is ignored (see text)

budget of global emission and deposition fluxes is balanced within the accuracy of these estimates.

REFERENCES

Adams, D. F., Farwell, S. O., Robinson, E., Pack, M. R. and Bamesberger, W. L. (1981a). Biogenic sulfur source strengths. *Environ. Sci. Technol.*, **15**, 1493–8.

Adams, D. F., Farwell, S. O., Pack, M. R. and Robinson, E. (1981b). Biogenic gas emissions from soils in Eastern and Southeastern United States. *J. Air Pollut. Control Assoc.*, **31**, 1083–9.

Almgren, T. and Hagstrom, I. (1974). The oxidation rate of sulphide in sea water. *Water Res.*, **8**, 395–400.

Anderson, J. W. (1980). Assimilation of inorganic sulfate into cysteine. In: Stumpf, P. K. and Conn, E. E. (Eds). *The Biochemistry of Plants*, Vol. 5, Academic Press, New York, pp. 203–23.
Andreae, M. O. (1979). Arsenic speciation in seawater and interstitial waters: the influence of biological–chemical interactions on the chemistry of a trace element. *Limnol. Oceanogr.*, **24**, 440–52.
Andreae, M. O. (1980). Dimethylsulfoxide in marine and freshwaters. *Limnol. Oceanogr.*, **25**, 1054–63.
Andreae, M. O. (1985). The emission of sulfur to the remote atmosphere. In: Galloway, J. N., Charlson, R. J., Andreae, M. O. and Rodhe, H. (Eds), *The Biogeochemical Cycling of Sulfur and Nitrogen in the Remote Atmosphere*, Reidel, Dordrecht, pp. 5–25.
Andreae, M. O. (1986). The ocean as a source of atmospheric sulfur compounds. In: Buat-Ménard, P. (Ed.). *The Role of Air–Sea Exchange in Geochemical Cycling*, Reidel, Dordrecht, pp. 331–62.
Andreae, M. O. (1991a). Biomass burning: Its history, use and distribution and its impact on environmental quality and global climate. In: Levine, J. S. (Ed.). *Global Biomass Burning*, MIT Press, Boston, Mass. (in press).
Andreae, M. O. (1991b). Photochemical production of carbonyl sulfide in open ocean waters. *J. Geophys. Res.* (in press).
Andreae, M. O. and Andreae, T. W. (1988). The cycle of biogenic sulfur compounds over the Amazon Basin. I. Dry season. *J. Geophys. Res.*, **93**, 1487–97.
Andreae, M. O. and Barnard, W. R. (1984). The marine chemistry of dimethylsulfide. *Mar. Chem.*, **14**, 267–79.
Andreae, M. O., Barnard, W. R. and Ammons, J. M. (1983). The biological production of dimethylsulfide in the ocean and its role in the global atmospheric sulfur budget. *Ecol. Bull. (Stockholm)*, **35**, 167–77.
Andreae, M. O. and Raemdonck, H. (1983). Dimethylsulfide in the surface ocean and the marine atmosphere: a global view. *Science*, **221**, 744–7.
Andreae, M. O., Ferek, R. J., Bermond, F., Byrd, K. P., Engstrom, R. T., Hardin, S., Houmere, P. D., LeMarrec, F., Raemdonck, H. and Chatfield, R. B. (1985). Dimethyl sulfide in the marine atmosphere. *J. Geophys. Res.*, **90**, 12 891–900.
Andreae, M. O., Berresheim, H., Andreae, T. W., Kritz, M. A., Bates, T. S. and Merrill, J. T. (1988). Vertical distribution of dimethylsulfide, sulfur dioxide, aerosol ions, and radon over the northwest Pacific Ocean. *J. Atmos. Chem.*, **6**, 149–73.
Andreae, M. O., Berresheim, H., Bingemer, H., Jacob, D. J. and Talbot, R. W. (1990a). The atmospheric sulfur cycle over the Amazon Basin, 2. Wet Season. *J. Geophys. Res.*, **95**, 16 813–24.
Andreae, M. O., Talbot, R. W., Berresheim, H. and Beecher, K. M. (1990b). Precipitation chemistry in central Amazonia. *J. Geophys. Res.*, **95**, 16 987–99.
Andreae, T. W., Andreae, M. O., Bingemer, H. G. and Leck, C. (1990). Measurements of DMS and H_2S over the western North Atlantic and the Tropical Atlantic. *Eos Trans. Am. Geophys. Union*, **71**, 1254.
Andreae, T. W., Cutter, G. A., Hussain, N., Radford-Knoery, J. and Andreae, M. O. (1991). Hydrogen sulfide and radon in and over the western North Atlantic Ocean. *J. Geophys. Res.* (in press).
Barnard, W. R., Andreae, M. O., Watkins, W. E., Bingemer, H. and Georgii, H. W. (1982). The flux of dimethylsulfide from the oceans to the atmosphere. *J. Geophys. Res.*, **87**, 8787–93.
Bates, T. S., Cline, J. D., Gammon, R. H. and Kelly-Hansen, S. R. (1987). Regional and seasonal variations in the flux of oceanic dimethylsulfide to the atmosphere. *J. Geophys. Res.*, **92**, 2930–8.

Berresheim, H. (1987). Biogenic sulfur emissions from the subantarctic and Antarctic Oceans. *J. Geophys. Res.*, **92**, 13 245–62.

Berresheim, H. and Jaeschke, W. (1982). Sulfur emissions from volcanoes. In: Georgii, H. W. and Jaeschke, W. (Eds), *Chemistry of the Unpolluted and Polluted Troposphere*. Reidel, Dordrecht, pp. 325–37.

Berresheim, H. and Jaeschke, W. (1983). The contribution of volcanoes to the global atmospheric sulfur budget. *J. Geophys. Res.*, **88**, 3732–40.

Bingemer, H. G. (1984). Dimethylsulfid in Ozean und mariner Atmosphäre. Experimentelle Untersuchung einer natürlichen Schwefelquelle für die Atmosphäre. *Ph.D. Dissertation*, J. W. Goethe Universität, Frankfurt-am-Main.

Bingemer, H. G., Andreae, M. O., Andreae, T. W., Artaxo, P. Helas, G., Mihalopoulos, N. and Nguyen, B. C. (1991). Sulfur gases and aerosols in and above the equatorial African rainforest. *J. Geophys. Res.* (in press).

Bowen, H. J. M. (1979). *Environmental Chemistry of the Elements*. Academic Press, New York.

Bremner, J. M. and Steele, C. G. (1978). Role of microorganisms in the atmospheric sulfur cycle. *Adv. Microb. Ecol.*, **2**, 155–201.

Brown, K. A. and Bell, J. N. B. (1986). Vegetation—the missing link in the global cycle of carbonyl sulphide (COS). *Atmos. Environ.*, **20**, 537–40.

Burnett, W. E. (1969). Air pollution from animal wastes. *Environ. Sci. Technol.*, **3**, 744–9.

Cadle, R. D. (1975). Volcanic emissions of halides and sulfur compounds to the troposphere and stratosphere. *J. Geophys. Res.*, **80**, 1650–2.

Casadevall, T. J., Johnston, D. A., Harris, D. M., Rose, W. I., Malinconico, L. L., Stoiber, R. E., Bornhorst, T. J., Williams, S. N., Woodruff, L. and Thompson, J. M. (1981). SO_2 emission rates at Mount St Helens from March 29 through December, 1980. *USGS Prof. Pap.*, **1250**, 193–200.

Challenger, F. (1951). Biological methylation. *Adv. Enzymol.*, **12**, 429–91.

Chen, K. Y. and Morris, J. C. (1972). Kinetics of oxidation of aqueous sulfide by O_2. *Environ. Sci. Technol.*, **6**, 529–37.

Cicerone, R. J., Delwiche, C. C., Harriss, R. and Dickinson, R. (1984). Critical processes affecting the distribution of chemical species: Biological and surface sources. In: *Global Tropospheric Chemistry: A Plan for Action*. National Academy Press, Washington, DC, pp. 55–68.

Cline, J. D. and Bates, T. S. (1983). Dimethyl sulfide in the equatorial Pacific Ocean: a natural source of sulfur to the atmosphere. *Geophys. Res. Lett.*, **10**, 949–52.

Cooke, R. B. and Kelly, C. A. Chapter 7 of this volume.

Cooper, J. A. (1980). Environmental impact of residential wood combustion emissions and its implications. *J. Air Pollut. Control Assoc.*, **30**, 855–61.

Cooper, D. J. and Saltzman, E. S. (1987). Uptake of carbonyl sulfide by silver nitrate impregnated filters: implications for the measurement of low level atmospheric H_2S. *Geophys. Res. Lett.*, **14**, 206–9.

Cox, R. A. and Sheppard, D. (1980). Reactions of OH radicals with gaseous sulphur compounds. *Nature*, **284**, 330–1.

Crafford, T. C. (1975). SO_2-emission of the 1974 eruption of Volcan Fuego, Guatemala. *Bull. Volcanol.*, **39**, 536–56.

Crutzen, P. (1976). The possible importance of CSO for the sulfate layer of the stratosphere. *Geophys. Res. Lett.*, **3**, 73–6.

Crutzen, P. J., Heidt, L. E., Krasnec, J. P., Pollock, W. H. and Seiler, W. (1979). Biomass burning as a source of the atmospheric gases CO, H_2, N_2O, CH_3Cl, and COS. *Nature*, **282**, 253–6.

Cullis, C. F. and Hirschler, M. M. (1980). Atmospheric sulfur: natural and man-made sources. *Atmos. Environ.*, **14**, 1263–78.
Delmas, R. (1982). On the emission of carbon, nitrogen and sulfur in the atmosphere during bushfires in intertropical savannah zones. *Geophys. Res. Lett.*, **9**, 761–764.
Delmas, R. and Servant, J. (1982). The origins of sulfur compounds in the atmosphere of a zone of high productivity (Gulf of Guinea). *J. Geophys. Res.*, **87**, 11 019–26.
Delmas, R. and Servant, J. (1983). Atmospheric balance of sulphur above an equatorial forest. *Tellus*, **35B**, 110–20.
Delmas, R. and Servant, J. (1988). The atmospheric sulfur cycle in the tropics. In: Rodhe, H. and Herrera, R. (Eds). *Acidification in Tropical Countries*. Wiley, New York, pp. 43–72.
Delmas, R., Baudet, J., Servant, J. and Baziard, Y. (1980). Emissions and concentrations of hydrogen sulfide in the air of the tropical forest of the Ivory Coast and of temperate regions in France. *J. Geophys. Res.*, **85**, 4468–74.
Fall, R., Albritton, D. L., Fehsenfeld, F. C., Kuster, W. C. and Goldan, P. D. (1988). Laboratory studies of some environmental variables controlling sulfur emissions from plants. *J. Atmos. Chem.*, **6**, 341–62.
Ferek, R. J. and Andreae, M. O. (1983). The supersaturation of carbonyl sulfide in surface waters of the Pacific Ocean off Peru. *Geophys. Res. Lett.*, **10**, 393–6.
Ferek, R. J. and Andreae, M. O. (1984). Photochemical production of carbonyl sulfide in marine surface waters. *Nature*, **307**, 148–50.
Filner, P., Rennenberg, H., Sekiya, J., Bressan, R. A., Wilson, L. G., Le Cureux, L. and Shimel, T. (1984). Biosynthesis and emission of hydrogen sulfide by higher plants. In: Koziol, M. J. and Whatley, F. R. (Eds). *Gaseous Air Pollutants and Plant Metabolism*. Butterworth, Stoneham, MA, pp. 291–312.
Friend, J. P. (1973). The global sulfur cycle. In: Rasool, S. I. (Ed.). *Chemistry of the Lower Atmosphere*. Plenum Press, New York, pp. 117–201.
Froelich, P. N., Kaul, L. W., Byrd, J. T., Andreae, M. O. and Roe, K. K. (1985). Arsenic, barium, geranium, tin, dimethylsulfide, and nutrient biogeochemistry in Charlotte Harbor, Florida, a phosphorus-enriched estuary. *Estuarine, Coastal, Shelf Sci.*, **20**, 239–64.
Galloway, J. N. (1985). The deposition of sulfur and nitrogen from the remote atmosphere. In: Galloway, J. N., Charlson, R. J., Andreae, M. O. and Rodhe, H. (Eds). *The Biogeochemical Cycling of Sulfur and Nitrogen in the Remote Atmosphere*. Reidel, Dordrecht, pp. 143–75.
Giblin, A. E. and Wieder, R. K. This volume.
Giovanelli, J., Mudd, S. H. and Datko, A. H. (1980). Sulfur aminoacids in plants. In: Stumpf, P. K. and Conn, E. E. (Eds), *The Biochemistry of Plants*, Vol. 5. Academic Press, New York, pp. 453–505.
Goldan, P. D., Kuster, W. C., Albritton, D. L. and Fehsenfeld, F. C. (1987). The measurement of biogenic sulfur emissions from soils and vegetation: three sites in the eastern United States revisited. *J. Atmos. Chem.*, **5**, 439–67.
Graedel, T. E. (1979). Reduced sulfur emission from the open oceans. *Geophys. Res. Lett.*, **6**, 329–31.
Guenther, J., Lamb, B. and Westberg, H. (1989). U. S. National biogenic sulfur emissions inventory. In: Saltzman, E. S. and Cooper, W. J. (Eds). *Biogenic Sulfur in the Environment*. American Chemical Society Symposium Series No. 393. American Chemical Society, pp. 14–31.
Hanst, P. L., Spiller, L. L., Watts, D. M., Spence, J. W. and Miller, M. F. (1975). Infrared measurement of fluorocarbons, carbontetrachloride, carrbonylsulfide and other atmospheric trace gases. *J. Air Pollut. Control Assoc.*, **25**, 1220–6.

Haulet, R., Zettwoog, P. and Sabroux, J. C. (1977). Sulphur dioxide discharge from Mount Etna. *Nature*, **268**, 715–17.

Heald, E. F., Naughton, J. J. and Barnes, J. L. (1963). The chemistry of volcanic gases. 2. Use of equilibrium calculations in the interpretation of volcanic gas samples. *J. Geophys. Res.*, **68**, 545–57.

Hegg, D. A., Radke, L. R., Hobbs, P. V., Brock, C. A. and Riggan, P. J. (1987). Nitrogen and sulfur emissions from the burning of forest products near large urban areas. *J. Geophys. Res.*, **92**, 14 701–9.

Herrmann, J. and Jaeschke, W. (1984). Measurements of H_2S and SO_2 over the Atlantic Ocean. *J. Atmos. Chem.*, **1**, 111–23.

Hill, F. B., Aneja, V. P. and Felder, R. M. (1978). A technique for measurements of biogenic sulfur emission fluxes. *J. Environ. Sci. Health A*, **13**, 199–225.

Hobbs, P. V., Radke, L. F., Eltgroth, M. W. and Hegg, D. A. (1981). Airborne studies of the emissions from the volcanic eruptions of Mount St Helens. *Science*, **211**, 816–18.

Howarth, R. W. (1984). The ecological significance of sulfur in the energy dynamics of salt marsh and marine sediments. *Biogeochemistry*, **1**, 5–27.

Howarth, R. W. and Stewart, J. W. B. This volume.

Ivanov, M. V., Lein, A. Yu., Reeburgh, W. S. and Skyring, G. W. (1989). Interaction of sulphur and carbon cycles in marine sediments. In: Brimblecombe, P. and Lein, A. Yu (Eds). *Evolution of the Global Biogeochemical Sulphur Cycle*. Wiley and Sons, Chichester, pp. 125–79.

Iverson, R. L., Nearhoof, F. L. and Andreae, M. O. (1989). Production of dimethylsulfonium propionate and dimethylsulfide by phytoplankton in estuarine and coastal waters. *Limnol. Oceanogr.*, **34**, 53–67.

Jaeschke, W. (1988). *Untersuchungen zur Chemie des Schwefels in der Atmosphäre*. Habilitationsschrift, Univ. Frankfurt, Ber. d. ZUF, 24pp.

Jaeschke, W., Berresheim, H. and Georgii, H. W. (1982). Sulfur emissions from Mt. Etna. *J. Geophys. Res.*, **87**, 7253–61.

Jaeschke, W., Claude, H. and Herrmann, J. (1980). Sources and sinks of atmospheric H_2S. *J. Geophys. Res.*, **85**, 5639–44.

Jaeschke, W. and Herrmann, J. (1981). Measurements of H_2S in the atmosphere. *Int. J. Environ. Anal. Chem.*, **10**, 107–20.

Jaeschke, W., Schmidt, R. and Georgii, H. W. (1976). Preliminary results of stratospheric SO_2 measurements. *Geophys. Res. Lett.*, **3**, 517–19.

Jaeschke, W., Georgii, H. W., Claude, H. and Malewski, H. (1978). Contributions of H_2S to the atmospheric sulfur cycle. *Pageoph.*, **116**, 465–75.

Johnson, J. E. (1981). The lifetime of carbonyl sulfide in the troposphere. *Geophys. Res. Lett.*, **8**, 938–40.

Jørgensen, B. B. (1982). The sulfur cycle of a coastal marine sediment (Limfjorden, Denmark). *Limnol. Oceanogr.*, **22**, 814–32.

Jørgensen, B. B., Hansen, M. H. and Ingvorsen, K. (1978). Sulfate reduction in coastal sediments and the release of H_2S to the atmosphere. In: Krumbein, W. E. (Ed.). *Environmental Biogeochemistry and Geomicrobiology*, Vol. 1. Ann Arbor Science, Michigan, pp. 245–53.

Jørgensen, B. B. and Revsbech, N. P. (1983). Colorless sulfur bacteria, *Beggiatoa* sp. and *Thiovulum* sp., in O_2 and H_2S microgradients. *Appl. Environ. Microbiol.*, **45**, 1261–70.

Kadota, H. and Ishida, Y. (1972). Production of volatile sulfur compounds by microorganisms. *Annu. Rev. Microbiol.*, **26**, 127–63.

Kellog, W. W., Cadle, R. D., Allen, E. R., Lazrus, A. L. and Martell, E. A. (1972). The sulfur cycle. *Science*, **175**, 587–96.

Kim, K. H. and Andreae, M. O. (1987). Carbon disulfide in seawater and the marine atmosphere over the North Atlantic. *J. Geophys. Res.*, **92**, 14 733–8.

Koblentz-Mishke, O. J., Volkevinsky, V. V. and Kabanova, J. G. (1970). Plankton primary production of the world ocean. in: Wooster, W. S. (Ed.). *Scientific Exploration of the South Pacific*. National Academy Science, Washington DC, pp. 183–93.

Lacaux, J. P., Servant, J., Huertas, M. L., Cros, B., Delmas, R., Loemba-Ndembi, J. and Andreae, M. O. (1988). Precipitation chemistry from remote sites in the African equatorial forest. *Eos Trans. Am. Geophys. Union*, **69**, 1069.

Lamb, B., Westberg, H., Allwine, G., Bamesberger, L. and Guenther, A. (1987). Measurement of biogenic sulfur emissions from soils and vegetation: application of dynamic enclosure methods with Natusch filter and GC/FPD analysis. *J. Atmos. Chem.*, **5**, 469–91.

Liss, P. S. and Merlivat, L. (1986). Air–sea exchange rates: introduction and synthesis. In: Buat-Menard, P. (Ed.). *The Role of Air–Sea Exchange in Geochemical Cycling*. Reidel, Dordrecht, pp. 113–27.

Lovelock, J. E. (1974). CS_2 and the natural sulphur cycle. *Nature*, **248**, 625–6.

Lovelock, J. E., Maggs, R. J. and Rasmussen, R. A. (1972). Atmospheric dimethyl sulphide and the natural sulphur cycle. *Nature*, **237**, 452–3.

Luther, G. and Church, T. Chapter 6 of this volume.

MacDonald, G. A. (1955). Hawaiian volcanoes during 1952. *US Geol. Surv. Bull.*, **1021-B**, 15–108.

MacDonald, G. A. (1972). *Volcanoes*. Prentice-Hall, Englewood Cliffs, New Jersey.

McDowell, W. H. (1987). Potential effects of acidic atmospheric deposition on tropical terrestrial ecosystems. In: Rodhe, H. (Ed.). *Effects of Acidification on Tropical Ecosystems*. Wiley-Interscience, New York.

McElroy, M. B., Wofsy, S. C. and Sze, N. D. (1980). Photochemical sources for atmospheric H_2S. *Atmos. Environ.*, **14**, 159–63.

Malinconico, L. L. (1979). Fluctuations in SO_2 emissions during recent eruptions of Etna. *Nature*, **278**, 43–5.

Millero, J. J., Hubinger, S., Fernandez, M. and Garnett, S. (1987). Oxidation of H_2S in seawater as a function of temperature, pH, and ionic strength. *Environ. Sci. Technol.*, **21**, 439–43.

Naughton, J. J., Lewis, V. A., Thomas, D. and Finlayson, J. B. (1975). Fume compositions found at various stages of activity at Kilauea volcano, Hawaii. *J. Geophys. Res.*, **80**, 2963–6.

Nguyen, B. C., Bergeret, C. and Lambert, G. (1984). Exchange rates of dimethyl sulfide between ocean and atmosphere. In: Brunsaert, W. and Jirka, G. H. (Eds). *Gas Transfer at Water Surfaces*. Reidel, Dordrecht, pp. 539–45.

Nriagu, J. O., Holdway, D. A. and Coker, R. D. (1987). Biogenic sulfur and the acidity of rainfall in remote areas of Canada. *Science*, **237**, 1189–92.

Okita, T. (1971). Detection of SO_2 and NO_2 gas in the atmosphere by Barringer spectrometer. ASCO, Pep. 8/7.

Okita, T. and Shimozuru, D. (1975). Remote sensing measurements of mass flow of sulfur dioxide gas from volcanoes. *Volcanol. Soc. Jpn. Bull.*, **19-3**, 153–7.

Peng, T. H., Broecker, W. S., Mathieu, G. G. and Li, Y. H. (1979). Radon evasion rates in the Atlantic and Pacific Oceans as determined during the Geosecs program. *J. Geophys. Res.*, **84**, 2471–86.

Rasmussen, R. A., Khalil, M. A. K. and Hoyt, S. D. (1982). The oceanic source of carbonyl sulfide (COS). *Atmos. Environ.*, **16**, 1591–4.

Rose, W. I., Bonis, S., Stoiber, R. E., Keller, M. and Bickford, T. (1973). Studies of volcanic ash from two recent Central American eruptions. *Bull. Volcanol.*, **37**, 338–64.

Saltzman, E. S. and Cooper, D. J. (1988). Shipboard measurements of atmospheric dimethylsulfide and hydrogen sulfide in the Caribbean and Gulf of Mexico. *J. Atmos. Chem.*, **7**, 191–209.

Saltzman, E. S., Savoie, D. L., Prospero, J. M. and Zika, R. G. (1986). Methanesulfonic acid and non-sea-salt sulfate in Pacific air: Regional and seasonal variations. *J. Atmos. Chem.*, **4**, 227–40.

Sandalls, F. J. and Penkett, S. A. (1977). Measurements of carbonyl sulphide and carbon disulphide in the atmosphere. *Atmos. Environ.*, **11**, 197–199.

Sapper, K. (1927). *Vulkankunde*. Engelhorn Verlag. Stuttgart.

Schoenau, J. J. and Germida, J. J. Chapter 10 of this volume.

Shepherd, E. S. (1938). The gases in rocks and some related problems. *Am. J. Sci. 5th Ser.*, **235a**, 311–51.

Slatt, B. J., Natusch, D. F. S., Prospero, J. M. and Savoie, D. L. (1978). Hydrogen sulfide in the atmosphere of the northern equatorial Atlantic Ocean and its relation to the global sulfur cycle. *Atmos. Environ.*, **12**, 981–91.

Smethie, W. M., Jr, Takahashi, T., Chipman, D. W. and Ledwell, J. R. (1985). Gas exchange and CO_2 flux in the tropical Atlantic Ocean determined from ^{222}Rn and pCO_2 measurements. *J. Geophys. Res.*, **90**, 7005–22.

Staubes, R. (1986). *Untersuchungen der Bodenexhalation von Carbonylfulfid, Dimethylsulfid und Schwefelkohlenstoff*. Diplomarbeit, Institut fur Meterologie und Geophysik, Johann-Wolfang-Geothe-Universität, Frankfurt-am-Main.

Staubes, R., Ockelmann, G. and Georgii, H. W. (1986). Emissions of biogenic sulfur compounds from various soils. Paper presented at 2nd International Symposium on Biosphere–Atmosphere Exchange, Mainz, Federal Republic of Germany, 16–21 March, 1986.

Stephens, E. R. (1971). Identification of odors from cattle feed lots. *Calif. Agric.*, **25**, 10–11.

Steudler, P. A. and Peterson, B. J. (1984). Contribution of the sulfur form salt marshes to the global sulfur cycle. *Nature*, **311**, 455–7.

Stith, J. L., Hobbs, P. V. and Radke, L. F. (1978). Airborne particle and gas measurements in the emissions from six volcanoes. *J. Geophys. Res.*, **83**, 4009–17.

Stoiber, R. E. and Bratton, G. (1978). Airborne correlation spectrometer measurements of SO_2 in eruption clouds form Guatemalan Volcanoes. *Eos Trans. Am. Geophys. Union*, **59**, 1222.

Stoiber, R. E. and Jepsen, A. (1973). Sulfur dioxide contributions to the atmosphere by volcanoes. *Science*, **182**, 577–8.

Stoiber, R. E. and Malone, G. B. (1975). SO_2-emission at the crater of Kilauea, at Mauna Ulu and at Sulfur Banks, Hawaii. *Eos Trans. Am. Geophys. Union*, **56**, 461.

Taylor, P. S. and Stoiber, R. E. (1973). Soluble material on ash from active Central American volcanoes. *Geol. Soc. Am. Bull.*, **84**, 1031–42.

Tsuya, H. (1955). Geological and petrological studies of vulcano Fuji, 5. *Tokyo Daigaku Jinshin Kenkyusho Iho*, **33**, 341–182.

Turner, S. M. and Liss, P. S. (1985). Measurements of various sulphur gases in a coastal marine environment. *J. Atmos. Chem.*, **2**, 223–32.

Turner, S. M., Malin, G. and Liss, P. S. (1989). Dimethylsulfide and (dimethylsulfonio) propionate in European coastal and shelf waters. In: Saltzman, E. S. and Cooper, W. J. (Eds). *Biogenic Sulfate in the Environment*. ACS Symposium Series No. 393, American Chemical Society, Washington DC.

Turner, S. M., Malin, G., Liss, P. S., Harbour, D. S. and Holligan, P. M. (1988). The seasonal variation of dimethyl sulfide and dimethylsulfoniopropionate concentrations in nearshore waters. *Limnol. Oceanogr.*, **33**, 364–75.

Warneck, P. (1988). *Chemistry of the Natural Atmosphere.* Academic Press, San Diego, New York, p. 499.

Wilson, L. G., Bressan, R. A. and Filner, P. (1978). Light-dependent emission of hydrogen sulfide from plants. *Plant Physiol.*, **61**, 184–9.

Winner, W. E., Smith, C. L., Koch, G. W., Mooney, H. A., Bewley, J. D. and Krouse, H. R. (1981). Rates of emission of H_2S form plants and patterns of stable sulphur isotope fractionation. *Nature*, **289**, 672–3.

Zepp, R. G. and Andreae, M. O. (1989). Factors affecting the photochemical formation of carbonyl sulfide in sea water. *Eos Trans. Am. Geophys. Union*, **70**, 1023.

Zepp, R. G. and Andreae, M. O. (1990). Photosensitized formation of carbonyl sulfide in sea water. In: Blough, N. V. and Zepp, R. G. (Eds). *Effects of Solar Ultraviolet Radiation on Biogeochemical Dynamics in Aquatic Environments.* Woods Hole Oceanographic Inst, Woods Hole.

Zettwoog, P. and Haulet, R. (1978). Experimental results on the SO_2 transfer in the Mediterranean obtained with remote sensing devices. *Atmos. Environ.*, **12**, 795–6.

APPENDIX TO CHAPTER 3: SULPHUR ISOTOPE EVIDENCE FOR EMISSION OF BIOGENIC SULPHUR

J. O. NRIAGU
National Water Research Institute, Burlington, Canada

and

H. R. KROUSE
Department of Physics, University of Calgary, Calgary, Canada

The direct measurement of sulphur gas fluxes from waters and terrestrial ecosystems to the atmosphere has proven problematical (Andreae and Jaeschke, this volume). The fluxes themselves can be highly variable in space and time, and many approaches for measuring the fluxes, such as the use of chambers, may alter flux rates. Controversy over the magnitude of these emissions continues, but estimates have tended to decrease over the past decade due mainly to better instrumentation and measurement.

The analysis of the natural abundance of stable isotopes in gases and atmospheric particles provides another approach for assessing biogenic sulphur gas fluxes. The approach is discussed in detail in another SCOPE volume (Krouse and Gorlenko, 1990). Here we briefly outline the use of stable isotope data and illustrate the potential power of the approach.

Sulphur isotope abundances are expressed on a $\delta^{34}S$ scale defined by:

$$\delta^{34}S = \frac{[^{34}S/^{32}S]\ \text{sample}}{[^{34}S/^{32}S]\ \text{standard}} - 1 \times 10^3$$

where $^{34}S/^{32}S$ is the number of ^{34}S atoms to ^{32}S atoms in the sample. Meteoritic troilite from Canyon Diablo has been chosen as an international standard because of its isotopic homogeneity and composition close to that of the mean value for terrestrial sulphur.

A number of biological processes can affect the sulphur isotopic abundances in compounds, and the $\delta^{34}S$ ratio can be used to trace the origin of sulphur in these compounds, including atmospheric gases and particles. For example, sulphide generated by bacterial sulphate reduction is characterized by large kinetic isotope effects whereby $^{32}SO_4^{2-}$ is reduced faster than $^{34}SO_4^{2-}$. The product sulphide is relatively depleted in ^{34}S, which usually

Sulphur Cycling on the Continents
Edited by R.W. Howarth, J.W.B. Stewart and M.V. Ivanov

means that such sulphide has quite negative $\delta^{34}S$ values. This has been demonstrated in the laboratory for the classical sulphate-reducing bacteria such as *Desulphovibrio desulphuricans* (see review by Chambers and Trudinger, 1972) as well as for a novel sequential reduction by bacillus reducing sulphate to sulphite and a clostridium reducing sulphite to sulphide (Hunt, 1979). The isotopic selectivity is quite variable and tends to be larger at lower reduction rates. In nature, the sulphide may be depleted in ^{34}S by as much as 70‰ as compared to the initial sulphate (Weyer, Krouse and Horwood, 1979). The dissolved sulphides in salt marsh porewaters are depleted by some 30 to 40‰ compared to the sulphate (Peterson and Howarth, 1987).

Living plants under sulphur stress release H_2S and perhaps other reduced sulphur compounds. This process is considered to involve reduction. This is confirmed by sulphur isotope data from laboratory experiments. The evolved H_2S was found to be as much as 15‰ depleted in ^{34}S as compared to foliar sulphur and SO_4 available to the roots (Winner *et al.*, 1981).

Information on the natural abundance of oxygen-stable isotopes can also prove useful in studying the sulphur cycle. Dissolved oxygen in near-surface water re-oxidizes much of the aqueous sulphide that was generated in anoxic environments (Giblin and Wieder, this volume; Andreae and Jaeschke, this volume). The oxygen isotope composition of sulphate provides evidence of this process, particularly at higher latitudes (van Everdingen, Shakur and Krouse, 1982).

Two studies, one in the Great Lakes Basin of North America and the other near sour gas (H_2S-rich) processing plants in Alberta, Canada, illustrate the power of using the natural abundances of sulphur-stable isotopes to elucidate sources of atmospheric sulphur. In the Great Lakes Basin study, sulphur in bulk precipitation samples was examined (Nriagu and Coker, 1978). Whereas the mean $\delta^{34}S$ value for rural stations was about 2‰ lower than for urban centres, the striking observation was that at all sampling sites, the $\delta^{34}S$ values during winter were about 4‰ higher than during summer despite the fact that concentrations at remote locations were lower. Further, the seasonable isotopic trend did not correlate with concentration changes. Fossil fuel consumption in urban areas for winter heating cannot explain the trend in remote areas. Incorporation of sulphur and associated isotope fractionation during rainfall and snowfall are not well understood but preliminary assessment showed that these phenomena could not be the dominant mechanism which determined the isotope behaviour. Although other possibilities could not be entirely excluded, the most likely explanation was that the ratio of biogenic to anthropogenic emission was higher in the warmer months.

The sour gas processing industry of Alberta, Canada, emits sulphur oxides which have $\delta^{34}S$ values ranging from +15 to +30‰. In contrast, the pre-industrial soils of Alberta had $\delta^{34}S$ values ranging from −30 to 0‰. On the basis of this isotopic difference, it was demonstrated that epiphytic lichens aquired sulphur rather directly from the atmosphere. In contrast, leaves and needles incorporated sulphur from two sources, atmospheric sulphur oxides and SO_4 transported from the soil by the root system (Krouse, 1977).

In contrast to the Great Lakes wetfall study, sulphur isotope data for atmospheric SO_2 were obtained in the Alberta study. At high SO_2 concentrations near sour gas processing facilities, δ^{34} values converged towards that of the stack gas (Krouse, 1980). At lower concentrations, $\delta^{34}S$ values were interpreted as reflecting mixtures of emissions from the industry and background sources. The background may contain emissions from distant processing plants.

For data from a number of sites at different times around sour gas plant operations, two important observations have emerged. In the warmer months, the $\delta^{34}S$ values for ambient SO_2 are lower and SO_2 concentrations are higher, particularly at sites where the plume does not impact continuously. This must reflect an increase in biogenic emissions, but what are the sources? An answer seems evident in the second observation, namely that two peaks are found in the distribution of $\delta^{34}S$ values for SO_2. The major peak is that of the industrial plume whereas the other coincides with the $\delta^{34}S$ values found for vegetation. Data in Figure 3A.1 show how the secondary peak in the distribution of $\delta^{34}S$ values of SO_2 occurs near that found for

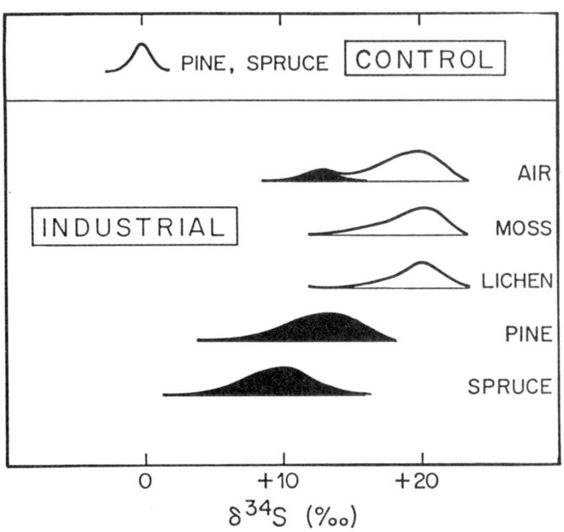

Figure 3A.1. Distributions in $\delta^{34}S$ values for atmospheric SO_2 and vegetation near the Ram River sour gas processing plant, Alberta, Canada. With the exception of the control area, each distribution curve is based upon 30 to 60 samples taken at various times and locations over a three-year period. The $\delta^{34}S$ value for the soil below a few centimetres depth is close to 0‰ at all sites whereas that of the emissions is near +22‰

conifer needles in a forested environment. Any oxidized H_2S from sulphate reduction should have possessed lower $\delta^{34}S$ values and a greater spread. It is more likely that this secondary peak reflects SO_2 generated by the oxidation of sulphide produced during decay of the litter, which mainly comprises fallen needles. Kinetic isotope effects during removal of sulphur from large organic molecules have not been measured but are expected to be relatively small, consistent with this interpretation.

In conclusion, stable sulphur isotope data with emphasis on the seasonal shifts in $\delta^{34}S$ values for wetfall and dryfall SO_2 appear to be sensitive indicators of biogenic sulphide emissions. The technique permits source assignments more readily than concentration measurements. Of course simultaneous measurements of both parameters is preferred along with meteorological parameters such as wind direction. In view of the range of estimates for biogenic sulphide fluxes, further sulphur isotope studies relevant to the problem are warranted.

REFERENCES

Andreae, M. O. and Jaeschke, W. This volume.

Chambers, L. A. and Trudinger, P. A. (1972). Microbiological fractionation of stable sulfur isotopes: A review and critique. *Geomicrobiol. J.*, **1**, 249–93.

Giblin, A. and Wieder, R. K. This volume.

Hunt, R. N. (1974). Oxygen isotope studies on sulphates. *Ph.D. Thesis*, University of Alberta, Edmonton.

Krouse, H. R. (1977). Sulphur isotope abundances elucidate uptake of atmospheric-sulphur emission by vegetation. *Nature*, **265**, 45–6.

Krouse, H. R. (1980). Sulphur isotopes in our environment. In: Fritz, P. and J. Ch. Fontes (Eds). *Handbook of Environmental Isotope Geochemistry*. Elsevier, Amsterdam, pp. 435–71.

Krouse, H. R. and Grinenko, V. A. (in press). *Stable Isotopes in the Assessment of Natural and Anthropogenic Sulphur in the Environment*. Wiley, Chichester.

Nriagu, J. O. and Coker, R. D. (1978). Isotopic composition of sulfur in precipitation within the Great Lakes Basin. *Tellus*, **30**, 365–75.

Peterson, B. J. and Howarth, R. W. (1987). Sulfur, carbon, and nitrogen isotopes used to trace organic matter flow in the salt-marsh estuaries of Sapelo Island, Georgia, *Limnology and Oceanography*, **32**, 1195–213.

van Everdingen, R. O., Shakur, M. A. and Krouse, H. R. (1982). Isotope geochemistry of dissolved, precipitated airborne and fallout sulfur species associated with springs near Paige Mountain, Norma Range, NWT. *Can. J. Earth Sci.*, **19**, 1395–407.

Weyer, K. O., Krouse, H. R. and Horwood, W. C. (1979). Investigation of regional geohydrology south of great Slave Lake, Canada, utilizing natural sulphur and hydrogen isotope variations. In: *Isotope Hydrology 1978*, International Atomic Energy Agency, Vienna, pp. 251–64.

Winner, W. E., Smith, C. L., Koch, G. W., Mooney, H. A., Bewley, J. E. and Krouse, H. R. (1981). Rates of emission of H_2S from plants and patterns of stable isotope fractionation. *Nature*, **289**, 672–3.

4 The Interactions of Sulphur with other Element Cycles in Ecosystems

R. W. HOWARTH
Section of Ecology and Systematics, Cornell University, Ithaca, NY, USA

and

J. W. B. STEWART
Department of Soil Science, University of Saskatchewan, Saskatoon, Canada

4.1 INTRODUCTION

The sulphur cycle is intimately connected with the cycles of all other major elements and many minor elements. The interactions of sulphur with these other element cycles has been discussed extensively in other SCOPE volumes (Likens, 1981; Ivanov and Freney, 1983; Bolin and Cook, 1983; Brimblecombe and Lein, 1989; Degens *et al.*, in press) and is an underlying theme in most of the chapters in this book. It is not our intention here to thoroughly review the interactions of sulphur with other element cycles or to repeat the information found in the other chapters. Rather, we wish to lay out a framework for understanding the diversity of sulphur–element interactions.

The ecological interest in nitrogen and phosphorus stems primarily from their role as plant nutrients. Sulphur also is a plant nutrient, although it is less often limiting than either nitrogen or phosphorus. But sulphur has numerous interactions with other elements beyond its role in plant nutrition. Below we first discuss the role of sulphur as an essential nutrient and then summarize these other types of element interactions.

4.2 THE CYCLING OF SULPHUR AS AN ESSENTIAL NUTRIENT

Sulphur is an essential element for all living plants, animals, and microorganisms. Three amino acids found in almost all proteins (cysteine, cystine, and

Sulphur Cycling on the Continents
Edited by R.W. Howarth, J.W.B. Stewart and M.V. Ivanov

methionine) contain sulphur. Sulphur is also found in sulpholipids, some vitamins, sulphate esters, and a variety of other compounds. Biogeochemists often distinguish between sulphate-ester sulphur, where the sulphur atom is bonded to oxygen, and carbon-bonded sulphur (which includes the sulphur in amino acids). Usually, most of the sulphur in vascular plants is in amino acids, resulting in a tight and predictable coupling between sulphur, carbon, nitrogen, and phosphorus in plant tissues (Johnson, 1984; Maynard, Stewart and Bettany, 1984; Mitchell, David and Harrison, this volume; Schoenau and Germida, this volume).

4.2.1 ACCUMULATION OF EXCESS SULPHUR AS SULPHATE AND SULPHATE ESTERS

The sulphur content of vascular plants is fairly plastic, and sulphur can accumulate in terrestrial plant tissues when soil sulphur is in excess compared to other nutrients such as nitrogen and phosphorus. Most of this sulphur which accumulates in plant tissues is in the form of sulphate esters and/or inorganic sulphate, and carbon-bonded sulphur is relatively non-plastic (Johnson, 1984; Maynard, Stewart and Bettany, 1984; Mitchell, David and Harrison, this volume). Although sulphate esters have not been specifically identified in trees (Mitchell, David and Harrison, this volume), sulphate esters definitely accumulate in crop plants subject to excess sulphate; currently available data do not distinguish between sulphate esters and inorganic sulphate in tree tissues (Johnson, 1984; Maynard, Stewart and Bettany, 1984).

If one considers only carbon-bonded sulphur in plants (that is, excluding inorganic sulphate and sulphate esters), then the molar N : S ratio varies little (N : S = 31 to 39) in a wide variety of vascular plants and plant tissues, including conifer needles (Kelly and Lambert, 1972), deciduous tree leaves (Johnson, 1984), graminous plants (Dijkshoorn and Van Wijk, 1967), and legumes (Dijkshoorn and Van Wijk, 1967). Again considering only carbon-bonded sulphur, the P : S ratio in vascular plants is typically near 1 : 1 (Maynard, Stewart and Bettany, 1984). Thus, we can define sulphur excesses in soils in terms of the ratios of available N : S and P : S. That is, we would expect that sulphur accumulates in vascular plants in the form of sulphate or sulphate esters when the ratio of available nitrogen to available sulphur in the soil is less than 30 or so (for a nitrogen limited system) or when the ratio of available phosphorus to available sulphur is less than 1 (for a phosphorus limited system).

4.2.2 SULPHUR LIMITATION IN TERRESTRIAL ECOSYSTEMS

In the absence of sulphur fertilization, the removal of sulphur in crop harvests can sometimes lead to sulphur limitation, both for rice crops (Lefroy *et al.*, this volume) and for upland agriculture (Maynard, Stewart and Bettany, 1984; Tabatabai, 1984; Schoenau and Germida, this volume). In forests, sulphur is not usually limiting, although sulphur deficiencies are found in some regions of low atmospheric deposition of sulphur and in some soils rich in nitrogen (Mitchell, David and Harrison, this volume). In addition to the potential to limit primary production, sulphur deficiencies in forests can increase the susceptibility of trees to fungal infections by increasing the concentration of non-sulphur-containing amino acids in leaves (Johnson, 1984; Mitchell, David and Harrison, this volume).

4.2.3 SULPHUR MINERALIZATION

Mineralization occurs when microbes catabolize organic molecules, respiring CO_2 and releasing sulphur. Carbon, N, S, and P cycles are interdependent because the elements are combined in organic matter in stoichiometric relationships. Although the N and S cycles are similar in many respects, mechanistic differences and imbalanced supplies can cause wide variations among the element ratios of nutrient pools in soils. Diverse communities of decomposers convert heterogeneous mixtures of organic N and S to simple inorganic forms. The chemical compositions of soil organic N and S, however, are largely undefined, and the degradation pathways are poorly understood.

The microbes responsible for mineralization are not confined to a particular group (Schoenau and Germida, this volume). Successional groups of organisms, rather than single species, frequently are implicated in multistep decomposition schemes. Soils are generally nutrient-poor environments, teeming with a diverse heterotrophic community consisting of bacteria, actinomycetes, fungi and protozoa that are in a state of starvation (Williams, 1985; Lynch, 1982). Comminution and mineralization of organic matter is also fostered by soil fauna (Swift, Heal and Anderson, 1979). Since the saprophytic organisms responsible for mineralization are non-specific, decomposer microbes are often studied collectively as 'functional groups' (Coleman, Reid and Cole, 1983). Faunal populations sometimes are better indicators of decomposition than is total microbial biomass, because fauna consume the microbes (Andren, Paustian and Rosswall, 1988).

During mineralization the molecular size of organic matter progressively decreases (depolymerization) before inorganic N and S are released. Organic matter must be broken down into its oligomeric or monomeric constituents before crossing the cell membranes of decomposers, although fauna may feed on particles (Swift, Heal and Anderson, 1979). Extracellular enzymes depolymerize polysaccharides, proteins, peptides, lipids, lignin, and other organic constituents. Aminopolysaccharides, such as the repeating units of *N*-acetylglucosamine in the chitin of microbial cell wall, typically account for 3 to 10% of soil organic N and likely contribute to N mineralization (Sowden, Chen and Schnitzer, 1977). Algae and bacteria may produce sulphated polysaccharides in soils, but these ester sulphates do not accumulate (Freney, 1986).

Proteins account for major proportions of soil organic N and S, and the amino acids released by proteases are important sources of mineralizable N and S. Most amino acids are bound in organic polymers or organo-mineral complexes, because free amino acids are rapidly taken up by decomposers and autotrophs (Loll and Bollag, 1983). Soil amino acids usually are determined after acid hydrolysis, but free amino acids are barely detectable in mild extracts (Freney, Stevenson and Beavers, 1982). Free amino acids, however, are the most likely source of mineralizable N and S, and production of NH_4^+ and H_2S during amino acid degradation are viewed as the terminal steps of mineralization.

Nitrogen mineralization is measured as NH_4^+ plus NO_3^- production, because NH_4^+ is often oxidized to NO_3^-. In contrast, H_2S is rarely detected during the degradation of amino-acid-S, and mineralization of soil organic S is measured as sulphate production. Cysteine-S may be oxidized while attached to the organic moiety, and organic S gases may be intermediaries in methionine degradation (Trudinger and Loughlin, 1981; Bremner, 1977). The S in ester sulphates is hydrolysed by sulphatase enzymes and released directly as inorganic sulphate (Fitzgerald, 1976). Sulpholipids in plant leaves account for major fluxes of S in terrestrial ecosystems, but the catabolic pathways of sulphonic acids (—C—SO_2—OH) in lipids are uncertain (Harwood and Nicholls, 1979).

4.2.4 MEASUREMENT OF SULPHUR FRACTIONS AND INTERPRETATION OF MINERALIZATION PROCESSES

Chemical fractionation of organic S identifies the major linkages between organic C and S. Organic sulphate is commonly defined as the fraction of organic S which is reduced by a mixture of hydriodic, formic, and hypophosphorous acids (Freney, 1961). The reducing mixture releases H_2S from organic S that is not directly bonded to C; therefore, 'organic sulphate' includes ester sulphate (—C—O—SO_2—O—), sulphamates (—C—N—SO_2—O—), and the second S in sulphocysteine (—C—S—S—). Organic sulphate typically accounts for 30 to 70% of organic S in soils (Freney, Melville and Williams, 1970). Carbon-bonded sulphur is reducible by Raney nickel. Some workers contend that Raney nickel reduces a well-defined fraction of organic S corresponding to amino-acid-S, and that the non-reducible C-bonded S is probably aliphatic sulphonates or heterocyclic S (Scott and Anderson, 1976; Scott, Bick and Anderson, 1981).

Chemical and physical separations expose general relationships between organic matter and soil environment, management history, and nutrient availability (Bettany, Stewart and Halstead, 1973; Bettany, Saggar and Stewart, 1980). Intensive extractions of organic matter with strong alkali and sonication suggested that organic sulphates were part of the aliphatic side chains of humic molecules, and fertility decreased when the side chains were eliminated (Bettany, Stewart and Saggar, 1979; Bettany, Saggar and Stewart, 1980). Anderson and co-workers (1981) observed that organic N and S were preferentially associated (narrow C–N–S) with the fine clay fractions, and suggested the clay-associated organic matter was a medium-term reservoir of available S.

Chemical separations distinguish between the bond classes of organic S, and some workers consider organic sulphate to be more 'susceptible to mineralization' or 'readily available' than C-bonded S (McLaren and Swift, 1977; Biderbeck, 1978). The possibility that organic sulphate may be more readily available, at least in the short term, is based on its chemical nature and physical accessibility: C—O—S linkages are easily hydrolysed by exocytoplasmic enzymes and chemical reagents (Houghton and Rose, 1976; Fitzgerald, 1978), and R—C—O—S is situated on aliphatic side chains rather than the condensed aromatic core of humus (Bettany, Stewart and Saggar, 1979). Chemical separations of organic matter, however, are too broad to generalize about the sources of mineralized N and S (McLachlan and De Marco, 1975). Incubation studies, aided by isotopic tracers, indicate that chemical fractions lack 'biological significance', because mineralizable N and S are derived from several different fractions (Chichester, Legg and Stanford, 1975; Freney, Melville and Williams, 1975). Care must therefore be taken with overall generalization of the processes (McGill and Cole, 1981), as mineralization of sulphate to obtain energy and enzymatic hydrolysation of sulphate can occur concurrently.

4.2.5 SULPHUR LIMITATION IN AQUATIC ECOSYSTEMS

In contrast to studies in terrestrial ecosystems, the role of sulphur in plant nutrition and the processes of organic sulphur decomposition have received fairly little study in aquatic ecosystems. Because of the high concentration of sulphate in seawater (28 mM; 0.9 g (S) l^{-1}), sulphur is probably never limiting to phytoplankton in marine ecosystems. In freshwater lakes, sulphur has not been found to limit primary production in any lake where it has been studied; however, the potential for sulphur limitation in lakes having extremely low sulphate concentrations has not been examined (Cook and Kelly, this volume). The ratio of C : P : S (molar) in freshwater phytoplankton is approximately 100 : 1 : 1 (Cook and Kelly, this volume). An appreciable amount of the sulphur in phytoplankton is probably present as sulphate esters (David and Mitchell, 1985). None the less, in contrast to plants in terrestrial ecosystems, the sulphur content of freshwater phytoplankton seems fairly constant and unresponsive to changes in ambient sulphate concentrations

(Cook and Kelly, this volume). Similarly, even though sulphate concentrations are much higher in seawater than in freshwaters, Cuhel and Waterbury (1984) have reported a C : S ratio for a marine phytoplankter which is very similar to that found for phytoplankton in freshwater lakes.

4.2.6 FORMATION OF DMS

One aspect of the sulphur metabolism of plants deserves special mention because of its atmospheric consequences: many species of plants, both algae and vascular plants, produce large amount of dimethylsulphonium propionate (DMSP) from sulphur-containing amino acids (Andreae and Jaeschke, this volume). This DMSP is converted to dimethyl sulphide (DMS), either within living plants or upon the death of plant tissues; the resulting production of DMS In the oceans is the largest source of biogenic sulphur to the atmosphere globally (Andreae and Jaeschke, this volume). Although the biochemical pathways of DMSP and DMS production are known, the physiological and ecological significance of this process remain unclear. For marine phytoplankton, DMSP and DMS production have been hypothesized to play a role in osmoregulation (Vairavamuthry, Andreae and Iverson, 1985; Andreae, 1986; Turner *et al.*, 1988) or in buoyancy control (Andreae, 1980; Barnard, Andreae and Iverson, 1984). It has also been postulated that DMSP production is indirectly related to the nitrogen nutrition of algae, with DMSP being a store for excess sulphate taken up while assimilating the molybdenum necessary to synthesize nitrate reductase or nitrogenase (Rudnick and Howarth, 1990; see also Section 4.6 below). This could also be true for vascular plants with high nitrate-reductase activities or with symbiotic nitrogen-fixing bacteria.

4.3 SULPHUR AS AN ELECTRON ACCEPTOR AND DONOR

4.3.1 SULPHATE REDUCTION

Sulphate can serve as a terminal electron acceptor for bacterial respiration in the absence of oxygen, and so dissimilatory sulphate reduction is coupled to carbon oxidation and nutrient mineralization in anoxic environments. Sulphate-reducing bacteria themselves metabolize only a limited range of relatively low-molecular-weight organic substrates (Sørensen, Christensen and Jørgensen, 1981; Ivanov *et al.*, 1989). Other bacteria ferment the complex organic matter in sediments, and their fermentation products become the substrates for the sulphate-reducing bacteria. Thus, it is a bacterial assemblage of fermenters and sulphate reducers which decompose organic carbon. Overall, two moles of carbon are oxidized to carbon dioxide for every mole of sulphate reduced (Jørgensen, 1977, 1982a; Howarth and Teal, 1979; Berner, 1984). The mineralization of other elements in organic matter is stoichiometrically related to the mineralization of carbon.

In coastal marine sediments and in marine wetlands, sulphate reduction is the major form of respiration; this process and the associated fermentation reactions which feed substrates to sulphate-reducing bacteria are responsible for most carbon and nutrient mineralization in these sediments (Jørgensen, 1982a; Howarth, 1984; Ivanov *et al.*, 1989; Giblin and Wieder, this volume). In freshwater environments, sulphate concentrations are much lower than in seawater, and sulphate often limits the rate of sulphate reduction in lake and freshwater–wetland sediments (Lovely and Klug, 1986; Cook and Kelly, this volume; Giblin and Wieder, this volume). None the less, sulphate reduction and associated fermentations frequently account for a significant percentage of the carbon and nutrient mineralization in freshwater sediments (Cook and Kelly, this volume; Giblin and Wieder, this volume). Dissimilatory sulphate reduction is not generally a major process in non-flooded, terrestrial ecosystems since soils tend to be oxic.

4.3.2 SULPHUR AND METHANE FLUXES

The formation of methane and fluxes of methane from sediments are intimately tied to the sulphur cycle. In the absence of sulphate, methanogenesis is the dominant process of organic matter degradation in anoxic sediments (Cook and Kelly, this volume; Giblin and Wieder, this volume). It was once thought that sulphate-reducing bacteria could completely out-compete methanogenic bacteria for substrates, and so methane should not be produced in sulphate-rich marine sediments. There is now abundant evidence that some methane is formed even in the presence of reasonably high concentrations of sulphate (see review by Ivanov *et al.*, 1989). However, methane is also rapidly oxidized in the presence of sulphate. Although the mechanism of this anoxic consumption of methane remains somewhat unclear, it appears to be coupled to sulphate reduction, either directly or indirectly (Ivanov *et al.*, 1989; Cook and Kelly, this volume; Giblin and Wieder, this volume). The net result is that methane fluxes from sediments are inversely correlated with sulphate concentrations in freshwater and brackish environments (King and Wiebe, 1980; Bartlett *et al.*, 1987; Yavitt, Lang and Wieder, 1987; Giblin and Wieder, this volume). Flooded rice paddies are thought to be major sources of methane to the world's atmosphere (Crutzen, 1987; Aselmann and Crutzen, 1989; Schutz, Seiler and Rennenberg, 1990). The addition of sulphur fertilizer to rice paddies may reduce the emission of methane from these systems.

4.3.3 CHEMOSYNTHESIS AND ANOXYGENIC PHOTOSYNTHESIS

Energy is conserved in the reduced sulphur compounds formed during sulphate reduction (Fenchel and Jørgensen, 1977; Howarth, 1984). Relative to aerobic respiration, approximately 75% of the potential energy in the organic matter consumed is transferred and stored in reduced sulphur during sulphate reduction (Howarth and Teal, 1980). When these reduced sulphur compounds are re-oxidized to sulphate, the energy is available to support bacterial growth. A variety of chemosynthetic bacteria can oxidize reduced sulphur compounds using oxygen or nitrate as electron acceptors and fixing carbon dioxide (Kelly, 1982; Jørgensen, 1982b). These chemosynthetic bacteria also assimilate other major and minor elements in stoichiometric proportion to carbon fixation, just as with primary production by plants. Chemosynthetic sulphur oxidizers are commonly found at the oxic/anoxic interface in sediments where both reduced sulphur and oxygen are abundant (Lein and Ivanov, this volume). Calculations based on energy flow suggest that chemosynthetic bacteria may fix enough carbon dioxide into organic carbon to support between 3 and 18% of the total rate of respiration in various nearshore marine sediments (Howarth, 1984).

The chemosynthetic bacteria require oxygen or nitrate to oxidize reduced sulphur compounds. However, sulphide can also be oxidized in the absence of oxygen and nitrate if light is present; a variety of photosynthetic bacteria, including purple and green sulphur bacteria and some cyanobacteria, use sulphide as an electron donor in anoxygenic photosynthesis (Pfennig, 1975; Gorlenko, 1979; Cohen, Gorlenko and Bonch-Osmolovskaya, 1989). Anoxygenic, photosynthetic bacteria occur in large numbers in the upper layers of anoxic water columns when enough light is present to support their photosynthesis. They are the dominant primary producers in some meromictic lakes, particularly those with relatively little iron to precipitate sulphides (Gorlenko, in press). Photosynthetic sulphur bacteria are also very active in bacterial mats on the surface of sediments receiving light (Cohen, Gorlenko and Bonch-Osmolovskaya, 1989). As with plants and chemosynthetic bacteria, carbon fixation by photosynthetic sulphur bacteria is stoichiometrically coupled to nutrient assimilation.

4.4 ACIDITY AND OXIDATION–REDUCTION CHEMISTRY OF SULPHUR

Most oxidation reactions of sulphur produce protons (acidity), while most reduction reactions consume protons. The process of sulphate reduction generates alkalinity and consumes protons in stoichiometric proportion as follows:

$$SO_4^{2-} + 2(CH_2O) + 2H^+ = H_2S + 2CO_2 + 2H_2O$$

Sulphate reduction in lakes helps mitigate the effects of acid deposition by generating alkalinity and consuming some of the acidity associated with sulphuric acid inputs (Schindler *et al.*, 1986; Cook *et al.*, 1986; Cook and Kelly, this volume). However, the mitigating effect persists only as long as the reduced sulphur formed from sulphate reduction is stored within the lake; if the reduced sulphur is re-oxidized to sulphate, acidity is produced and alkalinity consumed. For example:

$$H_2S + 2O_2 = SO_4^{2-} + 2H^+$$

Reduced sulphur formed during sulphate reduction can be stored in sediments either as metal sulphide minerals (principally iron sulphides) or as organic sulphur compounds formed by nucleophilic attack of sulphide on organic matter (Nriagu and Soon, 1985; Rudd, Kelly and Furitani, 1986; Cook and Kelly, this volume; Luther and Church, this volume). In some lakes, the availability of iron may limit the formation of iron sulphides, thereby limiting the storage of reduced sulphur and the extent of permanent alkalinity generation from sulphate reduction (Schindler, 1985; Carignan and Tessier, 1988; Giblin *et al.*, 1990).

Many wetland soils contain large amounts of pyrite (FeS_2) as a result of sulphate reduction and sulphide precipitation. Marine wetlands are particularly rich in pyrite (Howarth, 1984; Giblin and Wieder, this volume; Luther and Church, this volume). If these soils are drained, oxygen penetration increases, and pyrite is oxidized. The pyrite oxidation stoichiometrically creates protons:

$$FeS_2 + 15/4O_2 + 5/2H_2O = 2SO_4^{2-} + FeOOH + 4H^+$$

When marine and brackish wetland are drained, the soils can become so acidic as to be unable to support agricultural use or other plant growth (Van Breemen, 1982).

Coal seams also often contain large amounts of pyrite, a result of sulphate reduction in the ancient sediments from which the coal evolved. When the coal is mined, the pyrite is exposed to oxygen in air and oxygenated drainage waters. The pyrite is oxidized, and the resulting acid-mine drainage is a major environmental problem in many mining regions.

Elemental sulphur is often added to agricultural soils as a sulphur fertilizer (Schoenau and Germida, this volume). This sulphur is oxidized to sulphate, creating acidity:

$$S^o + 3/2O_2 + H_2O = SO_4^{2-} + 2H^+$$

The addition of sulphur as sulphate in fertilizers, such as in $CaSO_4$ or in superphosphate, will not acidify the soil.

4.5 PRECIPITATION, COMPLEXATION AND SORPTION CHEMISTRY OF SULPHUR

The inorganic chemistry of sulphur can affect the movement and availability of a variety of other elements. In many soils, sulphate is the most abundant divalent anion and is often the most abundant anion of any charge. As such, its abundance can influence the movement and adsorption of basic and acidic cations. In reducing environments, sulphides and polysulphides precipitate and complex with a variety of metals, often controlling the chemistry of the metals. Indirectly, precipitation of metal sulphides can control the availability of other elements, such as phosphorus. Below, we briefly present examples of some of these interactions.

4.5.1 SULPHATE AND THE MOBILITY OF CATIONS IN SOILS

In soils where sulphate is mobile, it is commonly the major anion serving as a 'carrier' of protons and other cations, determining their mobility (Johnson, 1984; Mitchell and Fuller, 1988; Mitchell, David and Harrison, this volume). The cation exchange capacity and base saturation of a soil determine which cations leach in association with the mobile sulphate; protons and the Al^{3+} are leached in those soils which have very low concentrations of basic cations and which have low cation exchange capacities (Mitchell, David and Harrison, this volume). In such soils, sulphate anions thus partially regulate the downstream export of acid and toxic aluminium.

The mobility of sulphate anions in soils is a function of interactions with other elements. The mechanisms of sulphate retention in soils are complex and are not completely understood, but adsorption on to aluminium and iron hydroxides appears to be an important mechanism in many soils (Rajan, 1978; Johnson, 1984; Mitchell, David and Harrison, this volume). Precipitation of aluminium sulphate minerals may also occur when the pH is low and aluminium concentrations are high (Matzner and Ulrich, 1981; Johnson, 1984; Mitchell, David and Harrison, this volume). Iron sulphate minerals can be important in retaining sulphate in drained (oxic) wetland soils (Van Breemen, 1982).

4.5.2 PRECIPITATION OF METAL SULPHIDES IN ANOXIC SOILS AND SEDIMENTS

In reducing sediments and flooded wetland soils, soluble sulphides (H_2S, HS^-, S^{2-}) are formed from sulphate reduction. Most of the sulphides are rapidly precipitated as metal sulphides, predominantly iron monosulphides and pyrite (FeS_2) (Berner, 1984; Howarth, 1984; Howarth and Jorgensen, 1984; Luther and Church, this volume; Giblin and Wieder, this volume). The precipitation keeps concentrations of soluble sulphides relatively low, which can be important since soluble sulphides are toxic to a variety of organisms. In many salt marshes, the rapid precipitation of sulphides in pyrite may minimize the toxic effects of sulphide and allow for the high plant productivity typically observed (Howarth and Teal, 1979; Chalmers, 1982). In some wetlands, however, a low availability of iron may limit pyrite formation, favouring higher concentrations of soluble sulphides and organic sulphur (Howarth, 1984).

4.5.3 IRON, SULPHIDES AND PHOSPHATE ADSORPTION

Most of the iron for precipitation of iron sulphides comes from reduction of iron(III) oxides and hydroxides within the sediment (Lord and Church, 1983; Berner, 1984; Canfield, 1989; Luther and Church, this volume). Significant amounts of oxidized iron persist even at depth in reducing sediments, but the most reactive oxidized iron is reduced and converted to iron sulphides (Berner, 1984). Since phosphate is highly sorptive on reactive iron(III) minerals, phosphate is released as the iron is reduced. This has long been recognized as a mechanism for returning phosphorus from sediments to the water column in lakes (Mortimer, 1941), although non-iron phases are also efficient adsorbers of phosphate in many lake sediments (Schindler, Hesslein and Kipphut, 1977; Bostrom, Jansson and Forsber, 1982; Wodka, Effler and Driscoll, 1985) and in calcaraeous marine sediments (Morse *et al.*, 1987). Caraco, Cole and Likens (1989) have proposed that the release of phosphorus from both freshwater and marine sediments is correlated with the sulphate concentration of the overlying water. This may explain why there tends to be less phosphorus adsorption in non-calcaraeous marine sediments than in freshwater lake sediments (Caraco, Cole and Likens, 1990).

4.5.4 OTHER METAL SULPHIDES AND POLYSULPHIDES

Although iron sulphides are the most abundant metal sulphides in most sediments, sulphides form highly insoluble minerals with a variety of other metals, including lead, copper, cadmium, zinc, silver, and mercury (Framson and Leckie, 1978). These trace metals can also co-precipitate in iron monosulphides and pyrite (Luther *et al.*, 1980). However, many of these metals also form stable, soluble complexes with sulphides and polysulphides (Boulegue *et al.*, 1982; Luther and Church, this volume). Thus, the mobility of many trace metals in anoxic sediments, soils, and waters is controlled by the balance between precipitation as metal sulphides and complexation with polysulphides (Boulegue *et al.*, 1982; Emerson, Jacobs and Tebo, 1982).

4.6 ELEMENT SUBSTITUTIONS AND CONFUSIONS

Many trace elements have biogeochemical cycles which closely parallel those of more abundant elements. For instance, the following analogues and substitutions are well known: strontium for calcium, arsenic for phosphorus, and germanium for silica. The cycles of two trace elements, selenium and molybdenum, are intimately related to the cycle of sulphur.

4.6.1 SULPHUR AND SELENIUM

Selenium is an essential trace element which is toxic if present in excessive concentrations (Fergusson, 1990). In natural systems, it exists as selenate (SeO_4^{2-}), selenite (SeO_3^{2-}), and organic and inorganic forms of selenide (Se^{2-}) (Cutter and Bruland, 1984; Fergusson, 1990). Chemically, these are all similar to forms of sulphur: sulphate, sulphite, and sulphides. In the selenide form, selenium substitutes for sulphur in proteins (Stadtman, 1974). Sulphate can inhibit the assimilation of selenate by some bacteria (Brown and Shrift, 1980; Bryant and Laishley, 1988).

Sulphate-reducing bateria are capable of reducing selenate to selenide (Zehr and Oremland, 1987). However, even low levels of sulphate inhibit selenate reduction by sulphate-reducing bacteria (Zehr and Oremland, 1987). Selenate reduction in sulphate-rich environments is probably mediated by other types of bacteria which are not inhibited by sulphate (Oremland *et al.*, 1989; Steinberg and Oremland, 1990).

4.6.2 SULPHUR AND MOLYBDENUM

Molybdenum is toxic to mammals if present in excessive concentrations (Fergusson, 1990) but is essential for the processes of nitrate reduction in plants and bacteria and of nitrogen fixation in bacteria (Fogg and Wolfe, 1954). Reduced forms of molybdenum bear little resemblance to sulphur compounds (Manheim and Landergren, 1978). However, the most oxidized form of molybdenum, molybdate (MoO_4^{2-}), is very similar in stereochemistry to sulphate (Howarth and Cole, 1985). As a result, sulphate can inhibit the assimilation of molybdate. This inhibition has been demonstrated in tomato plants (Stout and Meagher, 1948), bacteria (Elliott and Mortenson, 1975), mammalian intestines (Huising and Matrone, 1975; Cardin and Mason, 1976), and planktonic algae and cyanobacteria (Howarth and Cole, 1985; Cole *et al.*, 1986; Howarth, Marino and Cole, 1988). Sulphate is sometimes fed to sheep to reduce the toxic effects of excessive molybdenum in their food (Huising and Matrone, 1975). In oxic natural waters, molybdate is the thermodynamically stable form of molybdenum (Manheim and Landergren, 1978). Since the environmental abundance of sulphate in natural waters is thousands- to millions-fold higher than that of molybdenum, the inhibitory effect of sulphate can make molybdenum availability quite low. This may contribute to low rates of nitrogen fixation and nitrogen limitation in coastal marine ecosystems and in some saline lakes (Howarth and Cole, 1985; Howarth, Marino and Cole, 1988; Marino *et al.*, 1990).

REFERENCES

Anderson, D. W., Sagger, S., Bettany, J. R. and Stewart, J. W. B. (1981). Particle size fractions and their use in studies of soil organic matter: I. The nature and distribution of forms of carbon, nitrogen, and sulfur. *Soil Sci. Soc. Am. J.*, **45**, 767–72.

Andreae, M. O. (1980). The production of methylated sulfur compounds by marine phytoplankton. In: Trudinger, P. A. and Walter, M. R. (Eds). *Biogeochemistry of Ancient and Modern Environments*. Springer, NY, pp. 253–9.

Andreae, M. O. (1986). The ocean as a source of atmospheric sulphur compounds. In: Buat-Menard, P. (Ed.). *The Role of Air–Sea Exchange in Geochemical Cycling*. D. Reidel, Boston, pp. 331–62.

Andreae, M. O. and Jaeschke, W. A. This volume.

Andren, O., Paustian, K. and Rosswall, T. (1988). Soil biotic interactions in the functioning of agroecosystems. *Agriculture, Ecosystems and Environment*, **24**, 57–67.

Aselmann, I. and Crutzen, P. J. (1989). Global distribution of natural freshwater wetlands and rice paddies: their net primary productivity, seasonality and possible methane emissions. *J. Atmos. Chem.*, **8**, 307–58.

Barnard, W. R., Andreae, M. O. and Iverson, R. L. (1984). Dimethylsulfide and *Phaeocystics poucheti* in the southeastern Bering Sea. *Cont. Shelf Res.*, **3**, 103–13.

Bartlett, K. B., Bartlett, D. S., Harriss, R. C. and Sebacher, D. I. (1987). Methane emissions along a salt marsh salinity gradient. *Biogeochemistry*, **4**, 183–202.

Berner, R. A. (1984). Sedimentary pyrite formation: an update. *Geochim. Cosmochim. Acta*, **48**, 605–15.

Bettany, J. R., Saggar, S. and Stewart, J. W. B. (1980). Comparison of the amounts and forms of sulfur in soil organic matter fractions after 65 years of cultivation. *Soil Sci. Soc. Am. J.*, **44**, 70–5.

Bettany, J. R., Stewart, J. W. B. and Halstead, E. H. (1973). Sulfur fractions and carbon, nitrogen, and sulfur relationships in grassland, forest, and associated transitional soils. *Soil Sci. Soc. Am. Proc.*, **37**, 915–18.

Bettany, J. R., Stewart, J. W. B. and Saggar, S. (1979). The nature and forms of sulfur in organic matter fractions of soils selected along an environmental gradient. *Soil. Sci. Soc. Am. J.*, **43**, 981–5.

Biederbeck, V. O. (1978). Soil organic sulfur and fertility. In: Schnizer, A. and Khan, S. U. (Eds). *Soil Organic Matter. Development in Soil Science 8*. Elsevier Scientific Publishing Co., New York, pp. 273–310.

Bolin, B. and Cook, R. B. (1983). *The Major Biogeochemical Cycles and their Interactions*. Wiley, Chichester.

Bostrom, B., Jansson, M. and Forsber, C. (1982). Phosphorus release from lake sediments. *Arch. Hydrobiol. Beih. Ergebn. Limnol.*, **18**, 5–59.

Bremner, J. M. (1977). Role of organic matter in volatilization of sulphur and nitrogen from soils. In: *Soil Organic Matter Studies*, Vol. II. *Proceedings of an International Symposium, Braunschweig, Germany*. International Atomic Energy (IAEA), Vienna, Austria, pp. 229–39.

Brimblecombe, P. and Lein, A. Yu. (1989). *Evolution of the Global Biogeochemical Sulphur Cycle*. Wiley, Chichester.

Brown, T. A. and Shrift, A. (1980). Assimilation of selenate and selinite by *Salmonella typhimurium*. *Can. J. Microbiol.*, **26**, 671–5.

Bryant, R. D. and Laishley, E. J. (1988). Evidence for two transporters of sulfur and selenium oxyganions in *Clostridium pasteurianum*. *Can. J. Microbiol.*, **34**, 700–3.

Canfield, E. (1989). Reactive iron in marine sediments. *Geochim. Cosmochim. Acta*, **53**, 619–32.

Caraco, N., Cole, J. J. and Likens, G. E. (1989). Evidence for sulfate-controlled phosphorus release from sediments of aquatic systems. *Nature*, **341**, 316–18.

Caraco, N., Cole, J. J. and Likens, G. E. (1990). A comparison of phosphorus immobilization in sediments of freshwater and coastal marine systems. *Biogeochemistry*, **9**, 277–90.

Cardin, C. J. and Mason, J. (1976). Molybdate and tungstate transfer in rat ileum: competitive inhibition by sulphate. *Biochim. Biophys. Acta*, **455**, 937–46.

Carignan, R. and Tessier, A. (1988). The co-diagenesis of sulfur and iron in acid lake sediments of southwestern Quebec. *Geochim. Cosmochim. Acta*, **52**, 1179–88.

Chalmers, A. G. (1982). Soil dynamics and the productivity of *Spartina alterniflora*. In: Kennedy, V. S. (Ed.), *Estuarine Comparisons*. Academic Press, New York, pp. 231–42.

Chichester, F. W., Legg, J. O. and Stanford, G. (1975). Relative mineralization rates of indigenous and recently incorporated ^{15}N-labeled nitrogen. *Soil Sci.*, **120**, 455–60.

Cohen, Y., Gorlenko, V. M. and Bonch-Osmolovskaya, E. A. (1989). Interactions of sulphur and carbon cycles in microbial mats. In: Brimblecombe, P. and Lein, A. Yu. (Eds). *Evolution of the Global Biogeochemical Sulphur Cycle*. Wiley, Chichester, pp. 191–238.

Cole, J. J., Howarth, R. W., Nolan, S. S. and Marino, R. (1986). Sulfate inhibition of molybdate assimilation by planktonic algae and bacteria: some implications for the aquatic nitrogen cycle. *Biogeochemistry*, **2**, 179–96.

Coleman, D. C., Reid, C. P. P. and Cole, C. V. (1983). Biological strategies of nutrient cycling in soil systems. *Adv. Ecol. Res.*, **13**, 1–55.

Cook, R. B., Kelly, C. A., Schindler, D. W. and Turner, M. A. (1986). Mechanisms of hydrogen ion neutralization in an experimentally acidified lake. *Limnol. Oceanogr.*, **31**, 134–48.

Cook, R. B. and Kelly, C. A. This volume.

Crutzen, P. J. (1987). Role of tropics in atmospheric chemistry. In: Dickinson, R. E. (Ed.).*The Geophysiology of Amazonia. Vegetation and Climate Interactions*. Wiley, New York, pp. 107–30.

Cuhel, R. L. and Waterbury, J. B. (1984). Biochemical composition and short term nutrient incorporation patterns in a unicellular marine cyanobacterium, *Synechococcus* (WH7803). *Limnol. Oceanogr.*, **29**, 370–4.

Cutter, G. A. and Bruland, K. W. (1984). The marine biogeochemistry of selenium: a re-evaluation. *Limnol. Oceanogr.*, **29**, 1179–92.

David, M. B. and Mitchell, M. J. (1985). Sulfur constituents and cycling in waters, seston, and sediments of an oligotrophic lake. *Limnol. Oceanogr.*, **30**, 1196–207.

Degens, E. T., Kempe, S., Lein, A. Yu and Soroken, Y. (in press). *Interactions of Biogeochemical Cycles in Aqueous Ecosystems*. University of Hamburg.

Dijkshoorn, W. and Van Wijk, A. L. (1967). The sulphur requirements of plants as evidenced by the sulphur–nitrogen ratio in the organic matter. A review of published data. *Plant Soil*, **26**, 129–721.

Elliot, B. B. and Mortenson, L. E. (1976). Regulation of molybdate transport by *Clostridium pasteurianum*. *J. Bacteriol.*, **127**, 770–9.

Emerson, S., Jacobs, L. and Tebo, B. (1982). The behavior of trace metals in marine anoxic waters: solubilities at the oxygen–hydrogen sulfide interface. In Wong, C. S., Boyle, E., Bruland, K. W., Burton, J. D. and Goldberg, E. D. (Eds). *Trace Metals in Sea Water*. Plenum, New York, pp. 579–608.

Fenchel, T. and Jorgensen, B. B. (1977). Detritus food chains of aquatic ecosystems: the role of bacteria. *Adv. Microb. Ecol.*, **1**, 1–57.

Fergusson, J. E. (1990). *The Heavy Elements: Chemistry, Environmental Impact, and Health Effects*. Pergamon, Oxford.

Fitzgerald, J. W. (1976). Sulfate ester formation and jydrolysis: A potentially important yet often ignored aspect of the sulfur cycle of aerobic soils. *Bacteriolog. Rev.*, **49**, 698–721.

Fitzgerald, J. W. (1978). Naturally-occurring organosulfur compounds in soil. In: Nriagu, J. O. (Ed.). *Sulfur in the Environment, Part II—Ecological Impacts*. Wiley, New York, pp. 391–443.

Fogg, G. E. and Wolfe, M. (1954). Nitrogen metabolism of blue–green algae. *Symp. Soc. Gen. Microbiol.*, **4**, 99–125.

Framson, P. E. and Leckie, J. O. (1978). Limits of coprecipitation of cadmium and ferrous sulfides. *Am. Chem. Soc.*, **12**, 465–9.

Freney, J. R. (1961). Some observations on the nature of organic sulphur compounds in soil. *Aust. J. Agric. Res.*, **12**, 424–32.

Freney, J. R. (1986). Forms and reactions of organic sulfur compounds in soils. In: Tabatabai, M. A. (Ed.). *Sulfur in Agriculture*. *Agronomy No. 27*, American Society of Agronomy, Inc., Madison, WI, pp. 207–32.

Freney, J. R., Melville, G. E. and Williams, C. H. (1970). The determination of carbon bonded sulfur in soil. *Soil Sci.*, **109**, 310–18.

Freney, J. R., Melville, G. E. and Williams, C. H. (1975). Soil organic matter fractions as sources of plant-available sulphur. *Soil Biol. Biochem.*, **7**, 217–21.

Giblin, A. E. and Wieder, R. K. This volume.

Giblin, A. E., Likens, G. E., White, D. and Howarth, R. W. (1990). Sulfur storage and alkalinity generation in New England lake sediments. *Limnol. Oceanogr.*, **35**, 852–69.

Gorlenko, V. M. (1979). Phototrophic sulfur bacteria of salt meromictic lakes and their role in the sulfur cycle. In: Krumbein, W. E. (Ed.). *Environmental Biochemistry and Geomicrobiology*, Vol. 1, *The Aquatic Environment*. Ann Arbor Science, Ann Arbor, pp. 109–19.

Gorlenko, V. M. (in press). The role of purple and green bacteria in the carbon and sulfur cycles in stratified lakes. In: Degens, E. T., Kempe, S., Lein, A. and Soroken, Y. (Eds). *Interactions of Biogeochemical Cycles in Aqueous Ecosystems*. University of Hamburg.

Harwood, J. L. and Nicholls, R. G. (1979). The plant sulpholipid-a major component of the sulphur cycle. *Biochem. Soc. Trans.*, **7**, 440–47.

Houghton, C. and Rose, F. A. (1976). Liberation of sulfate from sulfate esters by soils. *Appl. Environ. Microbiol.*, **31**, 969–76.

Howarth, R. W. (1984). The ecological significance of sulfur in the energy dynamics of salt marsh and marine sediments. *Biogeochemistry*, **1**, 5–27.

Howarth, R. W. and Cole, J. J. (1985). Molybdenum availability, nitrogen limitation, and phytoplankton growth in natural waters. *Science*, **229**, 653–5.

Howarth, R. W. and Jørgensen, B. B. (1984). Formation of ^{35}S-labelled elemental

sulfur and pyrite in coastal marine sediments (Limfjorden and Kysing Fjord, Denmark) during short-term $^{35}SO_4^{2-}$ reduction measurements. *Geochim. Cosmochim. Acta*, **48**, 1807–18.
Howarth, R. W., Marino, R. and Cole, J. J. (1988). Nitrogen fixation in freshwater, estuarine, and marine ecosystems: 2. Biogeochemical controls. *Limnol. Oceanogr.*, **33**, 688–701.
Howarth, R. W. and Teal, J. M. (1979). Sulfate reduction in a New England salt marsh. *Limnol. Oceanogr.*, **24**, 999–1013.
Howarth, R. W. and Teal, J. M. (1980). Energy flow in a salt marsh ecosystem: the role of reduced inorganic sulfur compounds. *Am. Nat.*, **116**, 862–72.
Huising, J. and Matrone, G. (1975). Biological interactions of sulfate and molybdate. *Environ. Health Perspect.* **10**, 265.
Hunt, W. B., Stewart, J. W. B. and Cole, C. V. (1986). Concepts of sulfur, carbon, and nitrogen transformations in soils: evaluation by simulation modeling. *Biogeochemistry*, **2**, 163–77.
Ivanov, M. V. and Freney, J. R. (1983). *The Global Biogeochemical Sulphur Cycle*. Wiley, Chichester.
Ivanov, M. V., Lein, A. Yu., Reeburgh, M. S. and Skyring, G. W. (1989). Interaction of sulphur and carbon cycles in marine sediments. In: Brimblecombe, P. and Lein, A. Yu. (Eds). *Evolution of the Global Biogeochemical Sulphur Cycle*. Wiley, Chichester, pp. 125–79.
Johnson, D. W. (1984). Sulfur cycling in forests. *Biogeochemistry*, **1**, 29–43.
Jørgensen, B. B. (1977). The sulfur cycle of a coastal marine sediment (Limfjorden, Denmark). *Limnol. Oceanogr.*, **22**, 814–32.
Jørgensen, B. B. (1982a). Mineralization of organic matter in the sea bed: the role of sulfate reduction. *Nature*, **296**, 643–5.
Jørgensen, B. B. (1982b). Ecology of the bacteria of the sulphur cycle with special reference to anoxic–oxic interface environments. *Phil. Trans. R. Soc. Lond. B. Bio. Sci.*, **298**, 543–61.
Kelly, D. P. (1982). Biochemistry of chemolithotrophic oxidation of inorganic sulphur. *Phil. Trans. R. Soc. Lond. B. Bio. Sci.*, **298**, 499–528.
Kelly, J. and Lambert, M. J. (1972). The relationship between sulphur and nitrogen in the foliage of *Pinus radiata*. *Plant Soil*, **37**, 395–408.
King, G. M. and Wiebe, W. J. (1980). Regulation of sulfate concentrations and methanogenesis in salt marsh soils. *Estuarine Coastal Mar. Sci.*, **10**, 215–23.
Lefroy, R. D. B., Mamaril, C. P., Blair, G. J. and Gonzales, P. B. This volume.
Lein, A. Yu, and Ivanov, M. V. (This volume).
Likens, G. E. (1981). *Some Perspectives of the Major Biogeochemical Cycles*. Wiley, Chichester.
Loll, J. J. and Bollag, J. (1983). Protein transformation in soil. *Adv. Agron.*, **36**, 351–82.
Lord, C. J. and Church, T. M. (1983). The geochemistry of salt marshes: sedimentary ion diffusion, sulfate, reduction, and pyritization. *Geochim. Cosmochim. Acta*, **47**, 1381–91.
Lovely, D. R. and Klug, M. J. (1986). Model for the distribution of sulfate reduction and methanogenesis in freshwater sediments. *Geochim. Cosmochim. Acta*, **50**, 11–18.
Luther, G. W. and Church, T. M. (This volume).
Luther, G. W., Meyerson, A. L., Krajewski, J. J. and Hires, R. (1980). Metal sulfides in estuarine sediments. *J. Sediment. Petrol.*, **50**, 1117–20.
Lynch, J. M. (1982). Limits to microbial growth in soil. *J. Gen. Microbiol.*, **128**, 405–10.

McGill, W. G. and Cole, C. V. (1981). Comparative aspects of cycling of organic C, N, S, and P through soil organic matter. *Geoderma*, **26**, 267–86.

McLachlan, K. D. and DeMarco, D. G. (1976). Changes in soil sulphur fractions with fertilizer additions and cropping treatments. *Aust. J. Soil Res.*, **13**, 169–76.

McLaren, R. G. and Swift, R. S. (1977). Changes in soil organic sulphur fractions due to the long term cultivation of soils. *J. Soil Sci.*, **28**, 445–53.

Manheim, F. T. and Landergren, S. (1978). *Molybdenum. Handbook of Geochemistry*, sections 42 B-O. V. 2/5. Springer, New York.

Marino, R., Howarth, R. W., Shamess, J. and Prepas, E. (1990). Molybdenum and sulfate as controls on the abundance of nitrogen-fixing cyanobacteria in saline lakes in Alberta. *Limnol. Oceanogr.*, **35**, 245–59.

Matzner, E. and Ulrich, B. (1981). Balances of annual element fluxes within forest ecosystems in the Solling region. *Z. Pflanz. Bodenkd.*, **144**, 660–81.

Maynard, D. G., Stewart, J. W. B. and Bettany, J. R. (1984). Sulfur cycling in grassland and parkland soils. *Biogeochemistry*, **1**, 97–111.

Mitchell, M. J. and Fuller, R. D. (1988). Models of sulfur dynamics in forest and grassland ecosystems with emphasis on soil processes. *Biogeochemistry*, **5**, 133–63.

Mitchell, M. J., David, M. B. and Harrison, R. B. (This volume).

Morse, J. W., Zullig, J. J., Iverson, R. L., Choppin, G. R., Mucci, A. and Millero, F. J. (1987). The influence of seagrass beds on the chemistry of coarse-grained sediments in the Bahamas. *Mar. Chem.*, **22**, 71–83.

Mortimer, C. H. (1941). The exchange of dissolved substances between mud and water, I and II. *J. Ecol.*, **29**, 280–329.

Nriagu, J. O. and Soon, Y. K. (1985). Distribution and isotopic composition of sulfur in lake sediments of northern Ontario. *Geochim. Cosmochim. Acta*, **49**, 823–34.

Oremland, R. S., Hollibauh, J. T., Maest, A. S., Presser, T. S., Miller, L. G. and Culbertson, C. W. (1989). Selenate reduction to elemental selenium by anaerobic bacteria in sediments and culture: Biogeochemical significance of a novel, sulfate-independent respiration. *Appl. Environ. Microbiol.*, **55**, 2333–43.

Pfennig, N. (1975). The phototrophic bacteria and their role in the sulfur cycle. *Plant Soil*, **43**, 1–13.

Rajan, S. S. (1978). Sulfate adsorbed on hydrous alumina, ligands displaced, and changes in surface charge. *Soil Sci. Soc. Am. J.*, **42**, 39–44.

Rudd, J. W. M., Kelly, C. A. and Furitani, A. (1986). The role of sulfate reduction in the long term accumulation of organic and inorganic sulfur in lake sediments. *Limnol. Oceanogr.*, **31**, 1281–91.

Rudnick, D. T. and Howarth, R. W. (1990). Controls of dimethylsulfide production by marine phytoplankton. *Bull. Ecol. Soc. Am.*, **71**, (supplement), 309.

Schindler, D. W. (1985). The coupling of elemental cycles by organisms: evidence from whole-lake chemical perturbations. In: Stumm, W. (Ed.). *Chemical Processes in Lakes*. Wiley, New York, pp. 225–50.

Schindler, D. W., Hesslein, R. and Kipphut, G. (1977). Interactions between sediments and overlying waters in an experimentally eutrophied Precambrian shield lake. In: Golterman, H. L. (Ed.). *Interactions Between Sediments and Freshwater*. Junk, The Hague.

Schindler, D. W., Turner, M. A., Stainton, M. P. and Linsay, G. A. (1986). Natural sources of acid neutralizing capacity in low-alkalinity lakes of the Precambrian shield. *Science*, **232**, 844–7.

Schoenau, J. J. and Germida, J. J. (This volume).

Schutz, H., Seiler, W. and Rennenberg, H. (1990). Soil and land use related sources and sinks of methane (CH_4) in the context of the global methane budget. In:

Boouwman, A. F. (Ed.). *Soils and the Greenhouse Effect*. Wiley, Chichester, pp. 269–85.

Scott, N. M. and Anderson, G. (1976). Organic sulphur fractions in Scottish soils. *J. Sci. Food Agric.*, **27**, 358–66.

Scott, N. M., Bick, W. and Anderson, H. A. (1981). The measurement of sulphur-containing amino acids in some Scottish soils. *J. Sci. Food Agric.*, **32**, 21–4.

Sørensen, J., Christensen, D. and Jørgensen, B. B. (1981). Volatile fatty acids and hydrogen as substrates for sulfate-reducing bacteria in anaerobic marine sediment. *Appl. Environ. Microbiol.*, **42**, 5–11.

Sowden, F. J., Chen, Y. and Schnitzer, M. (1977). The nitrogen distribution in soils formed under widely differing climatic conditions. *Geochim. Cosmochim. Acta*, **41**, 1524–6.

Stadtman, T. C. (1974). Selenium biochemistry. *Science*, **183**, 915–21.

Steinberg, N. A. and Oremland, R. S. (1990). Dissimilatory selenate reduction potentials in a diversity of sediment types. *Appl. Environ. Microbiol.*, **56**, 3550–57.

Stout, P. R. and Meagher, W. R. (1948). Studies of the molybdenum nutrition of plants with radioactive molybdenum. *Science*, **108**, 471–3.

Swift, M. J., Heal, O. W. and Anderson, J. M. (1979). *Decomposition in Terrestrial Ecosystems*. University of California Press, Berkeley, CA, 372pp.

Tabatabai, M. A. (1984). Importance of sulfur in crop production. *Biogeochemistry*, **1**, 45–62.

Trudinger, P. A. and Loughlin, R. E. (1981). Metabolism of simple sulphur compounds. In: Neuberger, A. (Ed.). *Comprehensive Biochemistry, Amino Acid Metabolism and Sulphur Metabolism*. Elsevier Scientific Publishing Co., New York, pp. 165–256.

Turner, S. M., Malin, G., Liss, P. S., Harbour, D. S. and Holligan, P. M. (1988). The seasonal variation of dimethyl sulfide and dimethylsulfoniumpropionate concentrations in nearshore waters. *Limnol. Oceanogr.*, **33**, 364–75.

Vairavamurthy, A., Andreae, M. O. and Iverson, R. L. (1985). Biosynthesis of dimethylsulfide and dimethylpropiothein by *Hymenomonas carterae* in relation to sulfur source and salinity variations. *Limnol. Oceanogr.*, **30**, 59–70.

Van Breemen, V. (1982). Genesis, morphology, and classification of acid sulfate soils in coastal plains. In: Kitrick, J. A., Fanning, D. S. and Hossner, L. R. (Eds). *Acid Sulfate Weathering*. Soil Science Society of America, Madison, Wisconsin, pp. 95–108.

Williams, S. T. (1985). Oligotrophy in soil: Fact or fiction? In: Fletcher, M. and Floodgate, G. D. (Eds). *Bacteria in their Natural Environments*. Special Publications of the Society for General Microbiology, No. 16. Academic Press, London, pp. 81–110.

Wodka, M. C., Effler, S. W. and Driscoll, C. T. (1985). Phosphorus deposition from the epilimnion of Onondaga Lake. *Limnol. Oceanogr.*, **30**, 833–43.

Yavitt, J. B., Lang, G. E. and Wieder, R. K. (1987). Control of carbon mineralization to CH_4 and CO_2 in anaerobic, *Sphagnum*-derived peat from Big Run Bog, West Virginia. *Biogeochemistry*, **4**, 141–57.

Zehr, J. P. and Oremland, R. S. (1987). Reduction of selenate to selenide by sulfate-respiring bacteria: Experiments with cell suspensions and estuarine sediments. *Appl. Environ. Microbiol.*, **53**, 1365–9.

5 Sulphur Cycling in Marine and Freshwater Wetlands

ANNE E. GIBLIN
Marine Biological Laboratory, Woods Hole MA, USA

and

R. KELMAN WIEDER
Villanova University, Villanova PA, USA

5.1 INTRODUCTION

Wetland ecosystems are distributed world-wide from boreal regions to the tropics. Total wetland area has been estimated as 5.8×10^8 ha, although this estimate should be regarded as quite rough (Table 5.1). Saline and/or brackish wetlands comprise approximately 9% of the total global wetland area, while freshwater wetlands occupy the remaining area.

Primary production is quite variable among wetland types. Salt marshes and mangroves are considered to be among the most productive ecosystems of the world, with estimates of above ground net primary production ranging from 125 to greater than 1500 g (C) m^{-2} a^{-1} and total net primary production (above ground + below ground) reaching more than 4000 g (C) m^{-2} a^{-1} (Table 5.1; reviews by Odum, McIvor and Smith, 1982; Schubauer and Hopkinson, 1984). Production in freshwater wetlands also spans a wide range (see review by Brinson, Lugo and Brown, 1981). Among inland *Sphagnum*-dominated wetlands, above ground net primary production ranges from less than 40 to 440 g (C) m^{-2} a^{-1} (Wieder *et al.*, 1989). While net primary production in tidal freshwater wetlands may exceed that of saline wetlands (Whigham *et al.*, 1978), in general net primary production in freshwater wetlands is less than saline wetlands.

Wetland ecosystems, whether salt marshes, brackish water marshes, freshwater bogs, fens, marshes or swamps, are all characterized as having soils that are frequently or continuously waterlogged and that are often high in organic matter content (Cowardin *et al.*, 1979). The duration and frequency of waterlogging, however, are quite varied. For example, many salt marshes and

Sulphur Cycling on the Continents
Edited by R.W. Howarth, J.W.B. Stewart and M.V. Ivanov

Table 5.1. Characteristics of the world's wetlands

	Area (10^6 ha)	Net primary production (g (C) m^{-2} a^{-1})
Freshwater wetlands		
Bogs	297[a]	126–840[d]
Swamps	210[a]	420–2940[d]
Alluvial wetlands	19[a]	420–1680[d]
Saline wetlands		
Mangroves	14[b]	920–4000[e]
Salt marshes	38[b]	462–3234[f]

Sources:
[a]Matthews and Fung (1987).
[b]Maltby (1988).
[c]Woodwell *et al.* (1973).
[d]Brinson, Lugo and Brown (1981) (assuming 42% carbon).
[e]Odum, McIvor and Smith (1982).
and [f]Schubauer and Hopkinson (1984) (assuming 42% carbon).

some freshwater marshes are regularly flooded with each tidal cycle. Bottomland hardwood swamps may be dry during most of the year, but inundated during periods of high rainfall and/or runoff as often occurs during the spring months. Freshwater bogs may be waterlogged year-round, although water table fluctuations influence the depth of the waterlogged soil. Under saturated conditions, wetland soils are typically oxic at the surface and anoxic at depth (Teal and Kanwisher, 1961; Ingram, 1978; Howest *et al.*, 1981), yet the distinctness of the transition between oxic and anoxic zones is quite variable depending on such factors as soil drainage, horizontal rates of water flow, and the presence or absence of vascular plants that may deliver oxidants to otherwise anoxic soil layers (Sparling, 1966; Armstrong and Boatman, 1967; Armstrong, 1979; Clymo and Hayward, 1982; King, 1983; Dacey and Howes, 1984).

There has been intense interest in understanding the sulphur cycle in wetland ecosystems because high inputs of organic matter into wetland soils, along with oxic surface and anoxic subsurface zones, potentially allow sulphur to play a critical role in the biogeochemistry of wetlands. Interest in understanding the sulphur cycle in wetland ecosystems has focused on a variety of issues:

(1) Microbial dissimilatory sulphate reduction may account for much of the total organic matter decomposition (i.e. terminal carbon mineralization) in some wetlands.
(2) Hydrogen sulphide may reach concentrations in soils that are inhibitory or toxic to wetland plants, thereby affecting net primary production.

(3) Reduced sulphur may be an important energy source for chemoautotrophic bacteria, leading to new carbon fixation and thus affecting net ecosystem production and carbon balance.
(4) Wetlands may be a significant source of sulphur gases to the atmosphere, some of which may be oxidized and redeposited to the earth as acid precipitation.
(5) Sulphate reduction in wetlands may control methanogenesis and therefore methane emissions from wetlands, of potential importance because methane is a greenhouse gas.
(6) The oxidation of reduced sulphur compounds in wetland soils produces acidity and therefore may affect the use of such drained soils for agriculture.
(7) Studying sulphur storage in contemporary wetland soils may provide insights into the origins of sulphur in coal.

This chapter reviews what is known about the sulphur cycle in saline and freshwater wetlands.

Conceptually, the major features of the sulphur cycle are similar in saline and freshwater wetlands. However, dissolved sulphate concentrations in wetlands span more than three orders of magnitude, from greater than 45 mM (1.44 g (S) l^{-1} in some salt marsh soils (Luther and Church, 1988), to less than 15 μM (0.48 mg (S) l^{-1}) in freshwater bogs remote from sources of atmospheric sulphur deposition (Gorham and Detenbeck, 1986). For this, and other less well understood reasons, the quantitative importance of specific parts of the sulphur cycle differ considerably between saline and freshwater wetlands. Most of the information presented in this chapter is derived from studies of salt marshes and freshwater *Sphagnum*-dominated wetlands, since the largest data base exists for these types of systems. Whenever possible, generalizations to other wetland types are made.

5.2 FORMS OF SULPHUR IN WETLAND SOILS

Sulphur exists in wetland soils in a variety of oxidation states and may be present in gaseous, soluble, and/or solid forms. A complete understanding of sulphur cycling in wetland soils has been hampered by the analytical difficulty in the unambiguous determination of specific sulphur forms. Although considerable progress has been made regarding methodology, assessment of published data on sulphur forms is made with caution, in light of the various approaches to sulphur fractionation used by different authors.

The total sulphur concentration in wetland soils ranges from 0.2 to 16% of the dry weight (0.06 to 5.0 mmol g^{-1} dry weight) (Howarth and Teal, 1979; Casagrande, Gronli and Sutton, 1980; Postma, 1982; Altschuler *et al.*, 1983;

Lowe and Bustin, 1985; Giblin, 1986; Wieder and Lang, 1986). In general, saline wetland soils contain considerably higher sulphur concentrations than freshwater wetlands. Sulphur concentrations in soils from low-salinity, brackish wetlands (Lowe and Bustin, 1986; Feijtel, 1986; Giblin, 1988) are as high as, or higher than, those reported for salt marshes with higher salinity porewaters, indicating that at salinities above a few parts per thousand, sulphate concentration is not a major factor in determining sulphur content. In freshwater wetlands, sulphur concentrations in peat may be related to sulphate inputs (Wieder and Lang, 1986; Ferguson, Robinson and Press, 1984; Brown, 1985).

The majority of sulphur in wetland soils is present as reduced inorganic sulphur minerals (pyrite, iron monosulphides, and elemental sulphur), organic forms, or sulphate. In general, sulphur gases (Wieder, 1985; Behr, 1985) and dissolved reduced sulphur compounds (Howarth *et al.*, 1983) contribute little to the total sulphur in wetland soils. Soluble sulphate concentrations are sometimes low but often account for 10 to 20% or more of the total sulphur even at depth in waterlogged soils (Casagrande *et al.*, 1977; Altschuler *et al.*, 1983; Howarth, 1984; Behr, 1985; Neue and Mamaril, 1985; Spratt, Morgan and Good, 1987; Urban, Eisenreich and Grigal, 1989; Wieder and Lang, 1988). When wetland soils are drained for agriculture, the large store of reduced sulphur in the sediments may be oxidized, producing acid and sulphate. High concentrations of sulphate minerals (ferrous sulphate, jarosite, and aluminium sulphates) may be present in these soils (see review by Van Breemen, 1982).

In salt marshes and brackish swamps, a majority of the total sulphur is often present as reduced inorganic sulphur minerals (Howarth, 1984; Postma, 1982; Giblin, 1988), although organic sulphur has been found to be dominant in some brackish water marshes and swamps (Casagrande *et al.*, 1977; Lowe and Bustin, 1985) and in the surface peat in some salt marshes (Giblin, 1988). In salt marshes the majority of reduced inorganic sulphur is present as pyrite (FeS_2), with elemental sulphur and FeS typically making up less than a few per cent of the reduced inorganic sulphur (Kaplan, Emery and Rittenberg, 1963; Howarth and Teal, 1979; Lord and Church, 1983; Howarth, 1984; Giblin, 1988). Pyrite is also the overwhelmingly dominant form of sulphur in many brackish swamps (Postma, 1982). In contrast to marine and brackish wetland soils, most of the sulphur in freshwater wetland soils is typically present as organic rather than inorganic sulphur, with carbon bonded sulphur (reduced organic sulphur) much more abundant than ester-sulphate sulphur (oxidized organic sulphur) (Casagrande *et al.*, 1987; Altschuler *et al.*, 1983; Behr, 1985; Lowe, 1986; Spratt, Morgan and Good, 1987; Urban, Eisenreich and Grigal, 1989; Wieder and Lang, 1988). Of the reduced inorganic sulphur in freshwater peats, pyrite is often the dominant form, although concentrations are much lower than in salt marshes (Behr, 1985; Spratt, Morgan and Good, 1987; Urban, Eisenreich and Grigal, 1989; Wieder and Lang, 1988).

5.3 DISSIMILATORY SULPHATE REDUCTION AND ITS MEASUREMENT

Of the many possible transformations of sulphur in wetland soils (see Luther and Church, this volume), the process of dissimilatory sulphate reduction has received the most attention. Dissimilatory sulphate reduction involves a number of steps in which fermentative microorganisms break down organic matter into small organic molecules (e.g. lactate, acetate, alcohols, etc.) that then are oxidized to CO_2 by sulphate-reducing bacteria using sulphate as the terminal electron acceptor. Overall, two moles of organic carbon are oxidized for every mole of sulphate reduced (Howarth and Teal, 1979; Jørgensen, 1982; Grinenko and Ivanov, 1983; Howarth and Stewart, this volume). Sulphate reduction has received attention in wetland soils because in anoxic waterlogged soils it is an important, and sometimes dominant, pathway of carbon mineralization. In addition, sulphate reduction generates alkalinity, which buffers porewater pH, and produces hydrogen sulphide, which may be deleterious to plants.

Experimental studies in subtidal marine sediments (Westrich and Berner, 1984) have shown that where high levels (> 5 mM) of sulphate typical of the marine environment occur, organic carbon availability most often limits sulphate reduction. Sulphate may be limiting in freshwater environments where sulphate concentrations are lower (Cook, 1981), or at depth in some highly organic marine and brackish sediments where sulphate is exhausted by sulphate-reducing bacteria.

Sulphate reduction rates in wetlands have been estimated using a variety of techniques, including mathematical modelling of sulphate and bicarbonate profiles (Lord and Church, 1983; Casey and Lasaga, 1987), measurement of sulphate disappearance in sealed cores (Jørgensen, 1977; Howarth and Teal, 1979; King and Wiebe, 1980), calculation of sulphur retention from whole watershed sulphur budgets (Hemond, 1980; Bayley, Behr and Kelly, 1986; Bayley *et al.*, 1987; Ogden, 1982; Wieder, Yavitt and Lang, this volume), and direct measurement using $^{35}SO_4^{2-}$ (Howarth, 1979; Howarth and Teal, 1979; Howarth and Giblin, 1983; King, 1988; Hines, Knollmeyer and Tagel, 1989). Skyring (1987) and Ivanov *et al.* (1989) provide comprehensive discussions of the advantages, disadvantages, and assumptions associated with each of these approaches to measuring sulphate reduction.

The most commonly used method of measuring sulphate reduction is a radiotracer method in which tracer quantities of $^{35}SO_4^{2-}$ are injected into sediments and the formation of reduced ^{35}S is measured. The basic method was first used in subtidal marine sediments by Ivanov (1956); Jørgensen (1978) described the approach in some detail and provided a theoretical framework for interpreting the method. The method as used by most early investigators working in subtidal marine systems assayed only for acid-volatile sulphides as products of sulphate reduction. These products include both the free sulphides in the porewater and sulphide which has precipitated with iron to form iron monosulphides.

In his original work, Ivanov (1956) pointed out that some of the ^{35}S reduced during sulphate reduction in marine sediments was incorporated into a non-acid-volatile pool, pyrite (FeS_2). Unfortunately, this early report received little attention in the Western literature. More than 20 years later, Howarth (1979) reported that pyrite was the major end-product of short-term $^{35}SO_4^{2-}$ reduction measurements in salt-marsh sediments in Massachusetts, and that measuring the incorporation of ^{35}S only into acid volatile sulphur substantially underestimated sulphate reduction rates because a considerable quantity of reduced ^{35}S was recovered from forms which were not acid volatile. Howarth and colleagues extracted these forms using aqua regia, finding 5 to 10 times more ^{35}S in pyrite and elemental sulphur (also a non-acid-volatile form) than in soluble sulphide and iron monosulphides (the acid-volatile forms) (Howarth, 1979; Howarth and Teal, 1979; Howarth and Giblin, 1983; Howarth and Marino, 1984; Howarth and Merkel, 1984).

The use of aqua regia to analyse for ^{35}S in pyrite and elemental sulphur during $^{35}SO_4^{2-}$ reduction measurements can overestimate rates of reduction if care is not taken to fully rinse away unreduced $^{35}SO_4^{2-}$. Although this has caused some concern in salt-marsh studies (King, 1983; Howes, Dacey and Teal, 1983), other investigators find little or no unreduced $^{35}SO_4^{2-}$ remains after careful rinsing of marsh sediments (Howarth and Merkel, 1984) or lake sediments (Rudd, Kelly and Furutani, 1986; Cook and Kelly, this volume).

Any ambiguity regarding the determination of reduced ^{35}S during the measurement of sulphate reduction rates has largely been resolved by the application of the chromium reduction technique, which uses an acid solution of Cr(II) to reduce pyrite and elemental sulphur to H_2S (Zhabina and Volkov, 1978). The H_2S that is produced can be distilled and trapped, assayed for radioactivity, and quantified. Because chromium does not reduce sulphate to sulphide (Zhabina and Volkov, 1978; Howarth and Jørgensen, 1984; Howarth and Merkel, 1974; Wieder, Lang and Granus, 1985; Canfield *et al.*, 1986), the possible presence of any remaining unreduced sulphate in the soil sample is not of concern. Moreover, tests with yeast and specific organic compounds (amino acids, ester sulphates) have indicated that organic sulphur also is not liberated by chromium reduction (Howarth and Jørgensen, 1984; Wieder, Lang and Granus, 1985; Canfield *et al.*, 1986).

Howarth and Merkel (1984) compared the use of the chromium reduction assay with the aqua regia extraction in measuring sulphate reduction rates in salt marshes in Massachusetts and Georgia. They found excellent agreement between the two methods and confirmed that pyrite was the major short-term product of sulphate reduction. Using the chromium reduction assay, other workers have confirmed that pyrite is often a major product of sulphate reduction in marshes (Howes, Dacey and King, 1984; King *et al.*, 1985;

Hines, Knollmeyer and Tugel, 1989), although the extent of pyrite formation can vary in space and time even within a marsh (Hines, Knollmeyer and Tugel, 1989). The formation of ^{35}S-labelled pyrite during sulphate reduction measurements has also been noted in subtidal marine sediments (Ivanov, Lein and Kashparova, 1976; Ivanov *et al.*, 1980; Lein *et al.*, 1982; Howarth and Jørgensen, 1984). The incorporation of the radiolabel into pyrite is probably a result of rapid formation of pyrite, as originally proposed by Howarth (1979), since isotopic exchange between soluble sulphides and pyrite is not observed within the time scale of a day (Fossing and Jørgensen, 1990). See Luther and Church (this volume) for further discussion of the mechanisms involved.

Chromium reduction is now the most commonly used method of determining the reduced inorganic ^{35}S end-product of sulphate reduction in studies in salt marshes, but rates of sulphate reduction still can be underestimated if a substantial quantity of ^{35}S becomes incorporated into organic, carbon-bonded sulphur during incubation. Carbon-bonded sulphur can be formed either biotically through sulphate uptake and assimilatory sulphate reduction, or abiotically as H_2S reacts with organic matter via nucleophilic attack (Casagrande, Gronli and Sutton, 1980; Altschuler *et al.*, 1983; Brown, 1986; Francois, 1987; Luther and Church, this volume). There is no evidence for significant rates of formation of organic sulphur during short-term incubations with saline or brackish wetland soils with $^{35}SO_4^{2-}$ (Howarth and Merkel, 1984). However, in freshwater wetland soils, failure to account for the incorporation of ^{35}S into carbon-bonded S can lead to substantial underestimates of sulphate reduction (Wieder and Lang, 1988; Spratt and Morgan, submitted). The rapid incorporation of radiolabel into carbon-bonded sulphur has also been observed in lake sediments (Rudd, Kelly and Furutani, 1986).

5.3.1 SULPHATE REDUCTION IN SALTWATER WETLANDS

Much of the primary production in salt marshes takes place below ground (Valiela, Teal and Persson, 1976; Schubauer and Hopkinson, 1984; Good and Frasco, 1982). In studies where total below ground carbon budgets are available, sulphate reduction and associated fermentation reactions are the major anaerobic decomposition processes in soils; moreover, anaerobic decomposition equals or exceeds oxic respiration (Howarth and Teal, 1979; Howes, Dacey and King, 1984; King, 1988). In spite of the importance of sulphate reduction to carbon cycling in saltwater wetlands, there have been relatively few studies over annual cycles, and these largely have been confined to temperate salt marshes dominated by *Spartina alterniflora* (Table 5.2). To our knowledge, no seasonal studies have been carried out in tropical mangrove wetlands.

Table 5.2. Published annual sulphate reduction rates for saline marshes. The method used to make the measurement is also shown

Area	SO_4^{2-} reduction (mol (S) m^{-2} a^{-1})	Reference
Great Sippewissett Salt Marsh, MA		
	75	Howarth and Teal (1979)[a]
	<40	Howes, Dacey and Teal (1983)[b]
Sapelo Island, GA		
	40	Howarth (1984)[c]
Belle Baruch, Goat Island, NC		
	13.3	King (1988) short form[d]
	5.9	tall form[d]
Flax Pond, NY		
	13	Swider (1988) marsh flat[e]
Chapman's Marsh, NH		
	19	Hines, Knollmeyer and Tugel (1989) tall form[d]
	8	*S. patens*[d]
Netherlands		
	1–4	Oenema (1988)[f]
Clone Pt, UK		
	4.4, 9.0*	Nedwell and Abram (1978) creek[g]
	3.1, 4.1*	pan[g]

[a] ^{35}S Aqua regia digestion.
[b] CO_2 flux + single $^{35}SCr(II)$ digestion.
[c] ^{35}S Aqua regia + Cr(II) digestion.
[d] ^{35}S-S-Cr(II) digestion.
[e] Modelling and sealed jar.
[f] Modelling.
[g] ^{35}S AVS only.
* Correcting Nedwell and Abram (1978) using % CRS reported in Nedwell and Takii (1988).

Sulphate reduction rates in salt marshes are generally higher than rates in subtidal marine sediments (Howarth, 1984; Skyring, 1987). Variation in sulphate reduction rates over time, and between sites can be attributed to at least three factors: temperature, substrate availability, and partitioning between oxic and anoxic respirations. Howarth and Teal (1979) found a strong temperature response, but they also found a higher rate of sulphate reduction in the fall than in the spring at similar temperatures in a marsh in Massachusetts. They speculated that the higher autumn rates might be due to an increase in the amount of carbon available below ground when plants senesce. A similar increase during autumn was noted in soil CO_2 production in the same marsh (Howes, Dacey and Teal, 1983). King (1988) found that within a South Carolina salt marsh, temperature accounted for most of the variation in sulphate reduction rates over the season. However, Morris and Whitting (1986) observed a large difference between spring and fall rates of CO_2 production in South Carolina even though the temperatures were similar. Hines, Knollmeyer and Tugel (1989) found the seasonal pattern of sulphate reduction followed changes in temperature in a high marsh *Spartina patens* zone but not in a tall *Spartina alterniflora* zone; rates of sulphate reduction in the tall *S. alterniflora* zone were highest when plant growth was most rapid. Hines, Knollmeyer and Tugel (1989) suggested that the release of dissolved organic matter from plant roots during active growth may fuel sulphate reduction.

Patterns of primary production and biomass allocation seem to be responsible for many of the differences in the annual sulphate reduction rate reported for salt marshes. When a comparison is made among short *S. alterniflora* sites (Table 5.2), there does appear to be a relationship between sulphate reduction rate and below ground production. For example, higher sulphate reduction rates have been reported for soils from a Massachusetts marsh (Great Sippewissett) than more southern marshes where the growing season is longer and temperatures are higher, but where below ground production is lower (Table 5.2). Differences in sulphate reduction rates within sites in marshes both in Georgia and in South Carolina appear to be related to differences in below ground biomass and production (Howarth and Giblin, 1983; King, 1988). Sulphate reduction rates reported from salt marsh pans and creeks are generally much lower than those reported for vegetated areas (Table 5.2), suggesting that the roots, rhizomes, and perhaps exudates from the dominant macrophytes are fueling these high rates. Hartman (1984) compared sulphate reduction rates in areas of the same marsh where there were different amounts of above ground biomass, including an area which had been bare for more than three years, and a pan. Over the three-month study, sulphate reduction rates in undisturbed vegetated areas were considerably higher than in pans or areas of bare peat.

The degree of oxidation of marsh soils also affects sulphate reduction rates. Skyring, Oshrain and Wiebe (1979) initially reported that sulphate reduction rates were higher in more oxidized creekbank zones than in areas of short *S. alterniflora*. However, they did not include non-acid-volatile end-products of sulphate reduction in their measurements. Both Howarth and Giblin (1983) and King (1988) have reported that sulphate reduction rates are lower in the more oxidized and more productive creekbank stands of *S. alterniflora* than in interior marsh sites. Greater porewater drainage and air entry may allow more total decomposition via aerobic pathways in creekbank areas when compared to interior marsh soils. Oenema (1988) found very low sulphate reduction rates in a highly oxidized marsh in the Netherlands. Hines, Knollmeyer and Tugel (1989) found that sulphate reduction rates in the high marsh *S. patens* zone, where desiccation occasionally caused the soils to become oxidized to 11 cm, were lower than in the nearby *S. alterniflora* zone where redox potentials were more negative.

5.3.2 SULPHATE REDUCTION IN FRESHWATER WETLANDS

Sulphate reduction rates have been directly measured in only a few freshwater systems (Table 5.3). These rates are somewhat lower than reported for salt marshes, although the values overlap considerably. It should be emphasized that Cedar Swamp, Big Run Bog, and Buckle's Bog receive considerable inputs of SO_4^{2-} from sources other than directly incident precipitation. Wieder, Yavitt and Lang (1990) have argued that the advent of high SO_4^{2-} deposition via acid precipitation, coupled with minerotrophic drainage into Big Run Bog from the 276 ha of surrounding upland forest may have created conditions conducive to the dynamic sulphur cycling. A similar argument could be made for Cedar Swamp, with its substantial groundwater input. In the absence of further study, it presently is questionable whether or not the high rates of sulphate reduction reported in these studies are typical of all *Sphagnum*-dominated wetlands.

Very little is known about the controls on sulphate reduction rates in freshwater wetlands. Spratt and Morgan (1990) reported that sulphate reduction rates under mats of *Sphagnum* were up to five-fold lower than nearby areas where *Sphagnum* was absent in a New Jersey cedar swamp, possibly because photosynthesis from the moss oxidized the soil. The depth of the water table has been identified as another major factor controlling sulphate reduction rates (Spratt, Morgan and Good, 1987).

Traditionally, anaerobic carbon mineralization in *Sphagnum* wetlands was thought to be dominated by fermentation reactions, including methanogenesis. However, at Big Run Bog and Buckle's Bog, methane production accounted for only 3 to 12% of total anaerobic carbon mineralization of approximately 660 g (C) m^{-2} a^{-1} (55 mol m^{-2} a^{-1}) (Wieder, Yavitt and

Table 5.3. Sulphate reduction rates, sulphur inputs and sulphur outputs for selected freshwater wetlands

	Sulphate reduction (mol (S) m^{-2} a^{-1})	Sulphur inputs (mol (S) m^{-2} a^{-1})	Sulphur retention (mol (S) m^{-2} a^{-1})
Cedar Swamp, NJ			
Spratt and Morgan (1990) (Avg. of entire swamp)	2.64	0.058[a]	0.28
Big Run Bog, WV			
Wieder, Yavitt and Lang (1990)	17.2	0.057[b]	0.45
Buckle's Bog, MD			
Wieder, Yavitt and Lang (1990)	10.5		
Tillingbourne, UK			
Brown and MacQueen (1985)	0.1–0.25		0.15–0.2
Marcell Bog, MN			
Urban, Eisenrich and Grigal (1989)		0.023	0.008
Thoreau's Bog			
Hemond (1980)		0.04	0.031
ELA Bog 239			
Bayley, Behr and Kelly (1986) 1981–82		0.04	0.015
Bayley *et al.* (1987) 1983 acid added		0.105	0.077

[a]Precipitation chemistry from Yuretich *et al.* (1982).
[b]Wetfall only.

Lang, 1990; Yavitt, Lang and Wieder, 1987). Carbon dioxide production, resulting substantially from sulphate reduction, dominated anaerobic carbon mineralization (Wieder, Yavitt and Lang, 1990). While oxic respiration and methanogenesis are probably more important than sulphate reduction in decomposition in most freshwater wetlands, sulphate reduction rates may be very important in wetlands receiving sulphate from sources such as acid deposition or mine drainage.

Some studies have reported sulphate reduction rates which are considerably higher than would be indicated by the net retention of sulphur (Table 5.3). Therefore, the importance of sulphur reduction to carbon cycling may not be apparent from sulphur input–output budgets. The dynamic cycling of sulphur within the peat, discussed below, allows for more carbon to be mineralized via sulphate reduction than would be calculated from net sulphur retention. Such recycling is also important in salt marshes (Howarth and Teal, 1979; Hines, *et al.*, 1988; Gardner, 1990; Luther and Church, this volume).

5.4 THE CYCLING OF SULPHUR WITHIN THE SOIL

5.4.1 SALINE WETLANDS

Rates of sulphur burial (permanent accretion in the sediment, or precipitation in excess of re-oxidation) in saltwater wetlands are usually much less than 1 mol m^{-2} a^{-1} (32 g (S) m^{-2} a^{-1}) and are small compared with sulphate reduction rates (Howarth, 1984; King, 1988). Therefore, nearly all of the sulphide produced by sulphate reduction must be lost from the soils by some other mechanism. Although some sulphide is lost by volatilization and advection (below), the majority is oxidized in situ.

Sulphide may be directly oxidized in the porewater or converted to sulphide minerals which are later oxidized. Howarth and Teal (1979) and Giblin and Howarth (1984) proposed that pyrite formation and oxidation occurred on a seasonal basis. They suggested that a significant portion of the sedimentary pyrite was oxidized in the spring when sulphate reduction rates were low, but grass growth was rapid. It now appears that the importance of sulphur cycling through solid phases of sulphur may vary considerably from marsh to marsh, and that weather and tides may help determine sediment oxidation rates (Giblin, 1988). King (1988) found no seasonal pattern in concentrations of reduced inorganic sulphur, although concentrations did fluctuate over the course of the year. In a Louisiana marsh, Feitjel (1986) found that maximal pyrite oxidation occurred during periods of low water typical of the winter and early summer months. Luther and Church (1988) observed a very strong seasonal cycle of pyrite oxidation in a Delaware marsh. Hines, Knollmeyer and Tugel (1989) found that tidal and rainfall events produced large changes in soil redox in *Spartina patens* soils which resulted in rapid changes in the precipitation and dissolution of iron and in the magnitude and spatial distributions of sulphate reduction.

The mechanism by which pyrite is oxidized is not fully understood. Drainage and evapotranspiration allow air entry into the sediments (Dacey and Howes, 1984), but it is unlikely that oxidation proceeds by purely chemical mechanisms. In coal surface mines, more than 75% of the pyrite oxidation is carried out by chemoautotrophic bacteria of the genus *Thiobacillus* (Taylor, Wheeler and Nordstrom, 1984). The very rapid oxidation observed by Luther and Church (1988) suggests that microbial processes are driving the reaction. Luther *et al.* (1986) have also observed large concentrations of thiols in porewater during pyrite oxidation events, which they suggested are products of pyrite oxidation. Thiols would not be products of the chemical oxidation of sulphur (Luther *et al.*, 1986; Luther and Church, this volume). Previously, the cycling of organic sulphur in saltwater wetlands had largely been ignored because the cycling of organic sulphur through the vegetation is much smaller than the amount of sulphur used for respiration. The production of thiols during pyrite oxidation suggests that there may be a link between organic and inorganic sulphur pools (Luther *et al.*, 1986) which should be explored further.

5.4.1.1 The Significance of the Sulphur Cycle to Energy Flow in Saline Wetlands

In oxic systems the energy trapped during the formation of organic matter is liberated during decay so the flow of energy follows the flow of carbon. However, when decomposition occurs anoxically, part of the energy present in the organic carbon is not released, but remains conserved in high energy reduced compounds such as hydrogen sulphide (see discussion in Howarth and Teal, 1980; Howarth, 1984). During dissimilatory sulphate reduction, only 25% of the energy present in the organic matter is available for microbial growth and respiration, while nearly 75% of the energy remains trapped as sulphides. Hence carbon and energy flow are decoupled.

The energy retained in reduced compounds is released when they are oxidized. If this oxidation is carried out chemically, the energy originally present as photosynthetically fixed carbon is lost from the ecosystem. However, a variety of chemoautotrophic bacteria are capable of using the energy released in the oxidation of sulphide and other reduced sulphur compounds to support new carbon fixation. In saline wetlands, where much of the decomposition is carried out by sulphate-reducing bacteria, the amount of energy carried in reduced sulphur compounds is very large (Howarth, 1984). If we assume that 75% of the total decomposition proceeds by sulphate

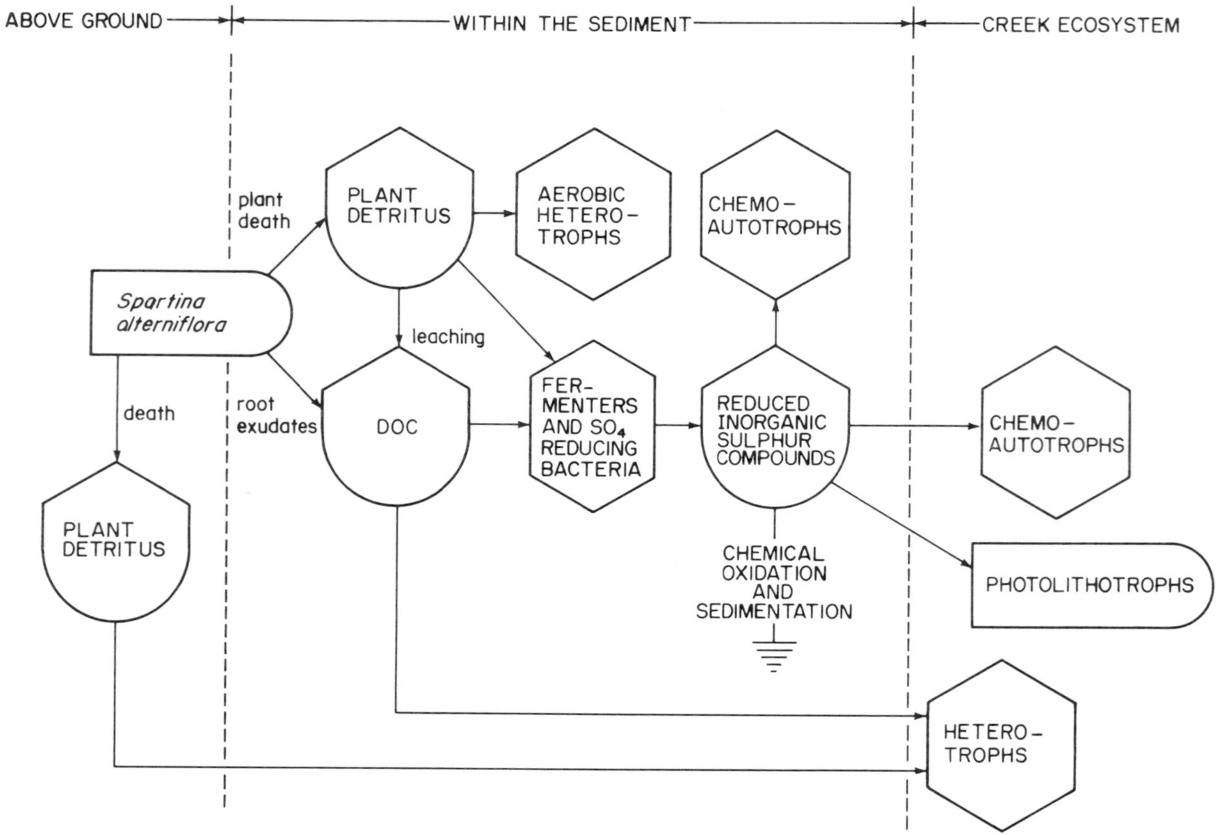

Figure 5.1. A model of energy flow in a salt marsh ecosystem. Adapted from Howarth and Teal (1980)

reduction, then more than half of the energy from photosynthesis is trapped, at least temporarily, as reduced sulphur.

A portion of the reduced sulphur produced each year is exported to creeks before being oxidized (Figure 5.1). Porewater fluxes to creeks may exceed 10 l m^{-2} day^{-1} (Howarth *et al.*, 1983), although these high rates may be restricted to areas within a few metres of creeks (Hemond and Fifield, 1982). Howarth *et al.* (1983) calculated that at most, 6 mol m^{-2} a^{-1} (192 g (S) m^{-2} a^{-1}) of reduced sulphur are exported to creeks in Great Sippewissett Marsh in such an advection of porewaters. It is not known how much of this energy is actually captured as opposed to being lost by chemical oxidation. At least a portion of the sulphide reaching creeks is used biologically, as evidenced by the presence of large mats of *Beggiatoa*, a chemoautotrophic bacterium, wherever porewater seeps occur. The majority of oxidation of reduced sulphur occurs within the soil, or at the soil surface. As discussed above, a large portion of this oxidation may be biologically catalysed. The fate of the large amount of energy released in this process is unknown but could potentially support CO_2 fixation by chemoautotrophic bacteria in the soil at rates as high as 275 to 500 g (C) m^{-2} a^{-1} (23 to 42 mol m^{-2} a^{-1}) (Howarth, 1984). However, a number of heterotrophic bacteria are also able to oxidize reduced sulphur compounds under oxic conditions, probably with little or no energy gain from the oxidation (Schoenau and Germida, this volume).

5.4.2 SULPHUR CYCLING IN FRESHWATER WETLANDS

In contrast to the extensive studies of sulphur cycling in saline and brackish wetlands, sulphur cycling in freshwater wetland systems has only recently generated much interest. One reason for the minimal effort in freshwater wetlands was the traditional belief that in freshwater wetlands dissimilatory sulphate reduction should be negligible, the process being limited by low sulphate concentrations (see review by Nedwell, 1984). However, the modern phenomenon of elevated atmospheric sulphur deposition in acid precipitation focused new efforts towards understanding sulphur cycling in freshwater wetland ecosystems (Gorham, Bayley and Schindler, 1984).

Most of the available information regarding sulphur cycling in freshwater wetlands comes from studies of *Sphagnum*-dominated wetland systems. In contrast to the high sulphate inputs to salt marshes, inputs of sulphur to freshwater wetlands via precipitation are quite low (Table 5.3). *Sphagnum*-dominated wetlands have consistently been identified as net sinks for atmospheric sulphur, rather than net sources. For truly ombrotrophic wetlands, receiving all of their water and nutrients from precipitation alone, the quantity of sulphur retained annually is less than input of sulphur from the atmosphere. However, in *Sphagnum* wetlands, with additional inputs from either groundwater or runoff from surrounding upland areas (i.e. minerotrophic sites), annual sulphur retention can exceed inputs from precipitation (Table 5.3).

Most of the sulphur in freshwater wetland soils is present as organic sulphur, but sulphur retention is not simply the result of slow incorporation of sulphur inputs directly into organic sulphur forms. Rather, both inorganic and organic sulphur fractions are involved in what appears to be a surprisingly rapid and dynamic cycling of sulphur, at least in *Sphagnum* wetlands. Radiotracer experiments have shown that in short-term incubations $^{35}SO_4^{2-}$ becomes incorporated into reduced inorganic sulphur, carbon-bonded sulphur, and ester-sulphate sulphur, with the majority of the reduced label being recovered from inorganic, rather than organic forms (Wieder and Lang, 1988; Spratt and Morgan, 1990). Apparently, the reduced inorganic sulphur pool turns over rapidly; rates of reduced sulphur oxidation are similar to rates of sulphur reduction (Wieder and Lang, 1988). In longer term studies following the fate of field-applied $^{35}SO_4^{2-}$, the majority of the reduced ^{35}S is recovered as carbon-bonded sulphur rather than reduced inorganic S (Behr, 1985; Brown and MacQueen, 1985; Brown, 1985). It appears that newly deposited SO_4^{2-} is initially converted to reduced inorganic sulphur, which turns over rapidly, on the time scale of days. A portion of this sulphide is incorporated into organic forms which turn over more slowly.

Although the number of sites examined thus far are few, in freshwater *Sphagnum* wetlands the process of sulphate reduction seems to be far more important than was previously imagined. Surprisingly, annual rates of sulphate reduction reported for freshwater *Sphagnum* wetlands overlap considerably with rates reported for saline wetlands, even though concentrations of dissolved sulphate in the *Sphagnum* wetland soils are typically less than 300 μM (9.6 mg (S) l^{-1} (Spratt, Morgan and Good, 1987; Wieder and Lang, 1988). Annual sulphate reduction far exceeds estimates of sulphur retention in Cedar Swamp and Big Run Bog (Table 5.3), attesting to the importance of reduced sulphur oxidation and rapid turnover of the reduced inorganic sulphur pool. This observation brings into question whether or not sulphate reduction in these *Sphagnum* wetlands is indeed limited by low sulphate concentrations. Radiotracer determination of sulphate reduction in short-term incubations of Cedar Swamp peat (Spratt, pers. comm.) and Big Run Bog peat (Wieder, Yavitt and Lang, 1990), in which background SO_4^{2-} concentrations were augmented up to 100-fold, did not result in a stimulation of sulphate reduction. In over 600 determinations of sulphate reduction in Big Run Bog and Buckle's Bog peat, rates of sulphate reduction were not positively correlated with the ambient sulphate pool size (Wieder, Yavitt and Lang, 1990).

Clearly, for sulphate reduction to continue, there must be an adequate supply of available SO_4^{2-}. However, in *Sphagnum* wetlands like Cedar Swamp, Big Run Bog, and Buckle's Bog, the size of the dissolved SO_4^{2-} pool apparently is not a good indicator of SO_4^{2-} availability. Rather, SO_4^{2-} availability is the result of dynamic cycling; the SO_4^{2-} pool turns over rapidly, being depleted by sulphate reduction and also being replenished by the oxidation of reduced inorganic sulphur.

The dynamic cycling of sulphur in these *Sphagnum*-dominated wetlands has implications for carbon balance and peat accumulation. Although most of the carbon mineralization in *Sphagnum*-dominated wetlands occurs aerobically, rates of peat accumulation are more affected by inputs of carbon into the anaerobic zone and by rates of anaerobic carbon mineralization (Clymo, 1984). At Buckle's Bog and Big Run Bog, net primary production is approximately balanced by carbon mineralized via oxic respiration and methanogenesis. However, with the addition of sulphate reduction as an anaerobic carbon mineralization pathway, these wetlands are presently in a state of negative carbon balance (Wieder, Yavitt and Lang, 1990).

The sulphur requirement for plant growth in freshwater wetlands has been estimated to be of the order of 0.1 to 1.0 mol m^{-2} a^{-1} (3 to 32 g m^{-2} a^{-1}) (calculated from Urban *et al.*, 1988 and production data in Table 5.1), which exceeds sulphur inputs in many locations (Table 5.3; Urban, Eisenreich and Grigal, 1989). Therefore, the recycling of sulphur within the bog is important for maintaining new primary production in addition to the role that sulphate plays in carbon decomposition. Sulphur appears to be mineralized more rapidly than C or N in aerobic portions of wetland soils (Urban, Eisenreich and Grigal, 1989).

5.5 FEEDBACKS BETWEEN THE SULPHUR CYCLE AND PRIMARY PRODUCTION

Sulphate reduction benefits plants by regenerating nutrients and buffering soil pH. However, the sulphides produced are potentially toxic and may inhibit nutrient uptake (Vamos and Koves, 1972; Goodman and Williams, 1961; Mitsui, 1965). It has been proposed that the concentration of sulphide in the sediment strongly influences both primary production (Howarth and Teal, 1979; Mendelssohn and Seneca, 1980; Howes *et al.*, 1981; King *et al.*, 1982) and plant species distribution (Nickerson and Thibodeau, 1985) in wetlands.

Wetland plants possess structural and metabolic adaptations to deal with the presence of sulphides (Crawford, 1978; Teal and Kanwisher, 1961; Thibodeau and Nickerson, 1986; Burdick and Mendelssohn, 1987). Some species of wetland plants are capable of lowering sulphide concentrations in sediments by allowing air to diffuse out of roots (Teal and Kanwisher, 1961) or by allowing sulphides to diffuse into roots where they can be oxidized or incorporated into organic compounds (Carlton and Forrest, 1982; Fry *et al.*, 1982). In spite of these adaptations roots and rhizomes may experience

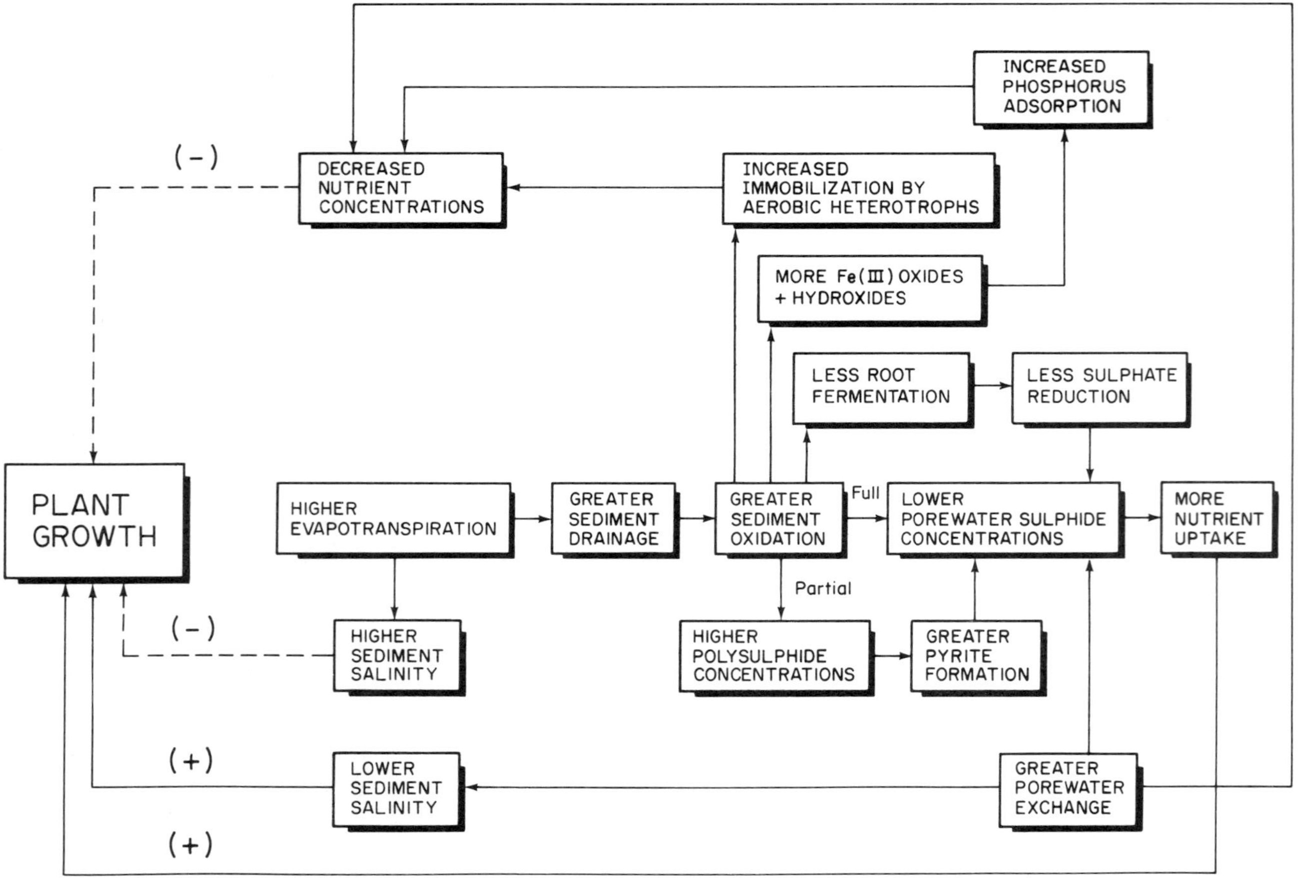

Figure 5.2. A model of the interaction of the sulphur cycle and the primary production in a salt marsh ecosystem

anoxia (Mendelssohn, McKee and Patrick, 1981). It has frequently been observed that wetland plants are more productive in areas of higher water movement (Sparling, 1966; references in Chalmers, 1982). A number of explanations for this phenomenon have been advanced including greater nutrient availability where water flow is rapid, removal of sulphides or other toxic metabolic end-products by flushing, and in saline wetlands, lower salinity stress. Research in salt marshes now suggests that primary production in these systems is regulated by a complex interaction between water flow and the sulphur cycle due to the dual role of sulphate reduction in generating both needed nutrients and metabolic poisons.

We have summarized the possible reactions between the sulphur cycle and marsh production in Figure 5.2. Some of these interactions are well known and the reader is referred to the summary by Chalmers (1982) as well as recent work by Wiegert, Chalmers and Randerson (1983), DeLaune, Smith and Tolley (1984), and Howes, Dacey and Goehringer (1986) for more details. The negative interactions on production postulated here under conditions of high drainage seem to be borne out by the long-term experiments of Chalmers, Wiegert and Wiebe (submitted), who observed that prolonged drainage ultimately resulted in declines in productivity in some sites and no changes in others.

Briefly, sulphate reduction generates sulphides which decrease plant nutrient uptake. Lowering dissolved sulphide concentration by oxidation or porewater exchange increases the ability of the plants to take up nutrients, and may reduce the excretion of root fermentation products, leading to lower rates of sulphate reduction and sulphide production, in a positive feedback loop. Water removal from evapotranspiration or porewater exchange also may lead to a negative feedback on production by increasing sediment salinity, or removing nutrients. Extensive oxidation will completely eliminate sulphide production and decomposition will proceed aerobically. This will further increase the plant's ability to take up nutrients by eliminating dissolved sulphides, but will promote nutrient immobilization by aerobic heterotrophs and phosphorus precipitation with iron oxides resulting in decreased nutrient concentrations.

5.6 SULPHUR GAS FLUXES TO THE ATMOSPHERE

A variety of volatile sulphur compounds are produced in wetlands including hydrogen sulphide (H_2S), methylmercaptan (MeSH), dimethyl sulphide (DMS), dimethyl disulphide (DMDS), carbonyl sulphide (COS), and carbon disulphide (CS_2). A portion of these biogenic sulphur gases are emitted to the atmosphere where they are eventually oxidized to SO_2 and SO_4^{2-}, contributing to the acidification of rainfall (Andreae and Jaeschke, this volume). Much

Table 5.4. Sulphur gas fluxes from wetlands

Area	Sulphur gas fluxes (μg (S) m^{-2} h^{-1})			Reference
	H_2S	DMS	other S gases	
Saline wetlands				
Mangrove swamps				
Florida				
red mangrove	12			Castro and Dierberg (1987)
black mangrove	0.7–7.9	0.3–10	<0.3–1.2	Cooper *et al.* (1987a)
Salt marshes				
Florida				
Distichlis spicata	0.1–4.7	0.7–1	0–1	Cooper *et al.* (1987a)
B. maritina	0.6–1.0	1–7	0.2–0.8	Cooper *et al.* (1987a)
Juncus	0.1–2.8	0.1–6.4	<0.1	Cooper *et al.* (1987a)
Juncus	0.9			Castro and Dierberg (1987)
mud flat	2.4			Cooper *et al.* (1987b)
sand flat	86	6.3	3.3	Cooper *et al.* (1987b)
Spartina alterniflora	2.9	51	5.0	Cooper *et al.* (1987b)
Juncus/Spartina				
(Cedar Island)	0.1–7570	4–54	1.3	Goldan *et al.* (1987)
(Cedar Island)	38	9.5	6.2	Lamb *et al.* (1987)
(Cedar Island)	2.3–6.9	0.8–179	2–10	Adams *et al.* (1981b)
(Cedar Island)	22	149		Aneja *et al.* (1979)
North Carolina				
Cox Hole				
vegetated marsh	<0.1	46	21	Aneja, Overton and Aneja (1981)
marsh flat	57	<1	3	Aneja, Overton and Aneja (1981)
Virginia				
Wallops Island	23–1495			Goldberg *et al.* (1981)

Delaware				
Canary Marsh				
frequently flooded	1	8	3	Adams *et al.* (1981a)
infrequently flooded	1.3	104	27	Adams *et al.* (1981a)
Massachusetts				
Great Sippewissett				
marsh (*Spartina* sp.)	234	328	100	Steudler and Peterson (1984)
creek	265	18	90	Steudler and Peterson (1984)
Delaware-Texas				
12 marshes	2–68 721	0.9–213	1–100	Adams *et al.* (1981b)
Freshwater wetlands				
Marshes				
Florida	9			Castro and Dierberg (1987)
Florida	0.1–4.7	0.7–1	0–1	Cooper *et al.* (1987a)
North Carolina	11	<1	<1	Aneja, Overton and Aneja (1981)
Temperate swamps				
Florida				
Cypress	0.5			Castro and Dierberg (1987)
New York				
stagnant	19	0.05	1	Adams *et al.* (1981a)
NY, NC, GA, Ontario				
9 areas		4–21		Nriagu, Holdway and Coker (1987)
Wet tropical swamps				
Ivory Coast	0.013–300			Delmas *et al.* (1980)
Bogs				
Ontario three areas		3–9		Nriagu, Holdway and Coker (1987)

of the research on sulphur gas emission in wetlands has focused on the relative importance of biogenic sulphur *vs* anthropogenically released sulphur as potential contributors to the acidification of rainfall. Measurements of sulphur gas fluxes have been complicated by analytical, methodological, and logistical difficulties (see Andreae and Jaeschke, this volume).

Most studies of sulphur gas fluxes from saline wetlands have been in marshes dominated by *Spartina alterniflora*, although several mangrove and *Juncus* sites have been examined. These studies have all shown that H_2S and DMS are the dominant gases being released to the atmosphere (Table 5.4). Some studies have demonstrated that emissions change on a diurnal basis (Cooper *et al.*, 1987a,b; DeMello *et al.*, 1987), or during the tidal cycle (Cooper *et al.*, 1987a; Steudler and Peterson, 1985). Estimates of emissions of both individual gases and the total gaseous sulphur flux span more than two orders of magnitude (Table 5.4). As discussed in Steudler and Peterson (1985) and Lamb *et al.* (1987) the cause for this variability has not been established. Methodological problems certainly contribute (Andreae and Jaeschke, this volume), but high spatial and temporal variability in fluxes should be expected in these ecosystems which have such dynamic cycles of sulphur.

The very high DMS fluxes which have been reported in areas colonized by *S. alterniflora* may not be representative of other saline wetlands. Steudler and Peterson (1984) found that DMS fluxes from unvegetated tidal creeks were only about 5% of those measured from sites vegetated by *S. alterniflora*, and DMS emission has been related more closely to *S. alterniflora* biomass than soil parameters (DeMello *et al.*, 1987). Dacey, King and Wakeham (1987) report finding high concentrations of dimethylsulphoniopropionate (DMSP) in the leaves of *S. alterniflora*; they speculated that DMSP may function in regulating osmotic pressure. DMSP concentrations were below detection levels in the leaves of seven other wetland species examined, including mangrove, although it has been reported in high concentrations in *S. anglia* leaves (Larher, Hamelin and Stewart, 1977). Dacey, King and Wakeham (1987) suggest that high fluxes of DMS occur when DMSP is degraded to DMS enzymatically in the leaves. It is not known whether DMS emissions are influenced by the chamber techniques used to measure gas fluxes. Most studies on sulphur gas emission have only been carried out for short time periods (Table 5.4; see also Andreae and Jaeschke, this volume), usually in the summer. If we extrapolate these rates to yearly fluxes (Figure 5.3) we find the loss of volatile sulphur gases represents only a small percentage of the total sulphur reduced by sulphate reduction (Steudler and Peterson, 1985) and is a negligible term in the total sulphur budget.

Sulphur gas fluxes from freshwater wetlands are generally lower than fluxes from saline wetlands (Table 5.4). These fluxes are discussed in detail elsewhere (Andreae and Jaeschke, this volume). Low rates of sulphur gas

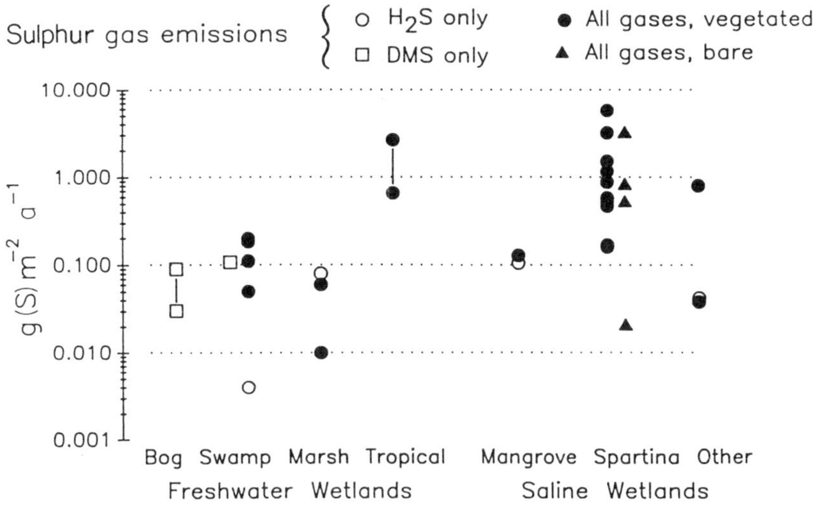

Figure 5.3. Sulphur gas emissions from fresh and saltwater wetlands. The values given in Table 5.4 were extrapolated to annual values. Values are reported separately for vegetated and bare areas in *Spartina marshes*

emission, and in particular H_2S, from freshwater wetlands may be a result of low rates of sulphate reduction. However, at Big Run Bog where 550 g (S) m^{-2} a^{-1} (17 mol m^{-2} a^{-1}) of sulphate is reduced (Table 5.3), H_2S emission is still minimal at 9.7 mg (S) (0.3 mmol m^{-2} a^{-1}) (Wieder, Yavitt and Lang, this volume). At Big Run Bog, sulphur gas flux does not substantially affect the wetland or watershed sulphur budget. In contrast, however, Nriagu, Holdway and Coker (1987) calculated dimethyl sulphide fluxes from bogs and swamps in northern Ontario to range from 25 to 184 mg (S) m^{-2} a^{-1} (0.78 to 5.8 mmol m^{-2} a^{-1}), based on porewater concentrations and application of a simple diffusion model. Although DMS emission was not directly measured, with sulphur inputs via precipitation of only 1 to 3 g (S) m^{-2} a^{-1} (30 to 100 mmol m^{-2} a^{-1}), a substantial fraction of the sulphur input could be returned to the atmosphere by DMS emission if the estimates of Nriagu, Holdway and Coker (1987) are correct (for a discussion of the assumptions used in making these calculations see Andreae and Jaeschke, this volume).

On a global basis, freshwater wetlands occupy about 10 times the area of saline wetlands (Table 5.1), but reported sulphur gas emissions from saline wetlands average about 1000 times greater than emissions from freshwater wetlands (Adams *et al.*, 1981a; Table 5.4; Figure 5.3). It seems likely that freshwater wetlands contribute substantially less sulphur to the global atmosphere than do saline wetlands.

5.7 EFFECT OF THE SULPHUR CYCLE ON METHANE FLUX

Wetlands make a significant contribution to the global methane budget (Matthews and Fung, 1987). It is well recognized that methane fluxes from freshwater wetlands are considerably greater than those from saline wetlands (Bartlett *et al.*, 1987 and references therein). This difference in methane emissions is due at least in part to changes in both the production and consumption of methane in the presence of sulphate-reducing bacteria. Sulphate reduction liberates more energy than methanogenesis, so sulphate reducers out-compete methanogens and dominate anaerobic processes when sulphate is available. At very low concentrations of sulphate, perhaps in the range of 10 to 30 μM SO_4^{2-} (0.3 to 1 μg (S) l^{-1}), methanogenesis becomes the dominant pathway for carbon decomposition. Methane is also anoxically oxidized to carbon dioxide in the presence of sulphate; this seems coupled to sulphate-reducing bacteria, although the mechanism remains unclear (see Ivanov, 1989; see also Howarth and Stewart, this volume). The anoxic consumption of methane obviously reduces the flux of methane to the atmosphere.

In saline and brackish water wetlands, a tight coupling between the sulphur cycle and methane release has been demonstrated in both field and laboratory experiments. Methane release from saltwater wetlands is correlated with sulphate depletion (King and Wiebe, 1980), and there is an inverse correlation of methane release and soil salinity across a broad range of saline and brackish wetland types (Bartlett *et al.*, 1987). The situation in freshwater *Sphagnum* wetlands appears to be more complex. The addition of high levels of sulphate to freshwater wetland soils decreases methane production (Yavitt, Lang and Wieder, 1987). However, at very low sulphate concentrations ($<$ 300 μM; 10 mg (S) l^{-1}), methane production does not appear to be closely related to sulphate concentration (Yavitt and Wieder, unpublished data).

5.8 SUMMARY

The major aspects of the sulphur cycle discussed above have been summarized for saline (Figure 5.4) and freshwater wetlands (Figure 5.5). The fluxes represent the range of values in literature and should only be taken as order of magnitude estimates. The fluxes and standing stocks are most representative of temperate salt marshes (Figure 5.4) and temperate bogs (Figure 5.5).

The tidal fluxes of sulphur through saline wetlands are usually very large in comparison to the rates of sulphur processing within the ecosystem (see Peterson *et al.*, 1983). Most of the sulphur reduced in the sediment is subsequently re-oxidized or exported to creeks, so only a small per cent of the reduced sulphur produced each year is buried. This rapid cycling occurs

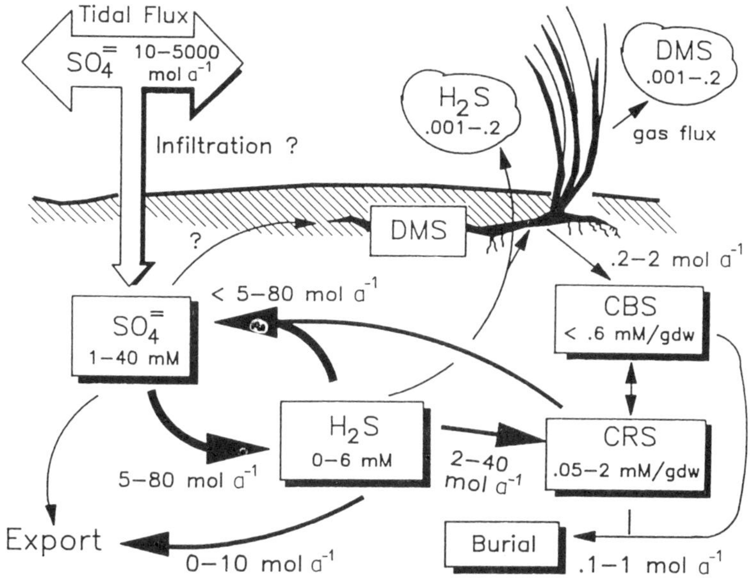

Figure 5.4. The sulphur cycle of saline wetlands. Fluxes and standing stocks, taken from the literature, are representative of temperate salt marshes. (See text for details)

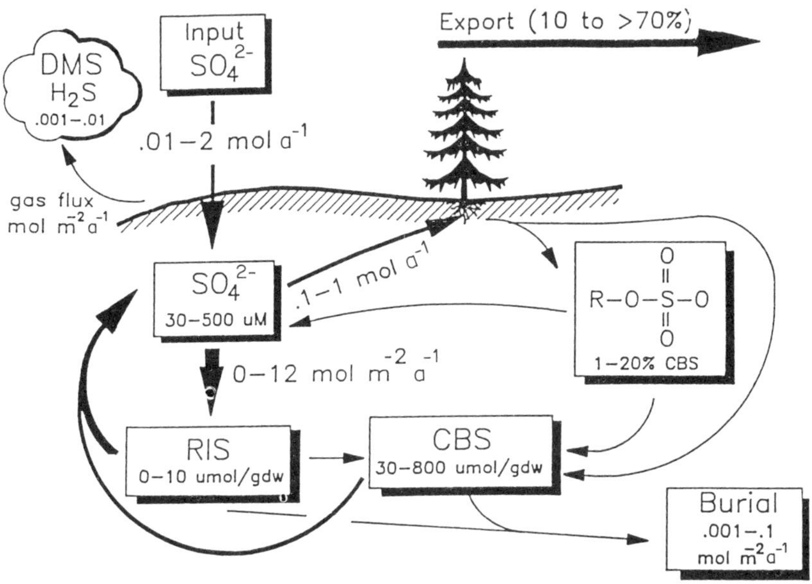

Figure 5.5. The sulphur cycle of freshwater wetlands. Fluxes and standing stocks, taken from the literature, are representative of temperate bogs and swamps. (See text for details)

primarily through inorganic reduced forms of sulphur. Gaseous fluxes are high in comparison to many other ecosystems but represent only a tiny fraction of the total cycled within the wetland.

Sulphur inputs to freshwater wetlands are much lower than saline wetlands. In contrast to saline wetlands, the amount of sulphur cycled within freshwater ecosystems equals or exceeds new inputs, perhaps by orders of magnitude. Another difference between fresh and saline marshes is the importance of sulphur uptake by the vegetation in the total sulphur budget. Gaseous losses are apparently lower than saline wetlands, but because of the low sulphur inputs, these losses may represent a significant fraction of the sulphur budget. The cycling of organic sulphur is much more important in freshwater wetlands than saline wetlands (at least as a percentage of total rate of sulphur cycling). Sulphur burial is lower in freshwater wetlands than in saline wetlands, but in general represents a greater percentage of the sulphur inputs.

REFERENCES

Adams, D. F., Farwell, S. O., Pack, M. R. and Robinson, E. (1981a). Biogenic sulfur gas emissions from soils in Eastern and Southeastern United States. *J. Air Poll. Control Assoc.*, **31**, 1083–9.

Adams, D. F., Farwell, S. O., Robinson, E., Pack, M. R. and Bamesberger, W. (1981b). Biogenic sulfur source strengths. *Environ. Sci. Technol.*, **15**, 1493–8.

Altschuler, Z. S., Schnepfe, M. M., Silber, C. C. and Simon, F. O. (1983). Sulfur diagenesis in Everglades peat and origin of pyrite in coal. *Science*, **221**, 221–7.

Andreae, M. O. and Jaeschke, W. This volume.

Aneja, V. P. (1986). Characterization of emissions of biogenic atmospheric hydrogen sulfide. *Tellus*, **38B**, 81–6.

Aneja, V. P., Overton, J. H. and Aneja, A. P. (1981). Emission survey of biogenic sulfur flux from terrestrial surfaces. *J. Air Poll. Control Assoc.*, **31**, 256–8.

Aneja, V. P., Overton, J. H., Cupitt, L. T., Durham, J. L. and Wilson, W. E. (1979). Direct measurements of emission rates of some atmospheric biogenic sulfur compounds. *Tellus*, **31**, 174–8.

Armstrong, W. (1979). Aeration in higher plants. *Adv. Bot. Res.*, **7**, 225–32.

Armstrong, W. and Boatman, D. J. (1967). Some field observations relating the growth of bog plants to conditions of soil aeration. *J. Ecol.*, **55**, 101–10.

Bartlett, K. B., Bartlett, D. S., Harriss, R. C. and Sebacher, D. I. (1987). Methane emissions along a salt marsh salinity gradient. *Biogeochemistry*, **4**, 183–202.

Bayley, S. E., Behr, R. S. and Kelly, C. A. (1986). Retention and release of S from a freshwater wetland. *Water Air Soil Pollut.*, **31**, 101–14.

Bayley, S. E., Vitt, D. H., Newberry, R. W., Beatty, K. G., Behr, R. and Miller, C. (1987). Experimental acidification of a *Sphagnum*-dominated peatland: First year results. *Can. J. Fish. Aquat. Sci.*, **44**, 194–205.

Behr, R. S. (1985). Sulfur dynamics in an experimentally acidified mire in northwestern Ontario. *M.S. Thesis*, University of Manitoba.

Brinson, M. M., Lugo, A. E. and Brown, S. (1981). Primary productivity, decomposition and consumer activity in freshwater wetlands. *Annu. Rev. Ecol. Syst.*, **12**, 123–61.

Brown, K. A. (1985). Sulphur distribution and metabolism in waterlogged peat. *Soil Biol. Biochem.*, **17**, 39–45.
Brown, K. A. (1986). Formation of organic sulphur in anaerobic peat. *Soil Biol. Biochem.*, **18**, 131–40.
Brown, K. A. and MacQueen, J. F. (1985). Sulphate uptake from surface water by peat. *Soil Biol. Biochem.*, **17**, 411–20.
Burdick, D. M. and Mendelssohn, I. A. (1987). Waterlogging responses in dune, swale and marsh populations of *Spartina patens* under field conditions. *Oecologia*, **74**, 321–9.
Canfield, D. E., Raiswell, R., Westrich, J. T., Reaves, C. M. and Berner, R. A. (1986). The use of chromium reduction in the analysis of reduced inorganic sulfur in sediments and shales. *Chem. Geol.*, **54**, 149–55.
Carlson, P. R. and Forrest, J. (1982). Uptake of dissolved sulfide by *Spartina alterniflora*: Evidence from natural sulfur isotope abundance ratios. *Science*, **216**, 633–5.
Casagrande, D. J., Gronli, K. and Sutton, N. (1980). The distribution of sulfur and organic matter in various fractions of peat: Origins of sulfur in coal. *Geochim. Cosmochim. Acta*, **44**, 25–32.
Casagrande, D. J., Siefert, K., Berschinski, C. and Sutton, N. (1977). Sulfur in peat-forming systems of the Okefenokee Swamp and Florida Everglades: Origins of sulfur in coal. *Geochim. Cosmochim. Acta*, **41**, 161–7.
Casey, W. H. and Lasaga, A. C. (1987). Modeling solute transport and sulfate reduction in marsh sediments. *Geochim. Cosmochim. Acta*, **51**, 1109–20.
Castro, M. S. and Dierberg, F. E. (1987). Biogenic hydrogen sulfide emissions from selected Florida wetlands. *Water Air Soil Pollut.*, **33**, 1–13.
Chalmers, A. G. (1982). Soil dynamics and the productivity of *Spartina alterniflora*. In: Kennedy, V. S. (Ed.). *Estuarine Comparisons*. Academic Press, New York, pp. 231–42.
Chalmers, A. G., Wiegert, R. G. and Wiebe, W. J. (submitted). Effects of increased drainage in a short *Spartina alterniflora* marsh.
Clymo, R. S. (1984). The limits to peat bog growth. *Phil. Trans. R. Soc. Lond. B Biol. Sci.*, **303**, 605–54.
Clymo, R. S. and Hayward, P. M. (1982). The ecology of *Sphagnum*. In: Smith, A. J. E. (Ed.). *Ecology of Byrophytes*. Chapman Hall, London, pp. 229–89.
Cook, R. B. (1981). The biogeochemistry of sulfur in two small lakes. *Ph.D. dissertation*, Columbia University, New York, 234pp.
Cook, R. B. and Kelly, C. A. This volume.
Cooper, D. J., De Mello, W. Z., Cooper, W. J., Zika, R. G., Saltzman, E. S., Prospero, J. M. and Savoie, D. L. (1987a). Short-term variability in biogenic sulphur emissions from a Florida *Spartina alterniflora* marsh. *Atmos. Environ.*, **21**, 7–12.
Cooper, W. J., Cooper, D. J., Saltzman, E. S., De Mello, W. Z., Savoie, D. L., Zika, R. G. and Prospero, J. M. (1987b). Emissions of biogenic sulphur compounds from several wetland soils in Florida. *Atmos. Environ.*, **21**, 1491–5.
Cowdardin, L. M., Cater, V., Gollet, F. C. and Laroe, E. T. (1979). *Classification of Wetlands and Deepwater Habitats of the United States*. FWS/OBS-79/31, US Fish and Wildlife Service.
Crawford, R. M. M. (1978). Metabolic adaptations to anoxia. In: Hook, D. D. and Crawford, R. M. M. (Eds). *Plant Life in Anaerobic Environments*. Ann Arbor, Mich, Ann Arbor Science, pp. 119–36.
Dacey, J. W. and Howes, B. L. (1984). Water uptake by roots controls water table movement and sediment oxidation in short *Spartina* marsh. *Science*, **224**, 487–9.

Dacey, J. W. H., King, G. M. and Wakeham, S. G. (1987). Factors controlling emission of dimethylsulphide from salt marshes. *Nature*, **330**, 643–5.

DeLaune, R. D., Smith, C. J. and Tolley, M. D. (1984). The effect of sediment redox potential on nitrogen uptake, anaerobic root respiration and growth of *Spartina alterniflora* Loisel. *Aquat. Bot.*, **18**, 223–30.

Delmas, R., Baudet, J., Servant, J. and Baziard, Y. (1980). Emissions and concentrations of hydrogen sulfide in the air of the tropical forest of the Ivory Coast and of temperate regions in France. *J. Geophys. Res.*, **85**, 4468–74.

De Mello, W. Z., Cooper, D. J., Cooper, W. J., Saltzman, E. S., Zika, R. G., Savoie, D. L. and Prospero, J. M. (1987). Spatial and diel variability in the emissions of some biogenic sulfur compounds from a Florida *Spartina alterniflora* coastal zone. *Atmos. Environ.*, **21**, 987–90.

Feijtel, T. C. J. (1986). Biogeochemical cycling of metals in Barataria Basin. *Ph.D dissertation*. Louisiana State University.

Ferguson, P., Robinson, R. N. and Press, M. C. (1984). Element concentrations in five *Sphagnum* species in relation to atmospheric pollution. *J. Byrol.*, **13**, 107.

Fossing, H. and Jorgensen, B. B. (1990). Measurement of bacterial sulfate reduction in sediments: Evaluation of a single-step chromium reduction method. *Biogeochemistry*, **9**, 223–46.

Francois, R. (1987). A study of sulphur enrichment in the humic fraction of marine sediments during early diagenesis. *Geochim. Cosmochim. Acta*, **51**, 17–27.

Fry, B., Scalan, R. S., Winters, J. K. and Parker, P. L. (1982). Sulphur uptake by salt grasses, mangroves, and seagrasses in anaerobic sediments. *Geochim. Cosmochim. Acta*, **46**, 1121–4.

Gardner, L. R. (1990). Simulation of the diagenesis of carbon, sulfur, and dissolved oxygen in salt marsh sediments. *Ecol. Monogr.*, **60** (1), 91–111.

Giblin, A. E. (1988). Pyrite formation during early diagenesis. *Geomicrobiol. J.*, **6**, 77–97.

Giblin, A. E. and Howarth, R. W. (1984). Porewater evidence for a dynamic sedimentary iron cycle in salt marshes. *Limnol. Oceanogr.*, **29**, 47–63.

Goldan, P. D., Kuster, W. C., Albritton, D. L. and Fehsenfeld, F. C. (1987). The measurement of natural sulfur emissions from soils and vegetation: Three sites in the Eastern United States revisted. *J. Atmos. Chem.*, **5**, 439–67.

Goldberg, A. B., Maroulis, P. J., Wilner, L. A. and Bandy, A. R. (1981). Study of H_2S emissions from a salt water marsh. *Atmos. Environ.*, **15**, 11–18.

Good, R. E., Good, N. F. and Frasco, B. R. (1982). A review of primary production and decomposition dynamics of the belowground marsh component. In: Kennedy, V. S. (Ed.). *Estuarine Comparisons*. Academic Press, New York, pp. 139–57.

Goodman, P. J. and Williams, W. T. (1961). Investigations into 'dieback' in *Spartina townsendii*. *J. Ecol.*, **49**, 391–8.

Gorham, E., Bayley, S. E. and Schindler, D. W. (1984). Ecological effects of acid deposition upon peatlands: A neglected field in 'acid-rain' research. *Can. J. Fish. Aquat. Sci.*, **41**, 1256–68.

Gorham, E. and Detenbeck, N. E. (1986). Sulfate in bog waters: A comparison of ion chromatography with Mackereth's cation-exchange technique and a revision of earlier views on cause of bog acidity. *J. Ecol.*, **74**, 899–903.

Grinenko, V. A. and Ivanov, M. V. (1983). Principal reactions of the global biogeochemical cycle of sulfur. In: Ivanov, M. V. and Freney, J. R. (Eds). *The Global Biogeochemical Sulfur Cycle*, SCOPE 19. Wiley, New York, pp. 1–23.

Hartman, J. M. (1984). The role of wrack in disturbance in the vegetation of a New England salt marsh. *Ph.D dissertation*, University of Connecticut, 130pp.

Hemond, H. F. (1980). Biogeochemistry of Thoreau's Bog, Concord, Massachusetts. *Ecol. Monogr.*, **50**, 507–26.

Hemond, H. F. and Fifield, J. L. (1982). Subsurface flow in salt marsh peat: A model and field study. *Limnol. Oceanogr.*, **27**, 126–36.

Hines, M. E. (in press). Emissions of sulfur gases from wetlands. In: Adams, D. D., Seitzinger, S. P. and Crill, P. M. (Eds). *Cycling of Reduced Gases in the Hydrosphere.*

Hines, M. E., Knollmeyer, S. L. and Tugel, J. B. (1989). Sulfate reduction and other sedimentary biogeochemistry in a northern New England salt marsh. *Limnol. Oceanogr.*, **34**, 578–90.

Howarth, R. W. (1979). Pyrite: Its rapid formation in a salt marsh and its importance in ecosystem metabolism. *Science*, **203**, 49–51.

Howarth, R. W. (1984). The ecological significance of sulfur in the energy dynamics of salt marsh and coastal marine sediments. *Biogeochemistry*, **1**, 5–27.

Howarth, R. W. and Giblin, A. E. (1983). Sulfate reduction in the salt marshes at Sapelo Island, Georgia. *Limnol. Oceanogr.*, **28**, 70–82.

Howarth, R. W. and Jørgensen, B. B. (1984). Formation of ^{35}S-labelled elemental sulfur and pyrite in coastal marine sediments (Limfjorden and Kysing Fjord, Denmark) during short-term $^{35}SO_4$ reduction measurements. *Geochim. Cosmochim. Acta*, **48**, 1807–18.

Howarth, R. W. and Marino, R. (1984). Sulfate reduction in salt marshes, with some comparisons to sulfate reduction in microbial mats. In: Cohen, Y., Castenholz, R. W. and Halvorson, H. O. (Eds). *Microbial Mats: Stromatolites.* Alan R. Liss, New York, pp. 245–63.

Howarth, R. W. and Merkel, S. (1984). Pyrite formation and the measurement of sulfate reduction in salt marsh sediments. *Limnol. Oceanogr.*, **29**, 598–608.

Howarth, R. W. and Stewart, J. W. B. This volume.

Howarth, R. W. and Teal, J. M. (1979). Sulfate reduction in a New England salt marsh. *Limnol. Oceanogr.*, **24**, 999–1013.

Howarth, R. W. and Teal, J. M. (1980). Energy flow in a salt marsh ecosystem: The role of reduced inorganic sulfur compounds. *Am. Nat.*, **116**, 862–72.

Howarth, R. W., Giblin, A., Gale, J., Peterson, B. J. and Luther, G. W. (1983). Reduced sulfur compounds in the pore waters of a New England salt marsh. In: Hallberg R. O. (Ed.). *Environmental Biogeochemistry. Ecol. Bull. (Stockholm)*, **35**, 135–52.

Howes, B. L., Dacey, J. W. H. and Goehringer, D. D. (1986). Factors controlling the growth form of *Spartina alterniflora*: Feedbacks between above-ground production, sediment oxidation, nitrogen and salinity. *J. Ecol.*, **74**, 881–98.

Howes, B. L., Dacey, J. W. H. and King, G. M. (1984). Carbon flow through oxygen and sulfate reduction pathways in salt marsh sediments. *Limnol. Oceanogr.*, **29**, 1037–51.

Howes, B. L., Dacey, J. W. H. and Teal, J. M. (1983). Annual carbon mineralization and belowground production of *Spartina alterniflora* in a New England salt marsh. *Ecology*, **66**, 595–605.

Howes, B. L., Howarth, R. W., Valiela, I. and Teal, J. M. (1981). Oxidation–reduction potentials in a salt marsh: Spatial patterns and interactions with primary production. *Limnol. Oceanogr.*, **26**, 350–60.

Ingram, H. A. P. (1978). Soil layers in mires: Function and terminology. *J. Soil Sci.*, **29**, 224–7.

Ivanov, M. V. (1956). Isotopes in the determination of the sulfate-reduction rate in Lake Belovod. *Microbiologia*, **25**, 305–9.

Ivanov, M. V., Lein, A. Yu and Kashparova, E. V. (1976). Intensity of formation and diagenetic transformation of reduced sulfur compounds in sediments of the Pacific Ocean. In: *The Biogeochemistry of Diagenesis of Ocean Sediments*. Mauka, Moscow, pp. 171–8 (in Russian).

Ivanov, M. V., Lein, A. Yu., Beylaev, S. S., Nesterov, A. I., Bonder, V. A. and Zhabina, N. N. (1980). Geochemical activity of sulfate reducing bacteria in benthal sediments of the North-West Indian Ocean. *Geokhim.*, **8**, 1238–49 (in Russian).

Ivanov, M. V., Lein, A. Yu., Reeburgh, M. S. and Skyring, G. W. (1989). Interaction of sulphur and carbon cycles in marine sediments. In: Brimblecombe, P. and Lein, A. Yu. (Eds). *Evolution of the Global Biogeochemical Sulphur Cycle*. Wiley, Chichester, pp. 125–79.

Jørgensen, B. B. (1977). The sulfur cycle of a coastal marine sediment (Limfjorden, Denmark). *Limnol. Oceanogr.*, **22**, 814–32.

Jørgensen, B. B. (1978). A comparison of methods for quantification of bacterial sulfate reduction in coastal marine sediments. I. Measurement with radiotracer techniques. *Geomicrobiol. J.*, **1**, 11–27.

Jørgensen, B. B. (1982). Mineralization of organic matter in the sea bed: The role of sulfate reduction. *Nature*, **296**, 643–5.

Kaplan, I. R., Emery, K. O. and Rittenberg, S. C. (1963). The distribution and isotopic abundance of sulfur in recent marine sediments off southern California. *Geochim. Cosmochim. Acta*, **27**, 297–331.

King, G. M. (1983). Sulfate reduction in Georgia salt marsh soils: An evaluation of pyrite formation using ^{35}S and ^{55}Fe tracers. *Limnol. Oceanogr.*, **28**, 987–95.

King, G. M. (1988). Patterns of sulfate reduction and the sulfur cycle in a South Carolina salt marsh. *Limnol. Oceanogr.*, **33**, 376–90.

King, G. M. and Wiebe, W. J. (1980). Regulation of sulfate concentrations and methanogenesis in salt marsh soils. *Estuarine Coastal Mar. Sci.*, **10**, 215–23.

King, G. M., Klug, M. J., Wiegert, R. G. and Chalmers, A. G. (1982). Relation of soil water movement and sulfide concentration to *Spartina alterniflora* production in a Georgia salt marsh. *Science*, **218**, 61–3.

Lamb, B., Westberg, H., Allwine, G., Bamesberger, L. and Guenther, A. (1987). Measurement of biogenic sulfur emissions from soils and vegetation: Application of dynamic enclosure methods with Natusch Filter and GC/FPD analysis. *J. Atmos. Chem.*, **5**, 469–91.

Larcher, F., Hamelin, J. and Stewart, G. R. (1977). L'acid dimethylsulfonium-3 propanique-3 de Spartina alglica. *Phytochemistry*, **16**, 2019–20.

Lein, A. Yu., Vaynshteyn, M. B., Namsarayev, B. B., Kashparova, E. V., Matrosov, A. G., Bondar, V. A. and Ivanov, M. V. (1982). Biogeochemistry of anaerobic diagenesis of recent Baltic Sea sediments. *Geochem. Int.*, **19**, 90–103.

Lord, C. J. and Church, T. M. (1983). The geochemistry of salt marshes: Sedimentary ion diffusion, sulfate reduction, and pyritization. *Geochim. Cosmochim. Acta*, **47**, 1381–91.

Lowe, L. E. (1986). Application of sequential extraction procedure to the determination of the distribution of sulfur forms in selected peat materials. *Can. J. Soil Sci.*, **66**, 337–45.

Lowe, L. E. and Bustin, R. M. (1985). Distribution of sulfur forms in six facies of peats of the Fraser River delta. *Can. J. Soil Sci.*, **65**, 531–41.

Luther, G. W. and Church, T. M. (1988). Seasonal cycling of sulfur and iron in porewaters of a Delaware salt marsh. *Mar. Chem.*, **23**, 295–309.

Luther, G. W. and Church, T. M. This volume.

Luther, G. W., Church, T. M., Scudlark, J. R. and Cosman, M. (1986). Inorganic and organic sulfur cycling in salt-marsh pore waters. *Science*, **232**, 746–9.

Maltby, E. (1988). Global wetlands—History, current status and future. In. Hook, D. D. (Ed.). *The Ecology and Management of Wetlands*. Timber Press, Portland, Oregon, pp. 3–14.

Matthews, E. and Fung, I. (1987). Methane emission from natural wetlands: Global distribution, area, and environmental characteristics of sources. *Global Biogeochemical Cycles*, **1**, 61–86.

Mendelssohn, I. A., McKee, K. L. and Patrick, W. H. (1981). Oxygen deficiency in *Spartina alterniflora* roots: Metabolic adaptation to anoxia. *Science*, **214**, 439–41.

Mendelssohn, I. A. and Seneca, E. D. (1980). The influence of soil drainage on the growth of salt marsh cordgrass *Spartina alterniflora* in North Carolina. *Estuarine Coastal Shelf Sci.*, **11**, 27–40.

Mitsui, S. (1965). Dynamic aspects of nutrient uptake. In: *The Mineral Nutrition of the Rice Plant. Symp. Proc. Int. Rice Res Inst.* John Hopkins Press, pp. 53–62.

Morris, J. T. and Whiting, G. J. (1986). Emission of gaseous CO_2 from salt marsh sediments and its relationship to other carbon losses. *Estuaries*, **9**, 9–19.

Nedwell, D. B. (1984). The input and mineralization of organic carbon in anaerobic aquatic sediments. In: Marshall, K. C. (Ed.). *Advances in Microbial Ecology*. Plenum Press, New York, pp. 93–131.

Nedwell, D. B. and Abram, J. W. (1978). Bacterial sulphate reduction in relation to sulphur geochemistry in two contrasting areas of saltmarsh sediment. *Estuarine Coastal Shelf Sci.*, **6**, 341–51.

Nedwell, D. B. and Takii, S. (1988). Bacterial sulphate reduction in sediments of a European salt marsh: Acid-volatile and tin-reducible products. *Estuarine Coastal Shelf Sci.*, **26**, 599–606.

Neue, H. W. and Mamaril, C. P. (1985). Zinc, sulfur, and micronutrients in wetland soils. In: *Proc. Wetland Soils Characterization, Classification and Utilization*. International Rice Research Institute, pp. 307–19.

Nickerson, N. H. and Thibodeau, F. R. (1985). Association between pore water sulfide concentrations and the distribution of mangroves. *Biogeochemistry*, **1**, 183–92.

Nriagu, J. O., Holdway, D. A. and Coker, R. D. (1987). Biogenic sulfur and the acidity of rainfall in remote areas of Canada. *Science*, **237**, 1189–92.

Odum, W. E., McIvor, C. C. and Smith, T. J. (1982). *The Ecology of the Mangroves of South Florida: A Community Profile*. FWS/OBS-81/24, Fish and Wildlife Service.

Oenema, O. (1988). Early diagenesis in recent fine-grained sediments in the Eastern Scheldt. *Ph.D dissertation*, University of Utrecht, Netherlands, 223pp.

Ogden, J. G. (1982). Seasonal mass balance of major ions in three small watersheds in a maritime environment. *Water Air Soil Pollut.*, **17**, 119–30.

Peterson, B. J., Steudler, P. A., Howarth, R. W., Friedlander, A. I., Juers, D. and Bowles, F. P. (1982). Tidal export of reduced sulfur from a salt marsh ecosystem. *Environ. Biogeochem. Ecol. Bull. (Stockholm)*, **35**, 153–65.

Postma, D. (1982). Pyrite and siderite formation in brackish and freshwater swamp sediments. *Am. J. Sci.*, **282**, 1151–83.

Rudd, J. W. M., Kelly, C. A. and Furutani, A. (1986). The role of sulfate reduction in long term accumulation of organic and inorganic sulfur in lake sediments. *Limnol. Oceanogr.*, **31**, 1281–91.

Schoenau, J. J. and Germida, J. J. This volume.

Schubauer, J. P. and Hopkinson, C. S. (1984). Above- and belowground emergent macrophyte production and turnover in a coastal marsh ecosystem, Georgia. *Limnol. Oceanogr.*, **29**, 1052–65.

Skyring, G. W. (1987). Sulfate reduction in coastal ecosystems. *Geomicrobiol. J.*, **5**, 295–374.

Skyring, G. W., Oshrain, R. L. and Wiebe, W. J. (1979). As assessment of sulfate reduction rates in Georgia marshland soils. *Geomicrobiol. J.*, **1**, 398–400.
Sparling, J. H. (1966). Studies on the relationship between water movement and water chemistry in mires. *Can. J. Bot.*, **44**, 747–58.
Spratt Jr, H. G. and Morgan, M. D. (1990). Sulfur cycling in a cedar dominated freshwater wetland. *Limnol. Oceanogr.*, **35**, 1586–93.
Spratt, Jr H. G., Morgan, M. D. and Good, R. E. (1987). Sulfate reduction in peat from a New Jersey pinelands cedar swamp. *Appl. Environ. Microbiol.*, **53**, 1406–11.
Steudler, P. A. and Peterson, B. J. (1984). Contribution of gaseous sulphur from salt marshes to the global sulphur cycle. *Nature*, **311**, 455–7.
Steudler, P. A. and Peterson, B. J. (1985). Annual cycle of gaseous sulfur emissions from a New England *Spartina alterniflora* marsh. *Atmos. Environ.*, **19**, 1411–16.
Swider, K. T. (1988). Transformations of sulfur compounds in salt marsh sediments. *M.S. Thesis*, SUNY at Stony Brook.
Tabatabai, M. A. This volume.
Taylor, B. E., Wheeler, M. C. and Nordstrom, D. K. (1984). Stable isotope geochemistry of acid mine drainage: Experimental oxidation of pyrite. *Geochim. Cosmochim. Acta*, **48**, 2669–78.
Teal, J. M. and Kanwisher, J. (1961). Gas exchange in a Georgia salt marsh. *Limnol. Oceanogr.*, **6**, 388–99.
Thibodeau, F. R. and Nickerson, N. H. (1986). Differential oxidation of mangrove substrate by *Avicennia germinans* and *Rhizophora mangle*. *Am. J. Bot.*, **73**, 512–16.
Urban, N. R. and Bayley, S. E. (1986). The acid–base balance of peatlands: A short-term perspective. *Water Air Soil Polut.*, **30**, 791–800.
Urban, N. R., Eisenreich, S. J. and Grigal, D. F. (1989). Sulfur cycling in a forested *Sphagnum* bog in northern Minnesota. *Biogeochemistry*, **7**, 81–109.
Valiela, I., Teal, J. M. and Persson, N. Y. (1976). Production and dynamics of experimentally enriched salt marsh vegetation: Belowground biomass. *Limnol. Oceanogr.*, **21**, 245–52.
Vamos, R. and Koves, E. (1972). Role of light in the prevention of the poisoning action of hydrogen sulfide in the rice plant. *Ecology*, **53**, 519–25.
Van Breemen, V. (1982). Genesis, morphology and classification of acid sulfate soils in coastal plains. In: Kittrick, J. A., Fanning, D. S. and Hossner, L. R. (Eds). *Acid Sulfate Weathering*. Soil Science Society of America, Madison, WI, pp. 95–108.
Westrich, J. T. and Berner, R. A. (1984). The role of sedimentary organic matter in bacterial sulfate reduction: The G model tested. *Limnol. Oceanogr.*, **29**, 236–49.
Whigham, D. F., McCormick, J., Good, R. E. and Simpson, R. L. (1978). Biomass and primary production in freshwater tidal marshes of the Middle Atlantic Coast. In: Good, R. E., Whigham, D. F. and Simpson, R. L. (Eds). *Freshwater Wetlands: Ecological Process and Management Potential*. Academic Press, New York, pp. 3–20.
Wieder, R. K. and Lang, G. E. (1986). Fe, Al, Mn, and S chemistry of *Sphagnum* peat in four peatlands with different metal and sulfur inputs. *Water Air Soil Pollut.*, **29**, 309–20.
Wieder, R. K. and Lang, G. E. (1988). Cycling of inorganic and organic sulfur in peat from Big Run Bog, West Virginia. *Biogeochemistry*, **5**, 221–42.
Wieder, R. K., Lang, G. E. and Granus, V. A. (1985). An evaluation of wet chemical methods for quantifying sulfur fractions in freshwater wetland peat. *Limnol. Oceanogr.*, **30**, 1109–15.
Wieder, R. K., Yavitt, J. B. and Lang, G. E. (1990). Methane production and sulfate reduction in two Appalachian peatlands. *Biogeochemistry*, **10** (2), 81–104.

Wieder, R. K., Yavitt, J. B. and Lang, G. E. This volume.
Wieder, R. K., Yavitt, J. B., Lang, G. E. and Bennett, C. A. (1989). Annual aboveground net primary production at Big Run Bog, West Virginia. Castanea (in press).
Wiegert, R. G., Chalmers, A. G. and Randerson, P. F. (1983). Productivity gradients in salt marshes: The response of *Spartina alterniflora* to experimentally manipulated soil water movement. *Oikos*, **41**, 1–6.
Yavitt, J. B., Lang, G. E. and Wieder, R. K. (1987). Control of carbon mineralization to CH_4 and CO_2 in anaerobic, *Sphagnum*-derived peat from Big Run Bog, West Virginia. *Biogeochemistry*, **4**, 141–57.
Yuretich, R. F., Crerar, D. A., Kinsman, D. J. J., Means, J. T. and Borsik, M. P. B. (1982). Hydrogeochemistry of the New Jersey Coastal Plain. I. Major element cycles in precipitation and rainwater. *Chem. Geol.*, **33**, 1–21.
Zhabina, N. N. and Volkov, I. I. (1978). A method of determination of various sulfur compounds in sea sediments and rocks. In: Krumbein W. E. (Ed.). *Environmental Biogeochemistry and Geomicrobiology: Methods, Metals and Assessment*. Ann Arbor, Ann Arbor Science Publishers, pp. 735–45.

APPENDIX TO CHAPTER 5: SULPHUR INPUTS MAY AFFECT ORGANIC CARBON BALANCE OF *SPHAGNUM*-DOMINATED WETLANDS

R. KELMAN WIEDER
Villanova University, Villanova, PA, USA

JOSEPH B. YAVITT
Cornell University, Ithaca, NY, USA

and

GERALD E. LANG
West Virginia University, Morgantown, WV, USA

Freshwater peatlands cover some 450 million ha of the earth's land surface (Kivinen and Pakarinen, 1981), mainly north of 49° N latitude. Large areas of peatland, especially in northeastern North America and in northern Europe, receive large inputs of sulphur in acid deposition. The effect of this deposition on peatland ecosystems has received much less study than have effects on forests and lakes (Gorham, Bayley and Schindler, 1984). We have hypothesized that one effect of sulphate deposition to peatlands may be to increase rates of anoxic decomposition and decrease the storage of organic carbon in the peat (Wieder, Yavitt and Lang, 1990). This hypothesis is based on our research on sulphur biogeochemistry in Big Run Bog, West Virginia. Big Run Bog receives a considerable loading of atmospheric sulphur, some 1.8 g (S) m^{-2} a^{-1} (0.056 mol m^{-2} a^{-1} in wet deposition alone).

Total sulphur concentration in Big Run Bog peat averages 3.9 mg g^{-1} dry mass (123 μmol g^{-1} dry mass), with 81% as carbon-bonded S, 10% as ester-sulphate S, and less than 10% as inorganic S, including pyrite, iron monosulphides, elemental sulphur, and sulphate (Wieder and Lang, 1988). Given that organic sulphur pools rather than inorganic sulphur pools dominate in Big Run Bog peat, one might suppose that sulphur fluxes would be mainly through the organic pools (via mineralization and immobilization) rather than

Sulphur Cycling on the Continents
Edited by R.W. Howarth, J.W.B. Stewart and M.V. Ivanov

through the inorganic pools (via oxidation and reduction). Also, traditional views of anaerobic carbon mineralization in freshwater environments held that sulphate reduction should be of minimal importance, being limited by a low sulphate concentration (see review by Nedwell, 1984). Thus, we were surprised to find that the major fate of $^{35}SO_4^{2-}$ in short-term incubations was the formation of reduced inorganic sulphur via the process of dissimilatory sulphate reduction (Table 5A.1; Wieder and Lang, 1988). Furthermore, rates of sulphate reduction were comparable to rates reported for coastal marine sediments, high sulphate environments (Giblin and Wieder, this volume). High rates of sulphate reduction in Big Run Bog apparently can persist despite a low instantaneous dissolved sulphate concentration because there is a continuous replenishment of the dissolved sulphate pool via the oxidation of reduced inorganic sulphur (Wieder and Lang, 1988; Table 5A.1).

High sulphur deposition at Big Run Bog, along with the unexpectedly high rates of sulphate reduction in Big Run Bog peat, may have important implications for carbon balance. Big Run Bog is in many ways structurally and functionally similar to more northern peatland ecosystems. However, one difference is that at Big Run Bog, a maximum of only 2.25 m of peat has accumulated over a relatively long period of time (approximately 13 000 years; Wieder, 1985). Such slow rates of peat accumulation are typical of Appalachian *Sphagnum*-dominated peatlands (Maxwell and Davis, 1982; Watts, 1979; Cotter, 1983). In contrast, in many northern peatlands considerably greater quantities of peat have accumulated over much shorter periods of time. The slow rate of peat accumulation at Big Run Bog is probably not

Table 5A.1. Summary of sulphur pool sizes and transformation rates in *Sphagnum*-derived peat from Big Run Bog. Values are means of 36 determinations on peat collected in March 1986, and are based on direct measurements on the short-term fate of $^{35}SO_4^{2-}$ injected into peat samples. Details are in Wieder and Lang (1988). Reduced inorganic S is equivalent to chromium-reducible S, and therefore includes H_2S, FeS, S^0, and FeS_2. The formation rate listed for sulphate is calculated as the negative value of the sum of the formation of reduced inorganic S, ester-sulphate S, and carbon-bonded S. Because formalin-killed controls exhibited essentially no conversion of added label into either organic or inorganic products, the transformations listed below are biologically mediated. Formation rates and pool sizes are one gram wet weight basis

Form of S	Formation rate (nmol g^{-1} h^{-1})	Pool size (μmol g^{-1} h^{-1})	Turnover time (days)
Reduced inorganic S	3.22	1.57	47.7
Ester sulphate S	0.36	1.12	195.8
Carbon bonded S	0.53	17.31	2154.1
Sulphate S	−4.11	0.07	1.1

because of low carbon inputs by net primary production. Present day net primary production at Big Run Bog is quite high (approximately 450 g (C) m^{-2} a^{-1}), and in general net primary production in peatlands increases with decreasing latitude (Wieder and Lang, 1983; Wieder *et al.*, 1989). Moreover, it does not appear to be the case that high atmospheric sulphur deposition has inhibited the growth of *Sphagnum* species at Big Run Bog, as has been suggested for peatlands in the southern Pennines of Great Britain (Austin and Wieder, 1987).

Historically, longer growing seasons and generally higher mean annual temperatures associated with decreasing latitude probably have led to net primary production being greater in southern peatlands than in northern peatlands (Damman, 1979). Southern peatlands like Big Run Bog would seem likely sites for high rates of carbon accumulation because of the high rate of net primary production. That this is not true and that rates of peat accumulation are low implies not only that carbon mineralization to CO_2 and CH_4 must be relatively high, but also that annual fluxes of carbon to and from the atmosphere must be high.

At Big Run Bog, anaerobic carbon mineralization in the top 35 cm of peat is dominated by sulphate reduction (Table 5A.2). Of the total annual anaerobic carbon mineralization to CO_2 and CH_4 in the top 35 cm of peat of 53 mol m^{-2} (640 g (C) m^{-2} a^{1}), 64% is attributable to sulphate reduction, 23% to methanogenesis, and 12% to other processes (Wieder, Yavitt and Lang, 1990). The dominance of sulphate reduction is apparently not at the expense of methanogenesis. Rates of sulphate reduction and methane production in

Table 5A.2 Summary of anaerobic carbon mineralization in the top 35 cm of Big Run Bog peat. Methane production and anaerobic carbon dioxide production were obtained by measurement of gas accumulation in the headspace of flasks containing peat during incubation in the laboratory. Sulphate reduction data were determined radiometrically. Values are based on 191 determinations for CH_4 production and CO_2 production and on 333 determinations of sulphate reduction in peat samples collected on eight dates throughout the course of an entire year. Data are from Wieder, Yavitt and Lang, 1990.

Anaerobic carbon mineralization process	Range of values (mmol m^{-2} day^{-1})	Mean value (mmol m^{-2} day^{-1})	Integrated total (mol m^{-2} a^{-1})
Methane production	0–366	63	6.2
Carbon diozide production	72–1238	406	46.6
Sulphate reduction	2.5–1568	146	17.0

replicate peat subsamples are not negatively correlated (Wieder, Yavitt and Lang, 1990). Moreover, sulphate additions to peat do not consistently inhibit methane production. Although sulphate additions to concentrations of 1 mM (32 mg l^{-1}) or greater sometimes inhibit methane production, this inhibition is not consistently accompanied by an increase in anaerobic CO_2 production, as would be expected if sulphate-reducing bacteria were out-competing methane producers at these enhanced sulphate levels (Yavitt, Wieder and Lang, 1988).

Most of the world's peatlands receive considerably less atmospheric sulphur input than that received by Big Run Bog. The high sulphate input in acid deposition to Big Run Bog, and the spiralling of added sulphur between oxidized and reduced inorganic states during its residence in the bog, have created conditions where sulphate reduction contributes much more to anaerobic carbon mineralization than the sulphate deposition rate alone might suggest. Sulphate reduction in the top 35 cm of Big Run Bog peat is 17 mol m^{-2} a^{-1} (540 g (S) m^{-2} a^{-1}), whereas annual sulphate input from wet and dry deposition is only 0.11 mol m^{-2} (3.6 g (S) m^{-2} a^{-1}) (Wieder, Yavitt and Lang, 1990; see Tables 5A.1 and 5A.2). Carbon-balance calculations for Big Run Bog indicate that in the absence of sulphate reduction, the Big Run Bog peat deposit would more than likely be either at steady state or in a condition of net peat accumulation. However, under extant conditions aerobic and anaerobic carbon mineralization to CO_2 and CH_4 exceeds carbon fixation by net primary production. The carbon balance is negative and the peat deposit is apparently being degraded (Wieder, Yavitt and Lang, 1990).

Our work at Big Run Bog has considerably broader implications. Although estimates vary, peatland ecosystems (including tundra, boreal forests, and wetlands) may contain as much as 34% of the global soil carbon pool (Schlesinger, 1977; Houghton *et al.*, 1985; Trabalka, 1985). Most of the carbon in peatland ecosystems is stored as peat that has accumulated since the most recent glaciation. As such, peatlands have served as global net sinks for atmospheric carbon over the past several thousand years, and continue to represent significant carbon reservoirs as well as potential sources and sinks for atmospheric carbon (Miller, 1981; Trabalka, 1985; Sebacher *et al.*, 1986; Matthews and Fung, 1987; Moore and Knowles, 1987). It is quite possible that in regions that are not exposed to high atmospheric sulphur deposition, anoxic carbon mineralization may be dominated by methanogenesis. However, our findings from Big Run Bog suggest that if such peatlands were to receive increased atmospheric sulphur deposition, the effect may be to enhance sulphate reduction without inhibiting methane production. As a result, CH_4 emissions to the atmosphere may remain unaffected, while CO_2 emissions may be stimulated, thereby augmenting the role of peatlands as sources of atmospheric carbon. According to such a scenario, peatland ecosystems may provide an important link between anthropogenically produced acid precipitation and global carbon cycling.

REFERENCES

Austin, K. A. and Wieder, R. K. (1987). Effects of elevated H^+, SO_4^{2-}, NO_3^-, and NH_4^+ in simulated acid precipitation on the growth and chlorophyll content of 3 North American *Sphagnum* species. *Bryologist*, **90**, 221–9.

Cotter, I. F. P. (1983). The minimum age of the Woodfordian deglaciation of northeastern Pennsylvania and northwestern New Jersey. Ph D dissertation, Lenigh University, Betheleham, PA, USA.

Damman, A. W. H. (1979). Geographic patterns in peatland development in eastern North America. *Proceedings of the International Symposium on Classification of Peat and Peatlands*. Hyytiala, Finland, pp. 42–57.

Giblin, A. E. and Wieder, R. K. (This volume).

Gorham, E., Bayley, S. E. and Schindler, D. W. (1984). Ecological effects of acid deposition upon peatlands: a neglected field in 'acid rain' research. *Can. J. Fish. Aquat. Sci.*, **41**, 1256–68.

Houghton, R. A., Schlesinger, W. H., Brown, S. and Richards, J. F. (1985). Carbon dioxide exchange between the atmosphere and terrestrial ecosystems. In: Trabalka, J. R. (Ed.). *Atmospheric Carbon Dioxide and the Global Carbon Cycle*. DOE/ER-0239 US Dept. of Energy, Off. Energy Research, Carbon Dioxide Research Div., Washington DC, pp. 113–40.

Kivinen, E. and Pakarinen, P. (1981). Geographical distribution of peat resources and major peatland complex types in the world. *Annales Academiae Scientiarum Fennicae*, Series A, III. *Geologica-Geographica*, 132.

Matthews, E. and Fung, I. (1987). Methane emission from natural wetlands: global distribution, area, and environmental characteristics of sources. *Global Biogeochem. Cycles*, **1**, 61–87.

Maxwell, J. A. and Davis, M. B. (1972). Pollen evidence of Pleistocene and Holocene vegetation of the Allegheny Plateau, Maryland. *Quaternary Research*, **2**, 506–30.

Miller, P. C. (1981). Carbon balance in northern ecosystems and the potential effect of carbon dioxide induced climate change, CONF-8003118, US Dept. of Energy, Off. Health Environ. Research, Washington DC.

Moore, T. R. and Knowles, R. (1987). Methane and carbon dioxide evolution from subarctic fens. *Can. J. Soil Sci.*, **67**, 321–43.

Nedwell, D. B. (1984). The input and mineralization of organic carbon in anaerobic aquatic sediments. *Adv. Microb. Ecol.*, **7**, 93–131.

Schlesinger, W. H. (1977). Carbon balance in terrestrial detritus. *Annu. Rev. Ecol. Syst.*, **8**, 51–81.

Sebacher, D. I., Harriss, R. C., Bartlett, K. B., Sebacher, S. M. and Grice, S. S. (1986). Atmospheric methane sources: Alaska tundra bogs, an alpine fen, and a subarctic boreal marsh. *Tellus*, **38B**, 1–10.

Trabalka, J. R. (1985). Atmospheric carbon dioxide and the global carbon cycle, DOE/ER-0239. US Dept. of Energy, Off. Energy Research, Carbon Dioxide Research Div., Washington DC.

Watts, W. A. (1979). Late quaternary vegetation of central Appalacia and the New Jersey coastal plain. *Ecol. Monog.*, **49**, 427–69.

Wieder, R. K. (1985). Peat and water chemistry at Big Run Bog, a peatland in the Appalachian Mountains of West Virginia. *Biogeochemistry*, **1**, 277–302.

Wieder, R. K. and Lang, G. E. (1983). Net primary production of the dominant bryophytes in a *Sphagnum*-dominated wetland in West Virginia. *Bryologist*, **86**, 280–6.

Wieder, R. K. and Lang, G. E. (1988). Cycling of inorganic and organic sulphur in peat from Big Run Bog, West Virginia. *Biogeochemistry*, **5**, 221–42.

Wieder, R. K., Yavitt, J. B. and Lang, G. E. (1990). Methane production and sulphate reduction in two Appalachian peatlands. *Biogeochemistry*, **10**, 81–104.
Wieder, R. K., Yavitt, J. B., Lang, G. E. and Bennett, C. A. (1989). Aboveground net primary production at Big Run Bog, West Virginia. *Castanea* (in press).
Yavitt, J. B., Wieder, R. K. and Lang, G. E. (1988). Control of carbon mineralization to CH_4 and CO_2 in anaerobic *Sphagnum*-derived peat from Big Run Bog, West Virginia. *Biogeochemistry*, **4**, 141–57.

6 An Overview of the Environmental Chemistry of Sulphur in Wetland Systems

GEORGE W. LUTHER, III and THOMAS M. CHURCH
College of Marine Studies, University of Delaware, Lewes DE, USA

6.1 INTRODUCTION

In this chapter, we show the complexity of and gaps in our knowledge of the sulphur cycle in soils and sediments. Much of our discussion focuses on salt marshes and other marine systems, but we also draw comparisons with other systems. There is a wealth of detailed information on the components of the sulphur cycle in marine environments as compared to freshwater environments, perhaps because marine sediments are dominated by inorganic sulphur chemistry while freshwater sediments are dominated by organic sulphur chemistry. Organic sulphur chemistry is more complex because of the wide variety of compounds which can be formed under natural conditions, and our knowledge and understanding of the chemistry of the organic sulphur cycle in both marine and freshwater environments is still in its formative stages. Clearly, new techniques to determine species and functional groups are needed to advance our knowledge of organic sulphur biogeochemistry.

6.2 AN OVERVIEW OF THE SULPHUR CYCLE

Figure 6.1 shows a comprehensive schematic representation of the sulphur cycle which includes: (1) sulphate reduction to sulphide and its subsequent oxidation; (2) the incorporation of sulphide into reduced inorganic and organic sulphur; and (3) the possible links between the inorganic and organic pools via sulphur compounds of intermediate oxidation states. The sulphur species may be formed via abiotic and/or biotic reactions. Tables 6.1 and 6.2 summarize some of the known laboratory organic sulphur reactions which have been documented to be, or may be, of environmental importance. A

Sulphur Cycling on the Continents
Edited by R.W. Howarth, J.W.B. Stewart and M.V. Ivanov

glossary of significant inorganic and organic sulphur compounds and ions is given in the appendix to this chapter.

Salt marshes provide excellent systems to understand the cycling of sulphur (and other elements) in the marine environment because of high sulphate concentrations derived from oceanic sources and because of high sulphate reduction rates (Howarth, 1984; Giblin and Wieder, this volume). These conditions lead to the dissimilatory production of hydrogen sulphide which in turn reacts with a variety of other inorganic and organic compounds in the sediments and porewaters. Sulphur compounds with intermediate oxidation states are quickly formed because the surface of the sediments is exposed to the atmosphere and because the rhizosphere of marsh grasses can oxidize sediments at depth. Thus, diffusive pathways for gaseous exchange (e.g. O_2

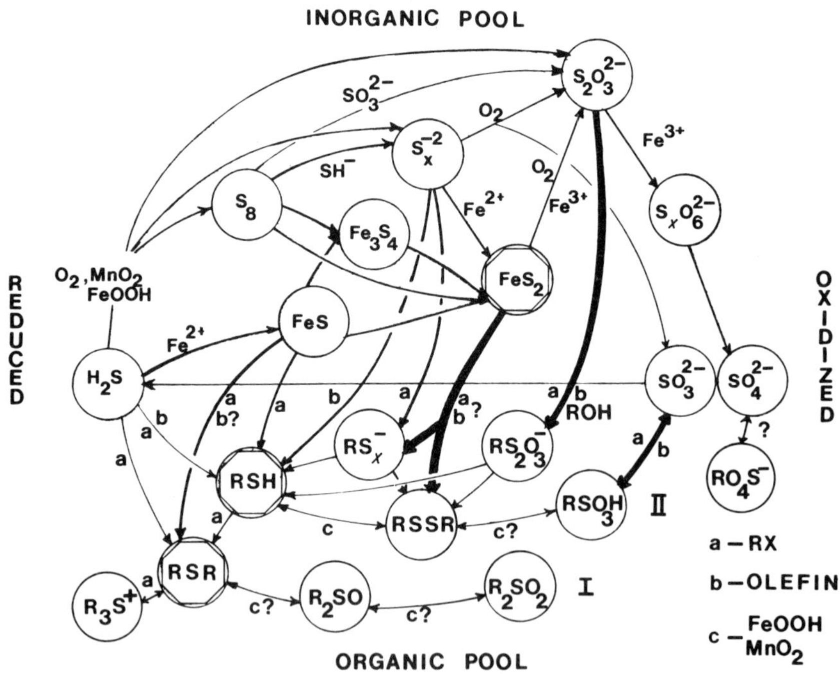

Figure 6.1. A schematic figure of the sulphur cycle. The upper half of the figure gives the major inorganic species whereas the lower half of the figure gives the principal organic functional groups. Oxidized species are on the right and more reduced species are to the left of the figure. The principal reduced sulphur compounds other than H_2S are considered to be FeS_2 (pyrite), organic sulphides (RSR), thiols (RSH) and organic disulphides (RSSR). Only the well documented stable and metastable sulphur compounds are provided in the figure. Organic pathways marked I and II are described in the text. Pathways marked with boldface lines and reactants with question marks are those that have not been documented in the marine environment

into and H_2S out of the sediments) are direct. Besides oxygen, other abundant oxidizing agents (e.g. FeOOH, MnO_2, NO_3^-) are available to oxidize reduced sulphur in marine sediments (Aller and Rude, 1988; Lord and Church, 1983; Howarth, 1984; Luther *et al.*, 1986; Luther, 1987) and in forest soils (Fuller *et al.*, 1986; Strickland *et al.*, 1987; Wieder and Lang, 1988). For this reason, abundant oxidized and highly reactive sulphur products coexist, chief in importance being polysulphides (S_x^{2-}) since they are a common reactant for both pyrite formation (Rickard, 1975; Howarth, 1979; Luther, 1991) and organic sulphur formation (LaLonde, Ferrara and Hayes, 1987; Oae, 1985; Vairavamurthy and Mopper, 1988; Kohnen *et al.*, 1989).

Table 6.1. Known organic sulphur reactions: formation of organic sulphur from inorganic sulphur

(1) Addition to olefins		
(a) $CH_2{=}CHCOO^- + HS_x^-$ $x \geqslant 1$ (bisulphide and polysulphide)	$\xrightarrow{H_2O}$	$(SH)CH_2CH_2COO^-$ (thiol)
(b) $CH_3CH{=}CH_2 + NaHSO_3$	$\longrightarrow$	$CH_3CH_2CH_2SO_3Na$ (sulphonate)
(c) $CH_3CH{=}CH_2 + S_2O_3^{2-}$	$\xrightarrow{H_2O}$	$CH_3CH_2CH_2S_2O_3^-$ (Bunte salt)
(2) Nucleophilic displacement of natural organohalides		
(a) $RCH_2CH_2X + HS_x^-$ $x \geqslant 1$ (bisulphide and polysulphide)	$\longrightarrow$	$RCH_2CH_2SH + X^-$ (thiol)
(b) $2\ RX + SH^-$	$\longrightarrow$	$R_2S + 2\ X^- + H^+$ (sulphide)
(c) $RX + S_x^{2-}$	$\longrightarrow$	$RS_x^- + X^-$ (polysulphide)
(d) $RX + S_2O_3^{2-}$	$\longrightarrow$	$RS_2O_3^- + X^-$ (Bunte salt)
(e) $RX + SO_3^{2-}$	$\longrightarrow$	$RSO_3^- + X^-$ (sulphonate)
(f) $RX + S_2O_3^{2-} + H_2O$	$\longrightarrow$	$RSH + HSO_4^- + X^-$ (thiol)
(3) Solid phase displacement of natural organohalides		
$2\ CH_3I + FeS$	$\longrightarrow$	$(CH_3)_2S + Fe^{2+} + 2\ I^-$

Table 6.2. Known organic sulphur reactions: oxidation of reduced organic sulphur compounds

(1) Thiol oxidation		
(a) RSH + MnO_2, FeOOH, O_2	⟶	RSSR (disulphide)
(b) RSSR + strong oxidants e.g. MnO_4^-	⟶	RSO_3H (sulphonate)
(2) Organic sulphide oxidation		
(a) RSR + O_2	⟶	R_2SO (sulphoxide)
(b) R_2SO + O_2	⟶	R_2SO_2 (sulphone)

For reactions 1(b), 2(a) and 2(b), metal oxide phases present in sediments may be effective oxidants of reduced organic sulphur.

However, other reactants (e.g. $S_2O_3^{2-}$, SO_3^{2-}, FeS_2) may be equally important, participating in reactions indicated with boldfaced lines in Figure 6.1.

6.3 INORGANIC SULPHUR GEOCHEMISTRY

Under anoxic conditions in salt marshes and other marine sediments, dissimilatory sulphate reduction is often the primary means of organic carbon mineralization (Jørgensen, 1982; Howarth, 1984; Ivanov *et al.*, 1989; Giblin and Wieder, this volume). The reduction of sulphate ion to bisulphide involves the transfer of eight electrons per S atom, and the process is dependent on microbial catalysis. However, under varying subsequent redox conditions, re-oxidation reactions serve to cycle sulphide back to sulphate (Howarth and Teal, 1979; Lord and Church, 1983; Giblin and Howarth, 1984). Besides dissimilatory sulphate reduction, assimilatory sulphate reduction can also produce H_2S which can be emitted by higher plants directly to the atmosphere (Rennenberg, 1989; Andreae and Jaeschke, this volume).

As detailed in Figure 6.1, a variety of partially oxidized species can be formed from bisulphide ion oxidation. Thiosulphate, elemental sulphur and polysulphides (Chen and Morris, 1982; Hoffman, 1977; Boulegue, Lord and Church, 1982; Howarth *et al.*, 1983; Luther, Varsolona and Giblin, 1985) are among the most important and are formed by abiotically or biotically catalysed oxidation reactions. Some of these partially oxidized sulphur species can result from microbial metabolism (Goldhaber and Kaplan, 1975) and/or from the reductive dissolution of metal oxide phases (Pyzik and Sommer, 1981).

Elemental sulphur is formed during the oxidation of bisulphide ion by a variety of oxidants. Hoffman (1977) showed elemental sulphur to be an important end-product in the reaction of bisulphide ion with peroxide. Pyzik and Sommer (1981) observed elemental sulphur in the reaction of iron(III) minerals with bisulphide ion, and Burdige and Nealson (1986) observed it in the reaction of MnO_2 with bisulphide ion. Elemental sulphur apparently results from the oxidation of metal monosulphides (Moses *et al.*, 1987) but does not result from the oxidation of pyrite (McKibben and Barnes, 1986; Moses *et al.*, 1987).

Polysulphides (S_x^{2-}) result from the same oxidation reactions which produce elemental sulphur. They are formed first from the reaction of bisulphide ion with zerovalent sulphur formed from bisulphide oxidation (Hoffman, 1977). As the chain length (x) increases, the polysulphides tend to decompose to elemental sulphur. Polysulphides can also be formed by the reaction of elemental sulphur and bisulphide ion (Giggenbach, 1972). Polysulphides are potentially important in the cycling of metals (Boulegue, Lord and Church, 1982). Polysulphides have been detected as an important component of dissolved sulphur in the interstitial waters of tidal wetland sediments (Boulegue, Lord and Church, 1982; Luther, Varsolona and Giblin; Luther *et al.*, 1986) and of subtidal sediments (Luther, Varsolona and Giblin, 1985). Polysulphides are also important intermediates for the formation of pyrite (Giblin, 1988; Howarth, 1979; Lord and Church, 1983; Luther *et al.*, 1982; Rickard, 1975). Recent work shows that polysulphides react with abundant lignin components from buried marsh grass material to produce organic sulphur (Francois, 1987a), and with activated olefins to produce thiols (Vairavamurthy and Mopper, 1988) and thiophenes (LaLonde, Ferrara and Hayes, 1987).

Both dissolved organic and inorganic sulphur compounds can provide powerful ligands for complexing and cycling trace metals. For example Cu(I) should be strongly complexed by species like $Cu(S_4)_2^{3-}$ and Cu(II) by species like $Cu(HS)_3^-$. Both complexes have stability constants (pK) of 18–19. Thiols such as cysteine can strongly complex Cu(I) with pK values of > 15 (Boulegue, Lord and Church, 1982). Recently, Shea and MacCrehan (1988b) have modelled Cu solubility in the anoxic sediments of the Chesapeake Bay and have shown that the major Cu complex is that of cysteine with a small contribution from S_x^{2-} complexes.

Thiosulphate ($S_2O_3^{2-}$) is the first product resulting from the decomposition of the solid sulphide precipitate, FeS_2 (Goldhaber, 1983; Moses *et al.*, 1987; Luther, 1987) and from the decomposition of polysulphides (Chen and Morris, 1972). Thiosulphate has been detected as a significant component of dissolved sulphur at all depths in salt marsh porewaters (Boulegue, Lord and Church, 1982; Howarth *et al.*, 1983; Luther, Varsolona and Giblin, 1985) and in subtidal porewaters (Luther, Varsolona and Giblin, 1985). Thiosulphate is an important nucleophile which can displace the halides and the hydroxide ion in organic compounds to produce thiols, acidity and sulphate (Roy and Trudinger, 1970; Oae, 1985). Luther *et al.* (1986) hypothesized that thiosulphate from pyrite oxidation can form organic sulphur (carbon-bonded sulphur) in salt marsh sediments.

Sulphite (SO_3^{2-}) and tetrathionate ($S_4O_6^{2-}$) have rarely been observed in the natural environment. Boulegue, Lord and Church (1982) and Luther, Varsolona and Giblin (1985) have reported sulphite as a minor component in a few salt marsh porewater samples, but sulphite was not detected by Howarth *et al.* (1983) in any samples over a year of sampling. Luther *et al.* (1986) reported tetrathionate in the porewaters from one salt marsh core.

Metal monosulphides, in particular FeS, are generally minor contributors to the total reduced inorganic sulphur pool in salt marsh sediments, and the major metal sulphide is pyrite, FeS_2 (Howarth and Teal, 1979; Cutter and Velinsky, 1988; Gardner, Wolaver and Mitchell, 1988; King, 1988; King, Howes and Dacey, 1985; Lord and Church, 1983; Swider and Mackin, 1989; Giblin and Wieder, this volume). Howarth and colleagues (Howarth, 1979; Howarth and Teal, 1979; Howarth and Giblin, 1983; Howarth and Merkel, 1984) using ^{35}S-sulphate incubations have shown that pyrite, not FeS, is the major short term end-product of sulphate reduction in many salt marsh sediments. Pyrite is commonly the most abundant metal sulphide in subtidal marine sediments as well (Berner and Westrich, 1985; Chanton, Martens and Goldhaber, 1987; Goldhaber and Kaplan, 1980; Sweeney and Kaplan, 1973). However, in contrast to marsh sediments, FeS is often the initial major end-product of sulphate reduction in subtidal coastal sediments (Howarth and Jørgensen, 1984; Westrich and Berner, 1988; Thode-Andersen and Jørgensen, 1989), although even in some subtidal sediments, pyrite rather than FeS dominates as the end-product in ^{35}S-sulphate incubations (Ivanov *et al.*, 1989; Christensen, 1989).

Pyrite provides an important sink for sulphur and trace metals in salt marsh sediments (Lord and Church, 1983) and in marine sediments in general (Berner, 1984; Berner and Raiswell, 1983; Morse and Cornwell, 1987; Raiswell and Berner, 1986). It is also an important energy component in the metabolism of tidal wetland ecosystems (Howarth and Teal, 1980; Howarth *et al.*, 1983; Howarth, 1984). Pyritization proceeds via the reaction of H_2S with detrital iron minerals also termed 'reactive iron' (Canfield, 1989). Pyritization can be modelled in sediments by the reaction between dissolved sulphide and iron oxide as goethite (Equation (6.1); Lord and Church, 1983 and references therein) or magnetite (Canfield and Berner, 1987):

$$FeOOH + 3/2\ HS^- + 1/16\ S_8 + 3/2\ H^+ = FeS_2 + 2H_2O \qquad (6.1)$$

This equation shows the need to have a source of oxidant to produce FeS_2, which is more oxidized than FeS.

Two general mechanisms have been proposed for the formation of FeS_2. Pyrite may form directly by reaction between polysulphide and Fe(II) (Rickard, 1975; Luther, 1991) or by the reaction of iron monosulphides with S_8 (e.g. Berner, 1970; Sweeney and Kaplan, 1973). The polysulphide reaction with Fe(II) is considered to be rapid and to result in single crystal morphology for FeS_2 whereas the S_8 reaction is thought to be slow and to result in framboidal morphology. Both morphologies have been observed in marine systems (Sweeney and Kaplan, 1973; Luther *et al.*, 1982; Giblin, 1988). The rapid formation from polysulphide reaction would seem to dominate in many saltmarsh sediments, where FeS is often rare (Howarth and Teal, 1979).

Although the rate of pyrite formation may be rapid in the surface zones of marsh sediments, on a net basis much of the pyrite is not preserved due to its oxidation to thiosulphate and sulphate (Boulegue, Lord and Church, 1982). Thus, a major portion of the pyrite which forms is quickly destroyed and recycled (Howarth and Teal, 1979; Lord and Church, 1983; Luther *et al.*, 1986; Luther and Church, 1988). Luther and Church (1988) have observed pyrite oxidation in salt marsh sediments on the time scale of a week or less. They have observed in the upper 10 cm of the sediment, large increased concentrations of porewater iron and thiol-like compounds concurrent with decreased pyrite concentrations before and after the spring tides. These time scales are shorter than recognized by previous studies in salt marsh sediments (Howarth and Teal, 1979; King *et al.*, 1982). Recently, Wieder and Lang (1988) have noted that rapid FeS_2 turnover also occurs in peat from a freshwater wetland. There is also a significant reoxidation of pyrite and other reduced iron compounds within subtidal marine sediments (Jørgensen, 1982; Howarth, 1984; Berner and Westrich, 1985). Recent laboratory studies by Kornicker and Morse (1991) and field studies (Huerta-Diaz and Morse, 1990) show extensive adsorption and co-precipitation of trace elements in pyrite. Thus, rapid FeS_2 turnover will likely affect the geochemical cycles of other trace elements.

At present, the oxidant(s) responsible for rapid pyrite oxidation in salt marsh sediments appears related to Fe(III) species. The oxidation of pyrite by oxygen has been documented as a slow process (Singer and Stumm, 1970; Moses *et al.*, 1987). In contrast, anaerobic oxidation by Fe(III) compounds is quite rapid (Moses *et al.*, 1987). Fe(III) may also be important in the oxidation of pyrite in anoxic, subtidal sediments (Howarth, 1984). Recently a molecular orbital theory (MOT) approach (Luther, 1987) has been used to explain the molecular basis for the reactivity of pyrite. Fe(III) can form a covalent bond to the surface of the pyrite through S, but molecular oxygen cannot. Thus, electron transfer (and oxidation of pyrite) is rapid with Fe(III). In pyrite oxidation, oxygen in the thiosulphate and sulphate formed comes mainly from water, not molecular oxygen (Taylor, Wheeler and Nordstrom, 1984a,b). Equations (6.2)–(6.4) describe the overall complete oxidation of

FeS_2 by Fe(III). Equation (6.4) is the sum of Equations (6.2) and (6.3)

$$FeS_2 + 6\ Fe^{3+} + 3\ H_2O = 7\ Fe^{2+} + S_2O_3{}^{2-} + 6\ H^+ \qquad (6.2)$$

$$S_2O_3{}^{2-} + 8\ Fe^{3+} + 5\ H_2O = 8\ Fe^{2+} + 2\ SO_4{}^{2-} + 10\ H^+ \qquad (6.3)$$

$$FeS_2 + 14\ Fe^{3+} + 8\ H_2O = 15\ Fe^{2+} + 2\ SO_4{}^{2-} + 16\ H^+ \qquad (6.4)$$

The MOT approach predicts that other oxide phases may also oxidize pyrite. Recently, Aller and Rude (1988) have shown that MnO_2 and other manganate solids oxidize sulphidic material in subtidal sediments. These oxidants are supplied by bioturbation (Aller and Rude, 1988; Berner and Westrich, 1985). The nature of this sulphidic material is unknown, but is assumed to be iron monosulphides or acid volatile sulphur (AVS). This oxidation can be mediated by chemolithotrophic microorganisms. Figure 6.2 describes the major processes that can occur in an estuarine ecosystem.

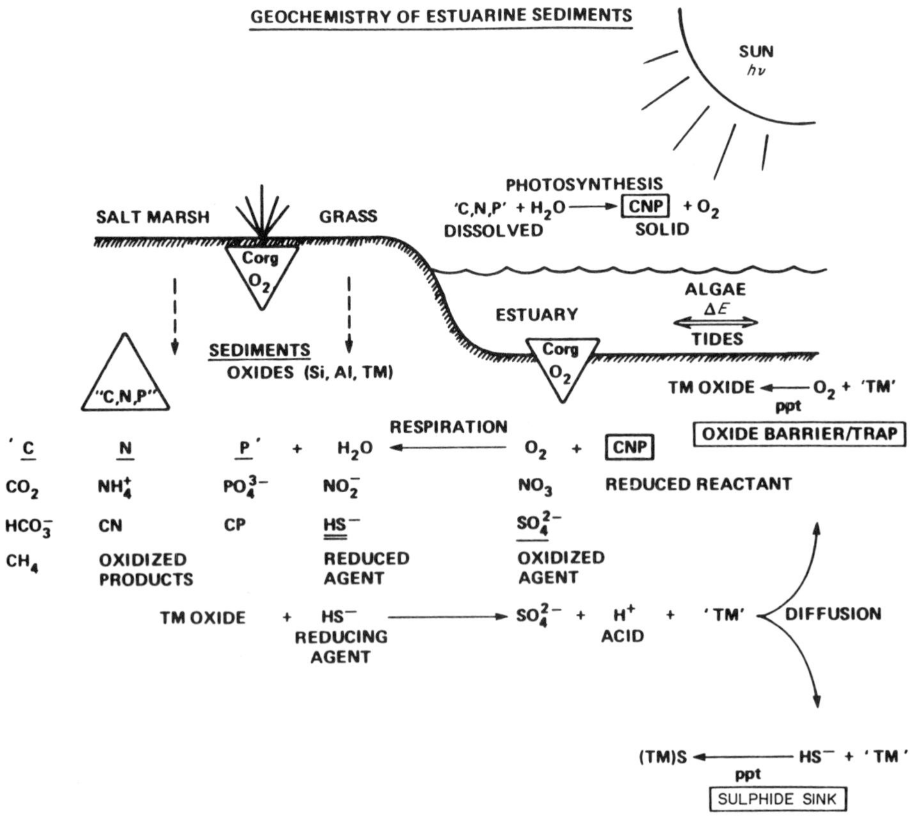

Figure 6.2. This schematic diagram depicts the major processes that can occur in an estuarine ecosystem

6.4 ORGANIC SULPHUR GEOCHEMISTRY

The role of organic sulphur in sulphur and trace element cycling is not well understood because of the diversity of organic sulphur compounds. Organic sulphur compounds in salt marsh environments are not well characterized, other than the volatile compounds dimethylsulphide (DMS), methanethiol (CH_3SH), and dimethyldisulphide (DMDS) (Adams *et al.*, 1981; Cooper *et al.*, 1989; Steudler and Peterson, 1984; Dacey, King and Wakeman, 1987; Andreae and Jaeschke, this volume). In particular, DMS results from the decomposition of dimethylsulphoniumproprionate (DMSP) in *Spartina alterniflora* (Dacey, King and Wakeman, 1987) and also phytoplankton (e.g. Charlson *et al.*, 1987; Nguyen *et al.*, 1988; Turner, Malin and Liss, 1989; Andreae and Jaeschke, this volume). Cooper *et al.* (1989) have found DMS, DMDS, and CS_2 to be produced in a freshwater *Cladium jamaicense* swamp in Florida, and Haines, Black and Bayer (1989) have found emission of CS_2, DMS, and ethanethiol (C_2H_5SH) from the forest litter, soil, and roots of *Stryphnodendron excelsum* in a central American rainforest. Biogenic sulphur emissions of these volatile compounds as well as COS and H_2S have been reviewed by Aneja and Cooper (1989), Rennenberg (1989), Andreae and Jaeschke (this volume), and Giblin and Wieder (this volume).

Non-volatile reduced organic sulphur materials are expected of the types R—S—H (thiols), R—S—R (organic sulphides), R—S—S—R (organic disulphides) with R being alkyl groups, or sulphur containing humic or fulvic compounds (Francois, 1987a; LaLonde, Ferrara and Hayes, 1987). Generally, such organic sulphur compounds could be diagenetic products of reactions under reducing conditions between high molecular weight lignin type materials and inorganic reduced sulphur (SH^-, FeS, FeS_2 and S_x^{-2}). These high molecular weight compounds can decompose to intermediate sulphur organic by-products which are soluble and perhaps volatile (Balzer, 1982).

In salt marshes and other sediments, partially to fully oxidized organic sulphur compounds are probably formed as well. These compounds should result from aerobic oxidation (by O_2) and from anoxic oxidation (by the metal oxides in reactions analogous to those described for inorganic reduced sulphur compounds) of reduced organic sulphur (Table 6.2). Soluble inorganic S compounds (e.g. $S_2O_3^{2-}$ and SO_3^{2-}) with sulphur in oxidation states ranging from −2 to +6 may react with organic compounds to form oxidized organic sulphur compounds. This range of oxidation states makes these compounds likely reactants for cycling between the inorganic and organic sulphur pools (Figure 6.1 and Table 6.1).

The organic sulphur compounds may be represented by two pathways: (1) sulphoxides and sulphones which are formed from R—S—R; and (2) sulphonates which can be produced by the oxidation of thiols and their disulphides (Oae, 1985). Within each pathway there is the possibility of reversibility which includes the reduction of RSO_3H to RSSR as shown in laboratory studies (Oae, 1985). Pathway 1 appears to be a process by which stable solid phase organic sulphur compounds such as thiopenes (cyclic R—S—R) can be buried (Brassell *et al.*, 1986; LaLonde, Ferrara and Hayes, 1987), whereas pathway 2 may be a process that serves to recycle sulphur through carbon compounds.

Organic sulphur pathways 1 and 2 are known to be linked at the RSH to RSR conversion, and recent work has shown that the link is reversible for methylated compounds via microbial processes (Kiene and Visscher, 1987). It is not clear that there is any other link between these two pathways. This work established that both pathways are occurring in the marine environment in the low molecular weight organic material.

Organic sulphur compounds that play important roles in the biochemical metabolism of plants, animals, and microbes include amino acid and peptide types (e.g. cysteine, cysteic acid, glutathione, taurine, methionine). These biochemically important sulphur-containing compounds may be transformed during microbially mediated decomposition processes (Balzer, 1982; Freney, 1986; Guerin and Braman, 1985; Kiene *et al.*, 1986; Kiene and Taylor, 1988), which follow reactions similar to those in pathways 1 and 2.

There has not been a detailed or comprehensive characterization of organic sulphur compounds in any waters and sediments, much less information on their means of formation or their fates. Much of the organic sulphur research has centred on freshwater wetlands, lake sediments, and forested mineral soils. Radiolabelled ^{35}S-sulphate incubations of soil and litter samples from a hardwood forest (Strickland *et al.*, 1987), from peats (Wieder and Lang, 1988), from cedar swamps (Spratt, Morgan and Good, 1987), and from lake sediments (Rudd, Kelly and Furutani, 1986) show that sulphate is incorporated into carbon-bonded S (C—S) and ester sulphate (C—OSO_3), with carbon-bonded sulphur being the major component or end-product. Using chemical extraction methods, several other groups have also noted the importance of carbon-bonded sulphur in freshwater peats (Bustin and Lowe, 1987; Lowe, 1986; Wieder and Lang, 1986; Wieder, Lang and Granus, 1987; Giblin and Wieder, this volume), in forest soils (Fitzgerald, Strickland and Ash, 1985; Fuller *et al.*, 1986; Mitchell and Fuller, 1988; Watwood *et al.*, 1988; Mitchell, David and Harrison, this volume), and in lake sediments (Nriagu and Soon, 1985; Rudd, Kelly and Furutani, 1986; Cook and Kelly, this volume). However, Tabatabai (1984) shows that organic sulphate esters predominate in some agricultural soils.

Lowe and Bustin (1985) have suggested that these two organic components (C—S and C—OSO_3) can be further subdivided into five other pools: hydrolysable C—S, hydrolysable HI—S forms, non-hydrolysable C—S forms, non-hydrolysable HI—S forms, and volatile forms lost on hydrolysis. Unfortunately, each of these pools may contain more than one functional group. For example, the non-hydrolysable C—S forms which they describe include sulphonates and sulphoxides. These compounds represent both of the organic sulphur pathways in Figure 6.1.

Although the long-term incorporation of sulphur is into the organic fraction, ^{35}S-sulphate studies show that inorganic sulphur species are the short-term end-products in peats (Wieder and Lang, 1988), in a bog (Urban, Eisenreich and Grigal, 1989), and in lake sediments (Rudd, Kelly and Furutani, 1986). These studies show that dissimilatory sulphate reduction is of importance in freshwater systems. These studies also support the earlier stable isotope data of Nriagu and Soon (1985) who showed that inorganic reduced sulphur transforms to organic sulphur in eutrophic lakes. In the Everglades, Altschuler *et al.* (1983) proposed that organic oxysulphur compounds are reduced in dissimilatory sulphate reduction.

Thiols (C—S—H) have been reported in salt marsh porewaters by Boulegue, Lord and Church (1982) and by Luther *et al.* (1986) using independent electrochemical techniques. Thiols complexed with iron (iron thiolates) have been reported using polarography (Luther and Church, 1988). In marine porewaters, Mopper and Delmas (1984) and Mopper and Taylor (1986) have reported measuring discrete thiols by derivatizing the thiols with *o*-phthalaldehyde followed by HPLC separation and fluorescent detection of the derivative. Using HPLC followed by direct electrochemical detection, Shea and MacCrehan (1988a) reported discrete thiols in Chesapeake Bay porewaters. Matrai and Vetter (1988) and Vetter *et al.* (1989) have measured thiols in sediment porewaters and coastal waters by derivatizing the thiols with monobromobimane followed by HPLC separation and fluorescent detection of the derivative.

Most recent work in the marine environment has centred on specific organic sulphur compounds or functional groups in subtidal sediments. Guerin and Braman (1985) have observed the production of 1–14 μmol CH_3SH g^{-1} dry weight in sediments from Tampa Bay during the analysis of sediments by the Cr(II) reduction method after the AVS had been removed. Their laboratory results show that Cr(II) solutions do not react with a variety of organic sulphur compounds. Thus, CH_3SH production comes from as yet unidentified solid inorganic or organometallic phases. Mopper and co-workers (1986, 1987, 1988) have reported on thiol compounds in shallow subtidal sediments and porewaters. On the addition of tributyl-phosphine (a S—S bond cleaving agent) to sediment slurries, up to nine times the porewater thiol concentrations were released from the sediments, demonstrating the importance of pathway II (Figure 6.1) in the marine environment.

Francois (1987a,b) observed sulphur enrichment of humic acids in the sediments of Jervis Inlet. Sulphur-stable isotope results and a peak of organic polysulphide (RS_x^-) sulphur at the oxic–anoxic interface suggests that active sulphidation of humic compounds occurs at this interface. Carbon to sulphur mole ratios below the redoxcline reached a low of 53 : 1. Francois (1987a) concluded that the depth of the redoxcline and iron limitation may be the important factors contributing to organic sulphur enrichment in the humic fraction. Canfield (1989) has noted the importance of iron speciation and limitation in determining whether organic or inorganic sulphur will be deposited in marine sediments. Francois's data on organic sulphur formation show similar characteristics to those observed in the Great Marsh, Delaware (Ferdelman, *et al.*, 1991) where the major zone of organic sulphur formation is in the upper 4–6 cm, which is the zone characterized by high rates of both sulphate reduction and pyrite oxidation.

It is unclear how all organic sulphur compounds or functional groups are formed in situ. However, there are several recent reports which shed some information on mechanisms. Activated olefins react with H_2S (Vairavamurthy and Mopper, 1987; Sinninghe Damste *et al.*, 1989) and inorganic polysulphides (Vairavamurthy and Mopper, 1988) to form thiols in marine porewaters (Table 6.1, reaction (1a)); polysulphides are more reactive. LaLonde, Ferrara and Hayes (1987) showed the basis for this reactivity using a molecular orbital approach and also showed that polysulphides react with conjugated olefin-carbonyls to form solid phase sulphur heterocyclic compounds (e.g. thiophenes). Isoprenoid thiophenes (cyclic C—S—C) are considered diagenetic products of chlorophyll-derived phytol or archaebacterial phytenes in marine sediments (Brassell *et al.*, 1986). Recently, Kohnen *et al.* (1989) have provided evidence for the direct incorporation of polysulphides into solid phase organic compounds to form six membered cyclic disulphides (C—C—S—S—C—C) and seven membered cyclic trisulphides (C—C—S—S—S—C—C).

Thayer, Olson and Brinckman (1984) reported on the direct reaction of FeS with organic halides to produce alkyl sulphides (R—S—R) and disulphides (RSSR) via a R—S—H intermediate. This reaction is analogous to the nucleophilic displacement of halides from organic halogen compounds by bisulphide ion which has been observed in polluted groundwaters (Schwarzenbach *et al.*, 1985). Thiosulphate formed from FeS_2 oxidation can oxidize further to sulphate or react with organic compounds (e.g. organic halides) to form Bunte salts ($RS_2O_3^-$) which lead to reduced organic sulphur formation (Oae, 1985). This reaction has not been documented in the marine environment, but has been hypothesized by Francois (1987c) and Luther *et al.* (1986). Recently, Barbash and Reinhard (1989) have reviewed the reactivity of various nucleophiles, including $S_2O_3^{2-}$, towards halogenated organic compounds and concluded that such reactions are feasible.

6.5 CONCLUSIONS

Although there have been numerous studies concerning the forms of sulphur in the environment, we are still far from a complete understanding of sulphur chemistry in natural ecosystems. This is due to the wide range of oxidation states of sulphur (−2 to +6), to sulphur's tendency to form a variety of compounds with carbon, oxygen and itself, and to methodological problems with measuring sulphur compounds. More data are needed on which specific organic sulphur compounds or functional groups are prevalent in the environment under varying redox conditions. In addition, there is a great need to identify and understand the mechanisms by which sulphur is incorporated into these organic compounds. Are the mechanisms abiotic, biotic or both, and are these mechanisms dependent upon the conditions characteristic of particular types of ecosystems? Clearly, more research is needed to understand and to predict accurately how ecosystems will respond to changes in sulphur inputs, to transformations, and to interactions with other element cycles.

REFERENCES

Adams, D. F., Farwell, S. O., Pack, M. R. and Robinson, E. (1981). Biogenic sulphur gas emissions from soils in eastern and southeastern United States. *APCA J.*, **31**, 1083–9.

Aller, R. C. and Rude, P. D. (1988). Complete oxidation of solid phase sulphides by manganese and bacteria in anoxic marine sediments. *Geochim. Cosmochim. Acta*, **52**, 751–65.

Altschuler, Z. S., Schneppe, M. M., Silber, C. C. and Simon, F. O. (1983). Sulphur diagenesis in Everglades peat and origin of pyrite in coal. *Science*, **221**, 221–7.

Andreae, M. O. and Jaeschke, W. A. (This volume).

Aneja, V. P. and Cooper, W. J. (1989). Biogenic sulphur emissions: A review. In: Saltzman, E. S. and Cooper, W. J. (Eds). *Biogenic Sulphur in the Environment*. American Chemical Society, Washington DC, pp. 2–14.

Balzer, W. (1982). Organic sulphur in the marine environment, Chapter 13 In: Duursma, E. K. and Dawson, R. (Eds). *Marine Organic Chemistry*, Elsevier, pp. 395–414.

Barbash, J. E. and Reinhard, M. (1989). Reactivity of sulphur nucleophiles toward halogenated organic compounds in natural waters. In: Saltzman, E. S. and Cooper, W. J. (Eds). *Biogenic Sulphur in the Environment*. American Chemical Society, Washington DC, pp. 101–38.

Berner, R. A. (1970). Sedimentary pyrite formation. *Am. J. Sci.*, **268**, 1–23.

Berner, R. A. (1984). Sedimentary pyrite formation: An update. *Geochim. Cosmochim. Acta*, **48**, 605–15.

Berner, R. A. and Raiswell, R. (1983). Burial of organic carbon and pyrite sulphur in sediments over Phanerozoic time: a new theory. *Geochim. Cosmochim. Acta*, **47**, 855–62.

Berner, R. A. and Westrich, J. T. (1985). Bioturbation and the early diagenesis of carbon and sulphur. *Am. J. Sci.*, **285**, 193–206.

Boulegue, J., Lord, C. J. and Church, T. M. (1982). Sulphur speciation and associated trace metals (Fe, Cu) in the porewaters of Great Marsh, DE. *Geochim. Cosmochim. Acta*, **46**, 453–64.

Brassell, S. C., Lewis, C. A., deLeeuw, J. W., deLange, F. and Sinninghe Damste, J. S. (1986). Isoprenoid thiophenes: novel products of sediment diagenesis. *Nature*, **320**, 160–2.

Burdige, D. J. and Nealson, K. H. (1986). Chemical and microbiological studies of sulphide mediated manganese reduction. *Geomicrobiol, J.*, **4**, 361–87.

Bustin, R. M. and Lowe, L. E. (1987). Sulphur, low temperature ash and minor elements in humid-temperate peat of the Fraser River Delta, British Columbia. *J. Geol. Soc. (London)*, **144**, 435–50.

Canfield, E. (1989). Reactive iron in marine sediments. *Geochim. Cosmochim. Acta*, **53**, 619–32.

Canfield, D. E. and Berner, R. A. (1987). Dissolution and pyritization of magnetite in anoxic marine sediments. *Geochim. Cosmochim. Acta*, **51**, 645–59.

Chanton, J. P., Martens, C. S. and Goldhaber, M. B. (1987). Biogeochemical cycling in an organic-rich coastal marine basin. 7. Sulphur mass balance, oxygen uptake and sulphide retention. *Geochim. Cosmochim. Acta*, **51**, 1187–99.

Charlson, R. J., Lovelock, J. E., Andreae, M. O. and Warren, S. G. (1987). Oceanic phytoplankton, atmospheric sulphur, cloud albedo and climate. *Nature*, **326**, 655–61.

Chen, K. Y. and Morris, J. C. (1972). Kinetics of oxidation of aqueous sulphide by oxygen. *Environ. Sci. Technol.*, **6**, 529–37.

Christensen, J. P. (1989). Sulphate reduction and carbon oxidation rates in continental shelf sediments, an examination of offshelf carbon transport. *Cont. Shelf Res.*, **9**, 223–46.

Cooper, D. J., Cooper, W. J., de Mello, W. Z., Saltzman, E. S. and Zika, R. G. (1989). Variability in biogenic sulphur emissions from Florida wetlands. In: Saltzman, E. S. and Cooper, W. J. (Eds). *Biogenic Sulphur in the Environment*. American Chemical Society, Washington DC, pp. 31–43.

Cutter, G. A. and Velinsky, D. J. (1988). Sulphur diagenesis in a coastal salt marsh. *Mar. Chem.*, **23**, 311–27.

Dacey, J. W. H., King, G. M. and Wakeman, S. G. (1987). Factors controlling emission of dimethylsulphide from salt marshes. *Nature*, **330**, 643–5.

Ferdelman, T. G., Church, T. M. and Luther, G. W. III. (1991). Sulfur enrichment of humic substances in a Delaware salt marsh. *Geochim. Cosmochim. Acta*, **55**, 979–88.

Fitzgerald, J. W., Strickland, T. C. and Ash, J. T. (1985). Isolation and partial characterization of forest floor and soil organic sulphur. *Biogeochemistry*, **1**, 155–67.

Francois, R. (1987a). A study of sulphur enrichment in the humic fraction of marine sediments during early diagenesis. *Geochim. Cosmochim. Acta*, **51**, 17–27.

Francois, R. (1987b). A study of the extraction conditions of sedimentary humic acids to estimate their true in situ sulphur content. *Limnol. Oceanogr.*, **32**, 964–72.

Francois, R. (1987c). The influence of humic substances on the geochemistry of iodine in nearshore and hemipelagic marine sediments. *Geochim. Cosmochim. Acta*, **51**, 2417–27.

Freney, J. R. (1986). Forms and reactions of organic sulphur compounds in soils. In: Tabatabai, M. A. (Ed.). *Sulphur in Agriculture, Agronomy Monograph No. 27*. American Society of Agronomy, Madison, Wisconsin, pp. 207–32.

Fuller, R. D., Driscoll, C. T., Schindler, S. C. and Mitchell, M. J. (1986). A simulation model of sulphur transformations in forested Spodosols. *Biogeochemistry*, **2**, 313–28.
Gardner, L. R., Wolaver, T. G. and Mitchell, M. (1988). Spatial variations in the sulphur chemistry of salt marsh sediments at North Inlet, South Carolina. *J. Mar. Res.*, **46**, 815–36.
Giblin, A. E. (1988). Pyrite formation in marshes during early diagenesis. *Geomicrobiol. J.*, **6**, 77–99.
Giblin, A. E. and Howarth, R. W. (1984). Porewater evidence for a dynamic sedimentary iron cycle in salt marshes. *Limnol. Oceanogr.*, **29**, 47–63.
Giblin, A. E. and R. K. Wieder. This volume.
Giggenbach, W. (1972). Optical spectra and equilibrium distribution of polysulphide ions in aqueous solution at 20° C. *Inorg. Chem.*, **11**, 1201–7.
Goldhaber, M. B. (1983). Experimental study of metastable sulphur oxyanion formation during pyrite oxidation at pH 6–9 and 30° C. *Am. J. Sci.*, **283**, 193–217.
Goldhaber, M. B. and Kaplan, I. R. (1975). The sulphur cycle. In: Goldberg, E. D. (Ed.). *The Sea*. Wiley, New York, pp. 569–655.
Goldhaber, M. B. and Kaplan, I. R. (1980). Mechanisms of sulphur incorporation and isotope fractionation during early diagenesis in sediments of the Gulf of California. *Mar. Chem.*, **9**, 95–143.
Guerin, W. F. and Braman, R. S. (1985). Patterns of organic and inorganic sulphur transformations in sediments. *Org. Geochem.*, **8**, 259–68.
Haines, B., Black, M. and Bayer, C. (1989). Sulphur emissions from roots of the rain forest tree *Stryphnodendron excelsum*. In: Saltzman, E. S. and Cooper, W. J. (Eds). *Biogenic Sulphur in the Environment*. American Chemical Society, Washington DC, pp. 58–69.
Hoffman, M. R. (1977). Kinetics and mechanisms of oxidation of hydrogen sulphide by hydrogen peroxide in acidic solution. *Environ. Sci. Technol.*, **11**, 61–6.
Howarth, R. W. (1979). Pyrite: Its rapid formation in a salt marsh and its importance to ecosystem metabolism. *Science*, **203**, 49–51.
Howarth, R. W. (1984). The ecological significance of sulfur in the energy dynamics of salt marsh and coastal marine sediments. *Biogeochemistry*, **1**, 5–27.
Howarth, R. W. and Giblin, A. (1983). Sulfate reduction in the salt marshes at Sapelo Island, Georgia. *Limnol. Oceanogr.*, **28**, 70–82.
Howarth, R. W. and Jorgensen, B. B. (1984). Formation of ^{35}S-labelled elemental sulfur and pyrite in coastal marine sediments (Limfjorden and Kysing Fjord, Denmark) during short-term $^{35}S_4^{2-}$ reduction measurements. *Geochim. Cosmochim. Acta*, **48**, 1807–18.
Howarth, R. W. and Merkel, S. (1984). Pyrite formation and the measurement of sulfate reduction in salt marsh sediments. *Limnol. Oceanogr.*, **29**, 598–608.
Howarth, R. W. and Teal, J. M. (1979). Sulfate reduction in a New England salt marsh. *Limnol. Oceanogr.*, **24**, 999–1013.
Howarth, R. W. and Teal, J. M. (1980). Energy flow in a salt marsh ecosystem; the role of reduced inorganic sulphur compounds. *Am. Nat.*, **116**, 862–72.
Howarth, R. W., Giblin, A., Gale, J., Peterson, J. and Luther, G. W. (1983). Reduced sulfur compounds in the pore waters of a New England salt marsh. *Environ. Biogeochem. Ecol. Bull.*, **35**, 135–52.
Huerta-Diaz, M. A. and Morse, J. W. (1990). A quantitative method for determination of trace metal concentrations in sedimentary pyrite. *Mar. Chem.*, **29**, 119–44.
Ivanov, M. V., Lein, A. Yu, Reeburgh, W. S. and Skyring. (1989). Interaction of sulphur and carbon cycles in marine sediments. In: Brimblecombe, P. and Lein, A.

Yu. (Eds), *Evolution of the Global Biogeochemical Sulphur Cycle*, Wiley, Chichester, pp. 125–179.

Jørgensen, B. B. (1982). Mineralization of organic matter in the sea bed: the role of sulfate reduction. *Nature*, **296**, 643–5.

Kiene, R. P. and Taylor, B. F. (1988). Biotransformations of organosulphur compounds in sediments via 3-mercaptopropionate. *Nature*, **332**, 148–50.

Kiene, R. P. and Visscher, P. T. (1987). Production and fate of methylated sulphur compounds in sediment via 3-mercaptopropionate. *Nature*, **332**, 148–50.

Kiene, R. R. and Visscher, P. P. (1987). Production and fate of methylated sulphur compounds from methione and dimethylsulphoniopropionate in anoxic salt marsh sediments. *Appl. Environ. Microbiol.*, **53**, 2426–34.

Kiene, R. P., Oremland, R. S., Catena, A., Miller, L. G. and Capone, D. G. (1986). Metabolism of reduced methylated sulphur compounds in anaerobic sediments and by a pure culture of an estuarine methanogen. *Appl. Environ. Microbial.*, **52**, 1037–45.

King, G. M. (1988). Patterns of sulphate reduction and the sulphur cycle in a South Carolina salt marsh. *Limnol. Oceanogr.*, **33**, 376–90.

King, G. M., Howes, B. L. and Dacey, J. W. J. (1985). Short term end products of sulphate reduction in a salt marsh: formation of acid volatile sulphides, elemental sulphur, and pyrite. *Geochim. Cosmochim. Acta*, **49**, 1561–6.

King, G. M., Klug, M. J., Wiegert, R. G. and Chalmers, A. G. (1982). Relation of soil water movement and sulfide concentration to *Spartina alterniflora* production in a Georgia salt marsh. *Science*, **218**, 61–3.

Kohnen, M. E. L., Sinninghe Damste, J. S., ten Haven, H. L. and de Leeuw, J. W. (1989). Early incorporation of polysulphides in sedimentary organic matter. *Nature*, **341**, 640–1.

Kornicker, W. A. and Morse, J. W. (1991). Interactions of divalent cations with the surface of pyrite. *Geochim. Cosmochim. Acta*, **55**, 2159–71.

LaLonde, R. T., Ferrara, L. M. and Hayes, M. P. 1987). Low temperature polysulphide reactions of conjugated ene carbonyls: A reaction model for the geologic origin of S-heterocycles. *Org. Geochim.*, **11**, 563–71.

Lord, C. J. and Church, T. M. (1983). The geochemistry of salt marshes: sedimentary ion diffusion, sulphate reduction, and pyritization. *Geochim. Cosmochim. Acta*, **47**, 1381–91.

Lowe, L. E. (1986). Application of a sequential extraction procedure to the determination of the distribution of sulphur forms in selected peat materials. *Can. J. Soil Sci.*, **66**, 337–45.

Lowe, L. E. and Bustin, R. M. (1985). Distribution of sulphur forms in six facies of peats of the Fraser River Delta. *Can. J. Soil Sci.*, **65**, 531–41.

Luther, G. W., III (1987). Pyrite oxidation and reduction: molecular orbital theory considerations. *Geochim. Cosmochim. Acta*, **51**, 3193–9.

Luther, G. W., III (1991). Pyrite synthesis via polysulfide compounds. *Geochim. Cosmochim. Acta,* **55**, 2839–49.

Luther, G. W., III and Church, T. M. (1988). Seasonal cycling of sulphur and iron in porewaters of a Delaware salt marsh. *Mar. Chem.*, **23**, 295–309.

Luther, G. W., III, Varsolona, R. and Giblin, A. E. (1985). Polarographic analysis of sulphur species in marine porewaters. *Limnol. Oceanogr.*, **30**, 727–36.

Luther, G. W., III, Giblin, A., Howarth, R. W. and Ryans, R. A. (1982). Pyrite and oxidized iron mineral phases formed from pyrite oxidation in salt marsh and estuarine sediments. *Geochim. Cosmochim. Acta*, **46**, 2665–9.

Luther, G. W., III, Church, T. M., Scudlark, J. R. and Cosman, M. (1986). Inorganic and organic sulphur cycling in salt marsh porewaters. *Science*, **232**, 746–9.

McKibben, M. A. and Barnes, H. L. (1986). Oxidation of pyrite in low temperature

Matrai, P. A. and Vetter, R. D. (1988). Particulate thiols in coastal waters: the effect acidic solutions: rate laws and surface textures. *Geochim. Cosmochim. Acta*, **50**, 1509–20.
of light and nutrients on their planktonic production. *Limnol. Oceanogr.*, **33**, 624–31.

Mitchell, M. J. and Fuller, R. D. (1988). Models of sulphur dynamics in forest and grassland ecosystems with emphasis on soil processes. *Biogeochemistry*, **5**, 133–63.

Mopper, K. and Delmas, D. (1984). Trace determination of biological thiols by liquid chromatography and precolumn labeling with ophthaldehyde. *Anal. Chem.*, **56**, 2557–60.

Mopper, K. and Taylor, B. (1986). Biogeochemical cycling of sulphur: Thiols in coastal marine sediments. In: Sohn, M. (Ed.). *Organic Marine Chemistry*. Amer. Chem. Soc., Washington DC, pp. 324–39.

Moses, C. O., Nordstrom, D. K., Herman, J. S. and Mills, A. L. (1987). Aqueous pyrite oxidation by dissolved oxygen and by ferric iron. *Geochim. Cosmochim. Acta*, **51**, 1561–71.

Nguyen, B. C., Belvino, S., Mihalopoulos, N., Gostan, J. and Nival, P. (1988). Dimethylsulphide production during natural phytoplanktonic blooms. *Mar. Chem.*, **24**, 133–41.

Nriagu, J. O. and Soon, Y. K. (1985). Distribution and isotopic composition of sulphur in lake sediments of northern Ontario. *Geochim. Cosmochim. Acta*, **49**, 823–34.

Oae, S. (1985). Historical development of sulphur bonding: A view of an experimental organosulphur chemist. In: Bernardi, F., Csizmalia, I. F. and Mangini, A. (Eds). *Organic Sulphur Chemistry*, Elsevier, New York, pp. 1–67.

Pyzik, A. J. and Sommer, S. E. (1981). Sedimentary iron monosulphides: Kinetics and mechanism of formation. *Geochim. Cosmochim. Acta*, **45**, 687–98.

Raiswell, R. and Berner, R. A. (1986). Pyrite and organic matter in Phanerozoic normal marine shales. *Geochim. Cosmochim. Acta*, **50**, 1967–76.

Rennenberg, H. (1989). Synthesis and emission of hydrogen sulphide by higher plants. In: Saltzman, E. S. and Cooper, W. J. (Eds). *Biogenic Sulphur in the Environment*. American Chemical Society, Washington DC, pp. 44–57.

Rickard, D. T. (1975). Kinetics and mechanisms of pyrite formation at low temperatures. *Am. J. Sci.*, **275**, 636–52.

Roy, A. B. and Trudinger, P. A. (1970). *The Biochemistry of Inorganic Compounds of Sulphur*. Cambridge University Press, London.

Rudd, J. W., Kelly, C. A. and Furutani, A. (1986). The role of sulphate reduction in long term accumulation of organic and inorganic sulphur in lake sediments. *Limnol. Oceanogr.*, **31**, 1281–91.

Schwarzenbach, R. P., Giger, W., Schaffner, C. and Wanner, O. (1985). Groundwater contamination by volatile halogenated alkanes: abiotic formation of volatile sulphur compounds under anaerobic conditions. *Environ. Sci. Technol.*, **19**, 322–7.

Shea, D. and MacCrehan, W. A. (1988a). Determination of hydrophilic thiols in sediment porewater using ion-pair liquid chromatography coupled to electrochemical detection. *Anal. Chem.*, **60**, 1449–1454.

Shea, D. and MacCrehan, W. A. (1988b). Role of biogenic thiols in the solubility of sulphide minerals. *Sci. Total Environ.*, **73**, 135–41.

Singer, P. C. and Stumm, W. (1970). Acid mine drainage—the rate limiting step. *Science*, **167**, 1121–3.

Sinninghe Damste, J. S., Rijpstra, I. C., Kock-van Dalen, A. C., de Leeuw, J. W. and Schenck, P. A. (1989). Quenching of labile functionalized lipids by inorganic sulphur species: Evidence for the formation of sedimentary organic sulphur compounds at the early stages of diagenesis. *Geochim. Cosmochim. Acta*, **53**, 1343–55.

Spratt, H. G., Morgan, M. D. and Good, R. E. (1987). Sulphate reduction in peat from a New Jersey Pinelands cedar swamp. *Appl. Environ. Microbiol.*, **53**, 1406–11.

Steudler, P. A. and Peterson, B. J. (1984). Contribution of gaseous sulphur from salt marshes to the global sulphur cycle. *Nature*, **311**, 455–7.

Strickland, T. C., Fitzgerald, J. W., Ash, J. T. and Swank, W. T. (1987). Organic sulphur transformations and sulphur pool sizes in soil and litter from a southern Appalachian hardwood forest. *Soil Sci.*, **143**, 453–8.

Sweeney, R. E. and Kaplan, I. R. (1973). Pyrite framboid formation: laboratory synthesis and marine sediments. *Econ. Geol.*, **68**, 618–34.

Swider, K. T. and Mackin, J. E. (1989). Transformations of sulphur compounds in marsh-flat sediments. *Geochim. Cosmochim. Acta*, **53**, 2311–23.

Tabatabai, M. A. (1984). Importance of sulphur in crop production. *Biogeochemistry*, **1**, 45–62.

Taylor, B. E., Wheeler, M. C. and Nordstrom, D. K. (1984a). Oxygen and sulphur compositions of sulphate in acid mine drainage: evidence for oxidation mechanisms. *Nature*, **308**, 538–41.

Taylor, B. E., Wheeler, M. C. and Nordstrom, D. K. (1984b). Stable isotope geochemistry of acid mine drainage: experimental oxidation of pyrite. *Geochim. Cosmochim. Acta*, **48**, 2669–78.

Thayer, J. S., Olson, G. J. and Brinckman, F. E. (1984). Iodomethane as a potential metal mobilizing agent in nature. *Environ. Sci. Technol.*, **18**, 726–9.

Thode-Andersen, S. and Jorgensen, B. B. (1989). Sulphate reduction and the formation of ^{35}S-labeled FeS, FeS_2, and S^0 in coastal marine sediments. *Limnol. Oceanogr.*, **34**, 793–806.

Turner, S. M., Malin, G. and Liss, P. S. (1989). Dimethyl sulphide and (dimethylsulphonio) propionate in European coastal and shelf waters. In: Saltzman, E. S. and Cooper, W. J. (Eds). *Biogenic Sulphur in the Environment*. American Chemical Society, Washington DC, pp. 183–200.

Urban, N. R., Eisenreich, S. J. and Grigal, D. F. (1989). Sulphur cycling in a forested *Sphagnum* bog in northern Minnesota. *Biogeochemistry*, **7**, 81–109.

Vairavamurthy, A. and Mopper, K. (1987). Geochemical formation of organosulphur compounds (thiols) by addition of H_2S to sedimentary organic matter. *Nature*, **329**, 623–5.

Vairavamurthy, A. and Mopper, K. (1988). Organosulphur formation in marine sediments: laboratory studies on the reactivity of sulphur nucleophiles with organic Michael acceptors. In: Saltzman, E. S. and Cooper, W. J. (Eds). *Biogenic Sulphur in the Environment*. American Chemical Society, Washington DC, pp. 231–42.

Vetter, R. D., Matrai, P. A., Javor, B. and O'Brien, J. (1989). Reduced sulphur compounds in the marine environment. In: Saltzman, E. S. and Cooper, W. J. (Eds). *Biogenic Sulphur in the Environment*. American Chemical Society, Washington DC, pp. 243–61.

Watwood, M. E., Fitzgerald, J. W., Swank, W. T. and Blood, E. R. (1988). Factors involved in potential sulphur accumulation in litter and soil form a coastal pine forest. *Biogeochemistry*, **6**, 3–19.

Westrich, J. T. and Berner, R. A. (1988). The effect of temperature on rates of sulphate reduction in marine sediments. *Geomicrobiol. J.*, **6**, 99–117.

Wieder, R. K. and Lang, G. E. (1986). Fe, Al, Mn, S chemistry of sphagnum peat in four peatlands with different metal and sulphur input. *Water Air Soil Pollut.*, **29**, 309–20.

Wieder, R. K. and Lang, G. E. (1988). Cycling of inorganic and organic sulphur in peat from Bog Run, West Virginia. *Biogeochemistry*, **5**, 221–42.

Wieder, R. K., Lang, G. E. and Granus, V. A. (1987). Sulphur transformations in sphagnum-derived peat during incubation. *Soil Biol. Biochem.*, **19**, 101–6.

APPENDIX TO CHAPTER 6: GLOSSARY OF ENVIRONMENTALLY IMPORTANT INORGANIC AND ORGANIC SULPHUR COMPOUNDS AND IONS (WITH SULPHUR OXIDATION STATES)

INORGANIC SULPHUR COMPOUNDS

H_2S	hydrogen sulphide; SH^- bisulphide ion [S(−2)]
SO_4^{2-}	sulphate ion [S(+6)]
$S_2O_3^{2-}$	thiosulphate ion; also SSO_3^{2-} where terminal [S(−2)] and central [S(+6)]
SO_3^{2-}	sulphite ion; HSO_3^- bisulphite ion [S(+4)]
$S_xO_6^{2-}$	polythionate ion, if $x = 4$ tetrathionate also $O_3S{-}S_{x-2}{-}SO_3^{2-}$ where SO_3 is [S(+5)] and central [S(0)]
S_x^{2-}	polysulphide ion of length x, if $x = 4$ tetrasulphide also $S{-}S_{x-2}{-}S^{2-}$ where terminal [S(−1)] and central [S(0)]
SO_2	sulphur dioxide [S(+4)]
COS	carbonyl sulphide [S(−2)]
CS_2	carbon disulphide [S(−2)]
FeS_2	pyrite (marcasite has the same formula but a different crystal structure) [S(−1)]
Fe_3S_4	greigite [S(−2)] with 2 Fe(III) and 1 Fe(II)
FeS	iron monosulphide (also part of the acid-volatile sulphide component in sediments) [S(−2)]

MAJOR ORGANIC FUNCTIONAL GROUPS

R = organic group, for R = methyl the name is in parentheses.

Accurate oxidation states are not as easily obtained or calculated for organic compounds. Thus, formal charges [fc] are given. These are calculated by the equation below which requires a correct Lewis electron dot structure:

[fc] = maximum valence electrons of the atom in its standard state − 1/2 (the number of electrons in bonds) − the number of electrons not in bonds

e.g. S in $R{-}\ddot{\underset{\cdot\cdot}{S}}{-}R$

$$[fc] = 6 - 1/2\ (4) - 4 = 0$$

'Reduced'

R—SH	thiol (methanethiol) [S(0)]
R—S—R	sulphide (dimethylsulphide) [S(0)]
R—S—S—R	disulphide (dimethyldisulphide) [S(0)]
$R{-}S_x^-$	polysulphide of length x (methyltrisulphide for $x = 3$) also $R{-}S_{x-1}{-}S^-$ where terminal [S(−1)] and central [S(0)]

Oxidized

R_2SO — sulphoxide (dimethylsulphoxide) [S(+1)]

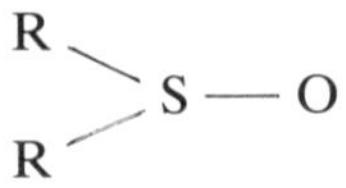

The S atom has a pyramidal geometry.

R_2SO_2 — sulphone (dimethylsulphone) [S(+2)]

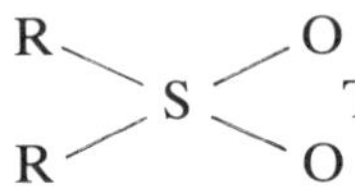

The S atom has a tetrahedral geometry.

$R—SO_3H$ — sulphonic acid (methanesulphonic acid) [S(+2)]
$R—SO_3^-$ sulphonate ion

$R—S_2O_3^-$ — thiosulphate or Bunte salt (methylthiosulphate) also $R—S—SO_3^-$ where [R—S(0)] and SO_3^- is [S(+2)]

R_3S^+ — sulphonium ion (trimethylsulphonium) [S(+1)]

*$R—O—SO_3^-$ — sulphate ester (methylsulphate) [S(+2)]

*Indicates no carbon to sulphur bond. All other organic compounds have at least one carbon to sulphur bond.

Special Cyclic Structures

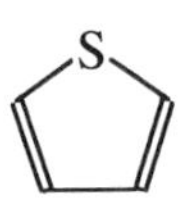

Thiophene

Thiolane

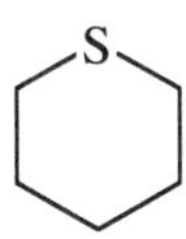

Thiane

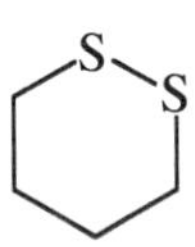

Cyclic disulphide

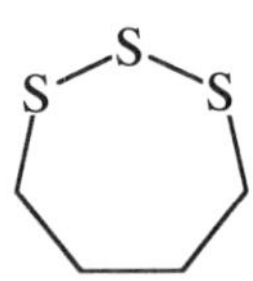

Cyclic trisulphide

7 Sulphur Cycling and Fluxes in Temperate Dimictic Lakes

ROBERT B. COOK
Oak Ridge National Laboratory, Oak Ridge TN, USA

and

CAROL A. KELLY
University of Manitoba, Winnipeg MB, USA

7.1 INTRODUCTION

Over the past 15 years, research impelled by the lake acidification issue has led to an improved understanding of the biogeochemistry of S in lakes. This research has shown that

(1) the S cycle in small lakes has been affected by increased atmospheric deposition (Rudd, Kelly and Furutani, 1986; Nriagu and Soon, 1985; Fry, 1986) with attendant effects on the acid–base chemistry of lakes (Landers *et al.*, 1988; Brakke, Landers and Eilers, 1988);
(2) under certain conditions biologically mediated SO_4^{2-} reactions, particularly bacterial SO_4^{2-} reduction, can significantly affect lakewater SO_4^{2-} chemistry (Kelly *et al.*, 1987; Baker and Brezonik, 1988);
(3) these reactions produce alkalinity and may counteract acidification (Cook *et al.*, 1986);
(4) historical changes in lakewater SO_4^{2-} concentrations may be recorded in sediments (Mitchell *et al.*, 1988);
(5) the diagenesis of S in freshwater sediments is more complex than previously thought (David and Mitchell, 1985; Rudd, Kelly and Furutani, 1986).

The purpose of this chapter is to examine the controls on biogeochemical reactions of S in the water column and sediments of temperate dimictic lakes. The processes causing fluxes of S between the water column and the sediments and the atmosphere also will be examined. Because the primary reactions that S undergoes are closely linked to organic matter, this chapter

Sulphur Cycling on the Continents
Edited by R.W. Howarth, J.W.B. Stewart and M.V. Ivanov

will also discuss the interactions of S with the elements predominant in organic matter (carbon, nitrogen, and phosphorus). We gratefully acknowledge Shirley Richards for allowing us to use her unpublished data. This is publication No. 3569 of the Environmental Sciences Division, ORNL.

7.2 REACTIONS AND FORMS OF SULPHUR

A schematic diagram of the S cycle in small lakes is presented in Figure 7.1, and the microorganisms performing the conversions of S in aquatic systems are presented in Table 7.1. Sulphur is supplied to lakes from atmospheric deposition in the form of SO_4^{2-} and SO_2 and from terrestrial portions of watersheds, primarily in the form of SO_4^{2-}, but also as organic S (David and Mitchell, 1985; Mitchell, David and Harrison, this volume). In northern temperate regions, the export of S from terrestrial ecosystems is in balance

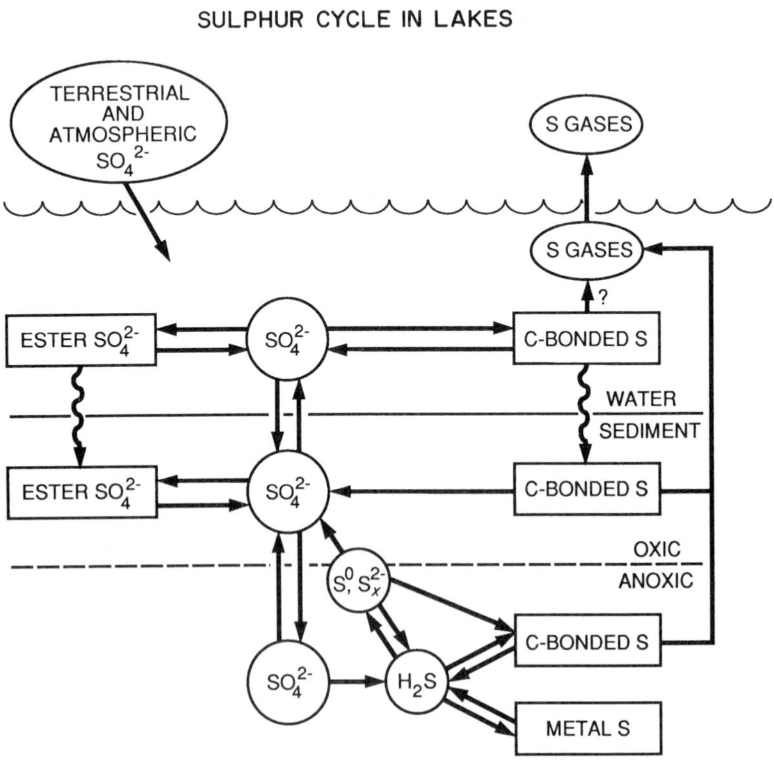

Figure 7.1. Schematic diagram of the sulphur cycle in small lakes. Polysulphides are indicated by S_x^{2-}

Table 7.1. Sulphur-using bacteria in aquatic systems

Group	S Conversion	Habitat requirements	Aquatic habitat example	Genera examples
Chemoorganotrophs that use oxidized S species as electron acceptors	$SO_4^{2-} \longrightarrow HS^-$ $S_2O_3^{2-} \longrightarrow HS^-$ or S^0 $S^0 \longrightarrow HS^-$ $SO_3^- \longrightarrow HS^-$	anaerobic; organic substrates available; light not required	anoxic sediments	*Desulfomonas*[a] *Desulfovibrio*[a] *Desulfotomaculum*[a] *Desulfuromonas*[b] *Campylobacter*[b]
	Note: SO_3^-, $S_2O_3^{2-}$, S^0 can be used by other heterotrophic genera			
Obligate and facultative chemoautotrophs that use reduced S as an energy source	$HS^- \longrightarrow S^0$ $S^0 \longrightarrow SO_4^{2-}$ $S_2O_3^- \longrightarrow SO_4^{2-}$	H_2S–O_2 interface; light not required	mud; hot springs; mine drainage	*Thiobacillus*[c] *Thiomicrospira*[c] *Achromatium*[d] *Beggiatoa*[e]
Phototrophs that use reduced S as an electron donor	$HS^- \longrightarrow S^0$ $S^0 \longrightarrow SO_4^{2-}$	anoxia; H_2S; light	shallow water anoxic sed; metalimnion or hypolimnion; anoxia water	*Chlorobium*[f] *Chromatrium*[f] *Ectothiorhodospira*[f] *Thiopedia*[f] *Rhodopseudomonas*[f]
Chemoorganotrophs that use organic-S compounds as energy sources or that hydrolyse esters	org S $\longrightarrow HS^-$ org S $\longrightarrow$ volatile org S ester $SO_4 \longrightarrow SO_4^{2-}$	source of organic-S compounds	sediments; water column	Broad variety
Microorganisms that use SO_4^{2-} or H_2S in biosynthesis	$SO_4^{2-} \longrightarrow$ protein $HS^- \longrightarrow$ protein $SO_4^{2-} \longrightarrow$ DMSP	non-specific	sediments; water column	Broad variety

[a]Pfennig *et al*. (1981); [b]Pfennig and Biebl (1981); [c]Kuenen and Tuovinen (1981); [d]La Riviere and Schmidt (1981); [e]Wiessner (1981); [f]Utanier *et al*. (1981).

with atmospheric inputs, so that the SO_4^{2-} concentration in lakes tends to be directly related to the rate of atmospheric deposition (Sullivan *et al.*, 1988). In the water column, SO_4^{2-} is taken up by plankton during assimilatory SO_4^{2-} reduction and converted into carbon-bonded S compounds (e.g. the S-containing amino acids cysteine, methionine, and cystine). Algae may release dimethylsulphoniumpropionate (DMSP), which is broken down to dimethyl sulphide (DMS), but this mechanism for producing DMS is presently well established only for marine algae (Andreae and Barnard, 1984; Andreae and Jaeschke, this volume). Ester sulphate (C—OSO_3) compounds also may be produced by plankton (Schiff and Hodson, 1973; David and Mitchell, 1985). The various organic-S compounds may be recycled in the water column or may settle out of the water column where they either accumulate or are recycled.

Lakewater SO_4^{2-} is conveyed by diffusion and bioturbation to the anoxic portions of the lake where it undergoes reactions, the most important of which is bacterial SO_4^{2-} reduction forming H_2S (Table 7.1). Some SO_4^{2-} is assimilated by known biosynthetic pathways into bacterial biomass. H_2S may either react to form metal sulphides, or diffuse to oxic zones where it is oxidized chemically or by chemoautotrophic or photoautotrophic S-oxidizing bacteria to S^0, S_x^{2-} (polysulphides) or SO_4^{2-} (Table 7.1; Luther and Church, this volume). Elemental sulphur may also serve as an electron acceptor in the anaerobic oxidation of organic matter (Table 7.1).

Reduced sulphur (H_2S, S^0, S_x^{2-}) may react with humic substances (Casagrande and Ng, 1979; Casagrande *et al.*, 1979; Vairavamurthy and Mopper, 1987, 1989; Francois, 1987; Luther and Church, this volume). The organic-S compounds formed are of unknown chemical make-up but may include thiols, mercaptans, and thioethers. Breakdown of organic-S compounds by heterotrophic bacterial metabolism may supply H_2S to the anoxic zones (Dunnette, 1973, 1989; Zinder and Brock, 1978; Jones, Simon and Roscoe, 1982).

Reduced S species may undergo chemical or biological oxidation (Wetzel, 1975; Luther and Church, this volume). Chemoautotrophic S-oxidizing bacteria are primarily aerobic taxa that can oxidize reduced S inside or outside of the cell (Table 7.1; Wetzel, 1975). The anaerobic phototrophic S-oxidizing bacteria are located on the sediments or in a narrow layer in the water column, where light, redox, and pH conditions are optimal (Table 7.1; Parkin and Brock, 1981).

Sulphur gases, such as dimethyl sulphide, carbonyl sulphide, methane thiol, and dimethyl disulphide, may be formed in the water column and sediments. The processes by which these gases are formed have not been adequately studied in freshwater systems, but they appear to be linked both to the decomposition of organic-S compounds in the sediments and to either decomposition or algal release in the water column (Richards, Kelly and Rudd, 1991).

The total amount of S in a particular sediment horizon is the net accumulation of particulate forms deposited at the time the sediment was laid down and S incorporated in the sediments following diffusion and reaction in the horizon. The palaeolimnological interpretation of sediment S profiles is confounded by the many mobile forms of S (SO_4^{2-}, H_2S, and soluble organic-S compounds) (Holdren *et al.*, 1984; Carignan and Tessier, 1988).

7.2.1 INCORPORATION OF SULPHUR BY BIOTA

All living organisms require S as a minor nutrient, in roughly the same atom proportion as phosphorus (Howarth and Stewart, this volume). Sulphur is present in freshwater algae at a ratio of about 1 S atom to 100 C atoms (0.15 to 1.96% by dry weight), and the S content varies with species, environmental conditions, and season (Baker *et al.*, 1989; King and Klug, 1982). Vascular plants, algae, and bacteria (except some anaerobes that require S^{2-}) have the ability to take up, reduce, and assimilate SO_4^{2-} into amino acids and convert SO_4^{2-} into ester sulphate compounds.

For most lakes, SO_4^{2-} concentrations in aerobic habitats are high enough to supply all of the S required by planktonic algae and bacteria. In anaerobic habitats, SO_4^{2-} is depleted but H_2S is usually present and can be used in biosynthesis. Because of this abundant supply, aquatic biota apparently do not take up S in excess of that required for growth, thereby maintaining relatively constant concentrations of S in organic matter. An exception to this generalization about storage is the phototrophic sulphur bacteria that store elemental sulphur (Table 7.1). This stored S^0 is for later use as an electron donor when other donors become unavailable, rather than for biosynthesis.

In the absence of changes in the major nutrients (P, N), changes in lakewater SO_4^{2-} concentrations will not likely lead to changes in algal S uptake and sedimentation, because the nutritional requirement of aquatic biota for S has not been affected. For example, experimental additions of SO_4^{2-} to lakes did not stimulate the algal uptake of S, as indicated by the similarity between C : S ratios in seston of acidified and control lakes at the Experimental Lakes Area (ELA), northwestern Ontario (J. Crusius, Columbia University, unpub. data) and by C : S ratios in seston of the control and treatment basins in Little Rock Lake, Wisconsin (molar C : S ranged from 99 to 144; Baker *et al.*, 1989). Knauss and Porter (1954) observed less than a doubling of SO_4^{2-} content in *Chlorella* when SO_4^{2-} concentrations were increased from 40 to 1400 mg (S) l^{-1} (1.3 to 44 mM). Also, examination of SO_4^{2-} uptake by a floating vascular plant over a 3000-fold variation in medium SO_4^{2-} concentration showed that total SO_4^{2-} uptake varied by only three-fold to four-fold, with the variation reflected in a change in inorganic SO_4^{2-} in the tissue (Datko *et al.*, 1978).

The lowest SO_4^{2-} concentration in the studies cited above is about 1 mg (S) l^{-1} (30 μM) (Crusius, unpub. data; Baker *et al.*, 1989). The algal uptake of SO_4^{2-} has not been studied in lakes having concentrations below this value. SO_4^{2-} concentrations of 0.1 mg (S) l^{-1} (3 μM) were reported for Lake Victoria (R. Hecky, Freshwater Institute, Winnipeg, unpub. data), and concentrations below 0.024 mg (S) l^{-1} (0.75 μM) have been observed in some lakes remote from human influence [e.g. in the mountainous western United States (Eilers *et al.*, 1989) and on the Kenai Peninsula, Alaska (Landers *et al.*, 1989)]. For lakes with low concentrations of SO_4^{2-}, the supply of SO_4^{2-} in the water column may limit the growth of aquatic biota, but this has not been demonstrated.

From the evidence on S uptake summarized above, the total assimilatory uptake of S by algae and bacteria and the accumulation of seston S in sediments can be expected to be related more to biomass or the trophic status of the lake than to the concentration of lakewater SO_4^{2-}. Increases in major nutrient (e.g. P) inputs to a lake may lead to an increased sedimentation of both organic C and organic S, without changing the C : S ratio of the sedimenting material.

Organic-S compounds may be recycled within the water column and the sediments. Ester sulphate compounds (C—OSO_3) may be hydrolysed to SO_4^{2-} (King and Klug, 1980; Luther and Church, this volume); S in amino acids and proteins may be broken down to form H_2S and other gaseous forms of S (Dunnette, 1973, 1989; Zinder and Brock, 1978; Jones, Simon and Roscoe, 1982; Andreae and Jaeschke, this volume), some of which may then undergo oxidation. The few studies of organic-S recycling have shown that between 25 and 80% of the seston S falling on the sediment–water interface is decomposed and re-enters the water column (Smith and Klug, 1981; King and Klug, 1982; David and Mitchell, 1985; Baker *et al.*, 1989). King and Klug (1982) estimated that less than 50% of the total S and 75% of the protein S was mineralized. David and Mitchell (1985) observed that half of the carbon-bonded S was recycled at the sediment–water interface (upper 1 cm). These studies assumed that the recycling could be determined from the comparison of rates of seston S inputs and total accumulation of organic S in the sediments. SO_4^{2-} reduction was assumed to be insignificant and, therefore, not a source of organic S formation (David and Mitchell, 1985; Baker *et al.*, 1989), or the possibility of this source was not considered (Smith and Klug, 1981; King and Klug, 1982). Because some of the organic S may be derived, not from seston, but from sulphides produced from bacterial SO_4^{2-} reduction (Rudd, Kelly and Furutani, 1986; Luther and Church, this volume), these recycling values may be underestimates. Measurement of seston S input and sediment accumulation of organic S in lakes with extremely low SO_4^{2-} concentrations is required to avoid this additional source of organic S.

7.2.1.1 Ester Sulphates

SO_4^{2-} may also be taken up by aquatic biota and converted into ester sulphate compounds (C—OSO_3) (Schiff and Hodson, 1973) in the water column (David and Mitchell, 1985). Ester sulphate compounds may also be formed within the sediment from SO_4^{2-} (Landers and Mitchell, 1988), although no evidence exists on whether the ester sulphate compounds are extracellular or intracellular. Measurements of the ester sulphate content of seston range from 32% (King and Klug, 1982) to 44–59% (David and Mitchell, 1985). The percentage of S in ester form in surficial sediments is similar or smaller than seston (35%, King and Klug, 1982; 22%, David and Mitchell, 1985). Ester hydrolase activity is present in sediments, showing that esters can be broken down biologically and can provide a source of SO_4^{2-} for SO_4^{2-} reduction (King and Klug, 1980). Mineralization of ester sulphate has been estimated by comparing seston input to sediment accumulation, yielding 43% (David and Mitchell, 1985) and $< 50\%$ (King and Klug, 1982) of the input ester sulphate being decomposed.

7.2.2 BACTERIAL SULPHATE REDUCTION

Dissimilatory SO_4^{2-} and S^0 reduction is the process by which bacteria use SO_4^{2-} and other oxidized forms of S (oxidation state of 0 or greater; e.g. $S_2O_3^{2-}$ and S^0) as terminal electron acceptors in the catabolism of organic matter (Table 7.1). The SO_4^{2-}- and S^0-reducing taxa are strictly anaerobes and require organic matter or hydrogen as the electron donors (Pfennig, Widdel and Truper, 1981).

Assimilatory SO_4^{2-} reduction is also performed by microorganisms in both the water column and the sediments for synthesis of S-containing compounds, primarily amino acids. In aerobic portions of the water column, assimilatory reduction will predominate over dissimilatory reduction, because the latter process is carried out only by obligate anaerobes. No measurements have been made of assimilatory reduction in sediments, but assimilatory reduction is probably minor compared to dissimilatory reduction. For example, in cultures of *Desulphovibrio*, only 1 mol of S was assimilated for every 400 mol of SO_4^{2-} reduced to H_2S for the purpose of meeting energy requirements (R. Cuhel and C. Taylor, pers. comm.). Of course, in sediments, non-SO_4^{2-}-reducing microorganisms are also assimilating S for cell synthesis. The energetic metabolism of these other bacteria is about four times that of the SO_4^{2-} reducers (Kelly, Rudd and Schindler, 1988), but their S assimilation would still be small compared to SO_4^{2-} reduction. Assimilatory reduction is linked to population size and cell growth. Thus, it should not respond to increases in SO_4^{2-} concentrations as does dissimilatory SO_4^{2-} reduction (as explained below).

The rate of bacterial SO_4^{2-} reduction is one of the better known processes in the S cycle in lakes. This rate has been measured directly by ^{35}S tracer (release of produced [^{35}S] H_2S or uptake of $^{35}SO_4^{2-}$) (Stuiver, 1967; Ivanov, 1968; Sorokin, 1975; Jones, Simon and Roscoe, 1982), depletion of SO_4^{2-} over time in the water column (Stuiver, 1967; Cook and Schindler, 1983; Kelly, Rudd and Schindler, 1988), and calculations of SO_4^{2-} fluxes from porewater gradients (Holdren *et al.*, 1984; Carignan, 1985; Rudd *et al.*, 1986). It is important to note that $^{35}SO_4^{2-}$ reduction rates represent a lower limit if

Table 7.2. Rates of bacterial sulphate reduction in lake sediments

Site	SO_4^{2-} (μmol l^{-1})	Reduction rate (mmol m^{-2} d^{-1})	Method[a]	Reference
Sagamore Lake	31	0.02	1	Rudd *et al.* (1986)
Little Rock Lake	27	0.02–0.05	1	Cook *et al.* (1987) Baker *et al.* (1989)
South Lake	50	0.03	1	Owen and Mitchell (1989)
Lake Gårdsjön	100	0.04	1	Andersson (1985)
McCloud Lake	35	0.04	1	Baker, Brezonik and Edgerton (1986)
Plastic Lake	70	0.04	1	Rudd *et al.* (1986)
Lake 239	31	0.05	2	Schindler *et al.* (1986)
Lake Tantare	50	0.05–0.2	1	Carignan (1988)
Dart's Lake	75	0.07	1	Rudd *et al.* (1986)
Red Chalk Lake	44	0.07	1	Rudd *et al.* (1986)
Langtjern	40	0.08	2	Wright (1983)
Lake 302 South	60	0.1–0.25	1	Rudd *et al.* (1986)
Lake 114	31	0.1	1	Rudd *et al.* (1986)
Twitchell Lake	23	0.1	1	Rudd *et al.* (1986)
Chubb Lake	43	0.14	1	Rudd *et al.* (1986)
Woods Lake	63	0.18	1	Rudd *et al.* (1986)
Lake 302 North	35	0.2	2	Kelly *et al.* (1987)
Lake 227	30	0.2–0.7	3	Cook (1981)
Big Moose Lake	37	0.3	1	Rudd *et al.* (1986)
Clearwater Lake	250	0.33	1	Carignan (1985)
Lake 223	30–100	0.3–1.2	3	Cook and Schindler (1983
Lake 226 North	30	0.5–1.7	3	Kelly *et al.* (1982)
Linsley Pond	25	2.1–2.8	4	Stuiver (1967)
Wintergreen Lake	50–200	3–15	4	Smith and Klug (1981)
Lake Mendota	200	7	3	Ingvorsen and Brock (198
Lake Holmsjon	1000–3000	31–44	4	Jensen and Andersen (198
Lake Anna	3000–20 000	0.2–450	4	Herlihy *et al.* (1987)

[a]Method: 1 = porewater gradient, 2 = loss of SO_4^{2-} from the water column, based on annual budgets, 3 = seasonal loss SO_4^{2-} from the anoxic hypolimnion, 4 = ^{35}S tracer.

the rates are measured without recovering all of the possible end-products (e.g. release of only [^{35}S] H_2S; see also Howarth and Teal, 1979; Giblin and Wieder, this volume). For the majority of lakes studied, the rates of bacterial SO_4^{2-} reduction range from 0.02 to 0.2 mmol m^{-2} day^{-1} (0.6 to 6 mg (S) m^{-2} day^{-1}) (Table 7.2). The lowest rates reported, 0.02 mmol m^{-2} day^{-1} (0.06 mg (S) m^{-2} day^{-1}), were measured in Little Rock Lake (Cook *et al.*, 1987; Baker *et al.*, 1989) and Sagamore Lake, and the highest (450 mmol m^{-2} day^{-1}; 14 g (S) m^{-2} day^{-1}) were measured in an impoundment with high concentrations of SO_4^{2-} (3 to 20 mM; 96 to 640 mg (S) l^{-1}) derived from acid mine drainage (Herlihy *et al.*, 1987).

Rates of bacterial SO_4^{2-} reduction are affected by organic carbon quantity and quality (Rudd, Kelly and Furutani, 1986; Sweerts, Rudd and Kelly, 1986; Giblin, Peterson and Fry, 1989) and also by the concentration of SO_4^{2-}(Cook and Schindler, 1983; Kelly and Rudd, 1984; Lovley and Klug, 1986). In the following two sections (7.2.2.1 and 7.2.2.2.), the influence of organic carbon supply and SO_4^{2-} concentrations on bacterial SO_4^{2-} reduction are described. Section 7.2.2.3 discusses the concept of the 'limiting' factor.

7.2.2.1 SO_4^{2-} Concentrations and Bacterial SO_4^{2-} Reduction

In marine systems, which have SO_4^{2-} levels as high as 28 mM (900 mg (S) l^{-1}), SO_4^{2-} concentrations are high enough that bacterial SO_4^{2-} reduction rates are independent of SO_4^{2-} (Boudreau and Westrich, 1984; Capone and Kiene, 1988; Giblin and Wieder, this volume). In contrast, in lakes with SO_4^{2-}concentrations below 1 to 2 mM (32 to 64 mg (S) l^{-1}), bacterial SO_4^{2-} reduction is typically first order with respect to SO_4^{2-} concentrations (Ramm and Bella, 1974; Kelly and Rudd, 1984; Lovley and Klug, 1986). For these systems, increases in SO_4^{2-} concentrations (caused, for example, by increases in atmospheric SO_4^{2-} deposition) will cause increases in the rates of SO_4^{2-} reduction. For example, in Lake 223, ELA, experimental additions of sulphuric acid caused lakewater SO_4^{2-} concentrations to increase three-fold, causing bacterial SO_4^{2-} reduction rates to increase both in the anoxic hypolimnion and in the sediments throughout the lake (Figure 7.2; Cook and Schindler, 1983; Cook *et al.*, 1986). In Lake 223, the accumulation of total S in the sediments increased primarily as a consequence of increased SO_4^{2-} reduction (Figure 7.3a). Similar findings on the relationship between SO_4^{2-} concentrations and SO_4^{2-} reduction rates have been reported in limnocorals (Perry, Baker and Brezonik, 1985; Schiff and Anderson, 1987) and in another whole-lake experiment (Rudd *et al.*, 1990).

Rates of bacterial SO_4^{2-} reduction for these systems can increase with increases in SO_4^{2-} concentrations, because the supply of organic carbon is large enough that only SO_4^{2-} is limiting the rate of SO_4^{2-} reduction in a particular lake. Prior to the experimental acid additions in Lake 223, organic matter decomposition in littoral sediments and in the hypolimnion occurred primarily by three processes: oxic decomposition, bacterial SO_4^{2-} reduction, and methanogenesis (Figure 7.4). In anoxic zones, decomposition was carried out primarily by methanogens (Kelly *et al.*, 1982; Kelly, Rudd and Schindler, 1988). With the SO_4^{2-} additions to Lake 223, bacterial SO_4^{2-} reduction was stimulated at the expense of methanogenesis (Cook and Schindler, 1983). This stimulation is well documented in anaerobic mixed-population systems and is related to the energy yield for bacterial SO_4^{2-} reduction being greater than that for methanogenesis (Lovley and Klug, 1986; Kuivila *et al.*, 1989;

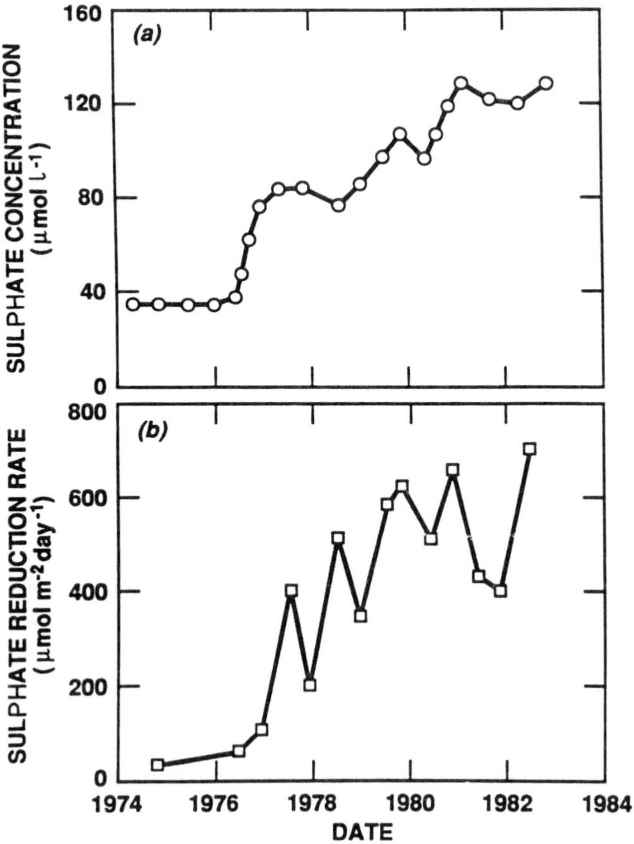

Figure 7.2. Lakewater SO_4^{2-} concentration and areal rates of bacterial sulphate reduction in the anoxic hypolimnion of Lake 223 in the Experimental Lakes Area of northwestern Ontario. Data from Cook and Schindler (1983) and Cook *et al.* (1986)

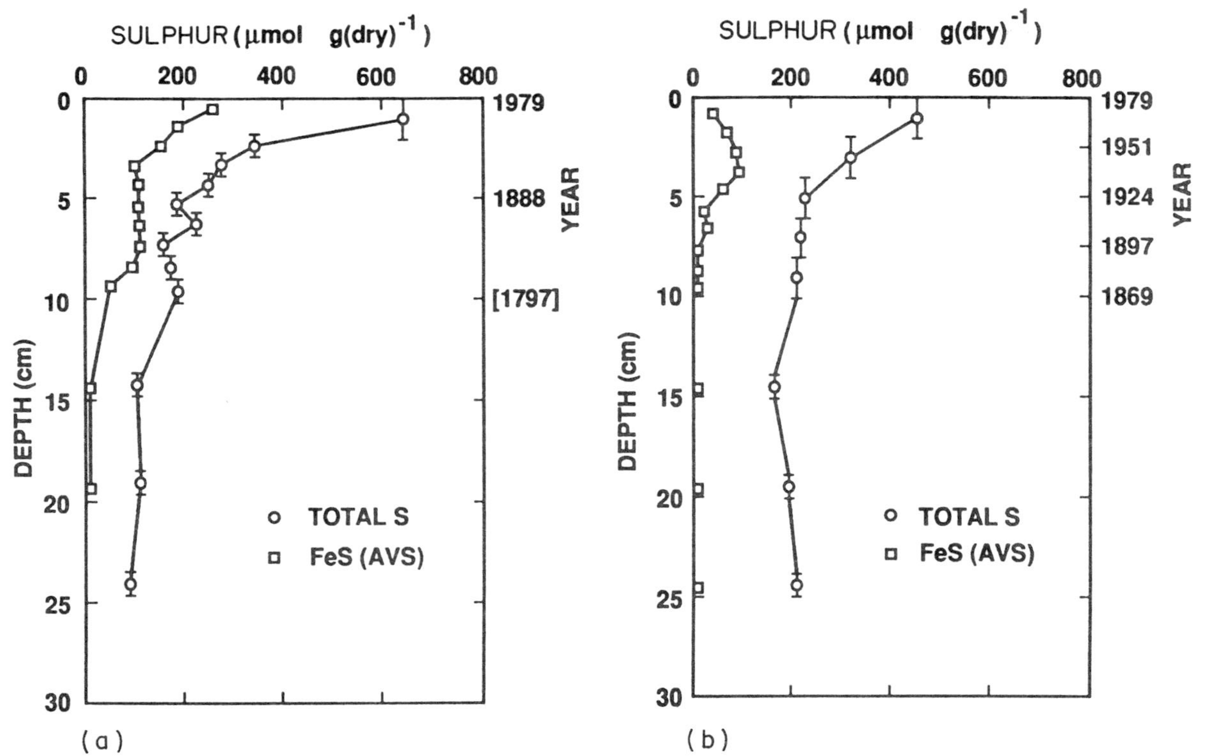

Figure 7.3. Acid volatile and total sulphur profiles in the sediments of (a) Lake 223 and (b) Lake 227. From Cook, 1981

Howarth and Stewart, this volume). As Lake 223 was acidified, a change occurred, not in the total amount of organic matter that decomposed, but in the fraction of decomposition performed by bacterial SO_4^{2-} reduction (Cook, 1981; Kelly *et al.*, 1982; Kelly, Furutani and Schindler, 1984). In the hypolimnion of Lake 223 during manipulation, the molar ratio of methanogenesis to SO_4^{2-} reduction was 3.2 : 1.0 (Kelly *et al.*, 1982; Kelly, Rudd and Schindler, 1988), suggesting that SO_4^{2-} reduction rates could increase further if more SO_4^{2-} were added. In many low-SO_4^{2-} freshwater lakes, most organic matter is decomposed in anoxic zones by methanogenic bacteria (Ingvorsen and Brock, 1982; Kelly *et al.*, 1982; Kelly, Rudd and Schindler, 1988; Kuivila *et al.*, 1989), so that organic matter will not generally limit bacterial SO_4^{2-} reduction in these systems. However, this does not mean that organic substrate availability plays no role in determining the rate of SO_4^{2-} reduction (see below).

The rate of supply of SO_4^{2-} to the zone of bacterial SO_4^{2-} reduction—typically the sediments—may limit the rate of bacterial SO_4^{2-} reduction. For most systems, SO_4^{2-} is supplied by molecular diffusion or bioturbation (see below). However, for Lake Anna, Virginia, which is affected by acid mine drainage, SO_4^{2-} is scavenged from the water column by iron oxyhydroxide particles, enhancing the rate of SO_4^{2-} supply to the reducing zone and

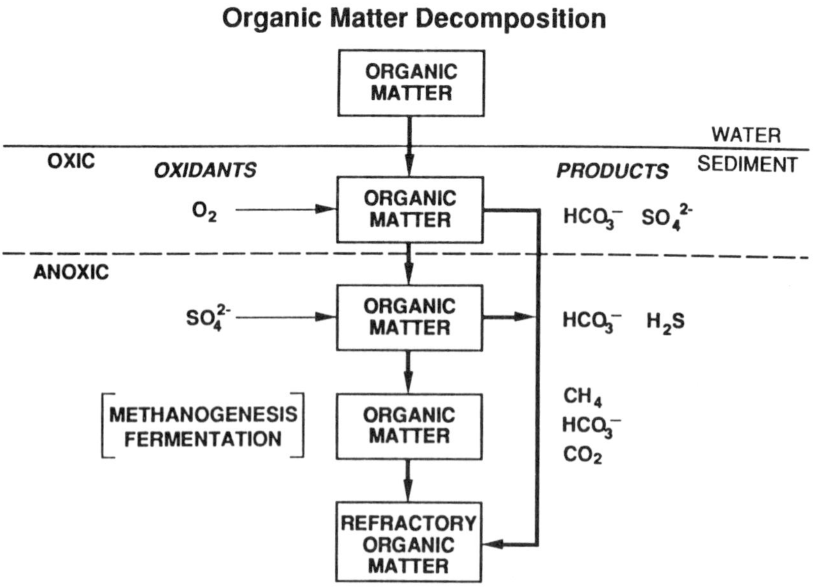

Figure 7.4. Schematic diagram of the primary reactions that decompose organic matter in lakes. Based on a figure from Martens and Jannasch (1983)

apparently increasing the rate of bacterial SO_4^{2-} reduction as measured by $^{35}SO_4^{2-}$ (Herlihy *et al.*, 1987; Mills, Bell and Herlihy, 1989). This system has the highest reported SO_4^{2-} reduction rate for any lake (Table 7.2).

The lowest concentration at which SO_4^{2-} reduction can occur is uncertain. At low SO_4^{2-} concentrations (< 20 to 40 μM; < 0.64 to 1.3 mg (S) l^{-1}), thermodynamic calculations suggest that methanogenic bacteria are energetically favoured over bacteria that reduce SO_4^{2-} (Cook, 1981). Rates of bacterial SO_4^{2-} reduction have been observed to be low at concentrations below 40 μM (1.3 mg (S) l^{-1}; Lovley and Klug, 1986), and the half-saturation constant for bacterial SO_4^{2-} reduction in cultures has been reported as ranging from 5 to 70 μM (0.16 to 2.2 mg (S) l^{-1}; Ingvorsen and Jørgensen, 1984). However, workers have reported SO_4^{2-} concentrations as low as 0.2 to 5 μM (0.006 to 0.16 mg (S) l^{-1}) in anoxic porewaters (Cook, 1981; Holdren *et al.*, 1984; Kelly and Rudd, 1984; Cook *et al.*, 1986, 1987; Rudd, Kelly and Furutani, 1986).

7.2.2.2 The Influence of Organic Carbon Supply on Bacterial SO_4^{2-} Reduction

The supply of organic matter to lake sediments influences the S reactions and retention in sediments. Gorham *et al.* (1974) found that the accumulation of S in sediments of lakes in the English Lake District was highly correlated to algal standing crop and the organic S content of the sediments. Although the forms of S accumulating and the mechanisms for their formation were not determined, Gorham *et al.* attributed this correlation to the occurrence of extensive anoxic zones that favour both preservation of seston S compounds and the reduction of SO_4^{2-} and subsequent retention of reduced S.

Sweerts, Rudd and Kelly (1986) observed that additions of flocculent organic-rich particles to surface sediments in situ caused an increase in the rates of oxygen consumption and bacterial SO_4^{2-} reduction. Because of the uptake of oxygen in the organic floc, the oxic–anoxic boundary moved closer to the surface of the consolidated sediment. The zone of SO_4^{2-} reduction also moved closer to the sediment surface, steepening the vertical SO_4^{2-} concentration gradient, thereby causing an increase in the diffusive flux and the rate of SO_4^{2-} reduction.

At Lake 227, ELA, the rate of bacterial SO_4^{2-} reduction was stimulated by experimental eutrophication (Cook, 1981, 1984). Nitrogen and phosphorus additions, which were made to simulate cultural eutrophication, caused primary production rates and concentrations of chlorophyll a to increase (Fee, 1979; Schindler, 1985). The total sedimentation rate increased from 45 g (dry wt) m^{-2} a^{-1} (Kipphut, 1978) to 210 g (dry wt) m^{-2} a^{-1} (Anderson, Schiff and Hesslein, 1987) as a result of the fertilization. Anoxic conditions in the water column increased and extended throughout the hypolimnion (Cook, 1984). These changes led to an increase in the habitat for bacterial reduction and an increase in rates of bacterial SO_4^{2-} reduction. The sediment accumulation of total S at the deepest point in Lake 227 increased from 9 to 20 mmol m^{-2} a^{-1} (0.3 to 0.64 g (S) m^{-2} a^{-1}; Figure 7.3b; Cook, 1984) as a consequence of both increased SO_4^{2-} reduction and sulphide retention and an increased sedimentation of organic S derived from algae. Although complete analyses for S species were not made, acid-volatile S and authigenic pyrite were among the forms accumulating (Cook, 1984).

Data from both eutrophic and oligotrophic lakes also illustrate the effect of organic carbon supply on bacterial SO_4^{2-} reduction. Eutrophic lakes (Mendota and Wintergreen) have much higher SO_4^{2-} reduction rates than oligotrophic lakes (Clearwater and Lake 223) even though the SO_4^{2-} concentrations in these four lakes are similar (Table 7.2).

A low rate of organic carbon supply may also limit the extent of bacterial SO_4^{2-} reduction. At McNearney Lake (northern Michigan, USA), Cook *et al.* (1990) hypothesize that the low supply of organic matter to the sediments causes the rate of bacterial SO_4^{2-} reduction and S accumulation to be low. The lake has been naturally acidic for the past several thousand years and is ultraoligotrophic, having very low levels of chlorophyll a (0.2 μg l^{-1}) and extremely low rates of bulk sedimentation (0.1 to 0.6 mm a^{-1}; Cook *et al.*, 1990). The SO_4^{2-} concentration in McNearney Lake is 75 μM (2.4 mg (S) l^{-1}); porewater concentrations in this lake have not been measured. The small concentration of S in the sediments has remained relatively constant over the past few hundred years (Figure 7.5) and exhibits no changes at the time of increases in the anthropogenic polycyclic aromatic hydrocarbons or lead (Cook *et al.*, 1990).

The stable S isotope content of sediments in McNearney Lake shows only a slight depletion of $\delta^{34}S$ (< 4‰), indicating a small amount of bacterial SO_4^{2-} reduction and of sulphide retention (Figure 7.5; Fry, 1990). Stable isotope data from other small oligotrophic lakes in which both high rates of bacterial SO_4^{2-} reduction and efficient retention of sulphides occur, indicate surface sediment $\delta^{34}S$ depletions of 10 to 15‰ compared with downcore $\delta^{34}S$ values (Nriagu and Soon, 1985; Fry, 1986; Fry, 1990). Additional evidence for low rates of bacterial SO_4^{2-} reduction in McNearney sediments is the absence of a recent increase in S concentrations and the relatively high molar C : S ratios (range: 111 to 162; Cook *et al.*, 1990). Retention of S by bacterial SO_4^{2-} reduction coupled with metal or organic S formation causes lower molar C : S ratios (< 75; Gorham *et al.*, 1974; Nriagu and Soon, 1985; Mitchell *et al.*, 1990). The sedimentary S accumulating in McNearney Lake is apparently derived from organic S formed in the water column, along with some limited amounts of S derived from bacterial SO_4^{2-} reduction. In

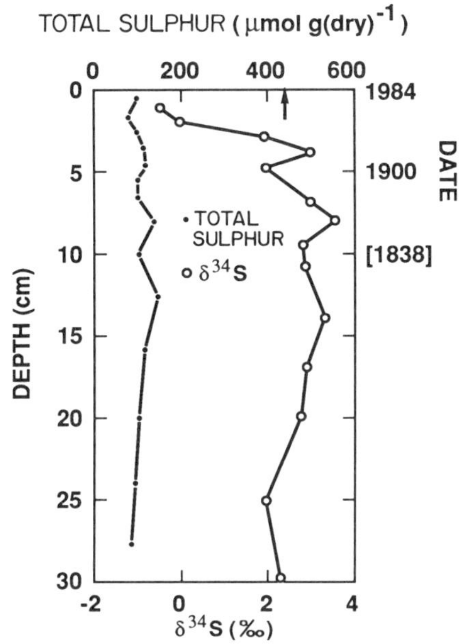

Figure 7.5. Total sulphur profile and stable sulphur isotope content ($\delta^{34}S$) in McNearney Lake, Michigan. (Fry, 1990). The anthropogenic Pb and polycyclic aromatic hydrocarbon rises occurred at 3 to 4 cm in this core (Cook *et al.*, 1990)

McNearney Lake, Cook *et al.* (1990) hypothesized that the low rates at which organic matter is supplied to the sediments may cause most of the organic matter to be oxidized aerobically at the sediment–water interface. Apparently, little of the organic matter in the anoxic portion of the sediments is suitable for bacterial SO_4^{2-} reduction, despite high lakewater SO_4^{2-} concentrations. Other workers have observed that bacterial SO_4^{2-} reduction rates in certain lakes were low and attributed these rates to low supplies of organic carbon (David and Mitchell, 1985; Rudd *et al.*, 1986).

7.2.2.3 Is Sulphate Concentration or Organic Carbon Supply Limiting Sulphate Reduction?

The limiting factor—the environmental parameter that is in least supply compared to the requirements of the organism—will determine the activity and growth rate of an organism. The previous two sections discuss evidence for both SO_4^{2-} and organic carbon being the factor limiting the activity of

SO_4^{2-} reducing bacteria. Recent models of SO_4^{2-} removal in acid-sensitive lakes assume that bacterial SO_4^{2-} reduction is first order with respect to SO_4^{2-} concentration (Kelly *et al.*, 1987; Baker and Brezonik, 1988) and thus consider SO_4^{2-} to be the limiting factor. Rate constants (which relate rates and concentrations) for the suite of lakes studied by Kelly *et al.* and Baker and Brezonik had a remarkably small range. However, as the model developers pointed out, this small range in rate constants occurs because the lakes to which the model was applied are oligotrophic and have similar rates of carbon supply to the sediments, as well as only a small component of SO_4^{2-} removal by algal uptake.

The Lake 227 experiment shows that when carbon supply changes, S retention in lake sediments also changes (Figure 7.3(b); Cook, 1981). The experiment by Sweerts, Rudd and Kelly (1986), in which organic-rich floc was added to the sediment surface of cores, partly explains this by showing that the increased organic carbon causes the oxic–anoxic interface within the sediment to move closer to the sediment surface. The location of this interface in the sediment, together with the SO_4^{2-} concentration, determines the concentration gradient and thus the flux of sulphate into the sediments. In a given lake, with a particular level of productivity, the average position of the oxic–anoxic boundary can be considered fixed at a certain depth by the supply of organic carbon. With increases in SO_4^{2-} concentration, both the concentration gradient and the rate of SO_4^{2-} reduction will increase because there is usually enough organic substrate to support increased SO_4^{2-} reduction. Therefore, SO_4^{2-} can only be considered limiting when the organic carbon supply is constant and at a high enough level.

The sediment depth of the oxic–anoxic boundary for ELA lakes is deeper in the winter than in the summer (C. A. Kelly, unpub. data), probably due to a combination of lower temperature and lower organic carbon supply. SO_4^{2-} reduction rates in summer are higher than average net annual rates, because of lower winter rates and also because some of the SO_4^{2-} reduced during the summer is apparently re-oxidized during the winter (Rudd, Kelly and Furutani, 1986). However, in the study cited, no sulphur was lost below the depth of 3 cm, which is probably the sediment depth below which oxygen does not penetrate in the winter. Even though the daily rate of SO_4^{2-} reduction varies throughout the year, and some reduced sulphur is re-oxidized, both of these phenomena are likely related to organic input to the sediments. For the suite of lakes examined by Kelly *et al.* (1987) and Baker and Brezonik (1988), the similar rate constants for in-lake SO_4^{2-} retention likely are a result of similar trophic status and similar organic matter sedimentation. These findings mean that neither the SO_4^{2-} concentration nor the organic carbon supply rate by itself is an absolute predictor of SO_4^{2-} reduction rates. Therefore, neither parameter strictly fits the definition of a limiting factor.

7.2.3 FATE OF SULPHIDE

After H_2S is produced in sediments, whether by SO_4^{2-} reduction or decomposition of organic sulphur, it either diffuses into the water column or remains in the sediments. H_2S accumulates in anoxic hypolimnetic water, but it is oxidized in aerobic waters and is present only in small concentrations in surface waters (Luther and Church, this volume). H_2S may escape to the atmosphere when the rate of gas exchange is high (e.g. fall overturn). Low pH conditions may also promote high rates of gas exchange, because at pH < 6 the solubility of $H_2S(g)$ is low (Cook, 1981) and the rates of chemical oxidation are slow (Chen and Morris, 1972; Millero *et al.*, 1987; Millero and Hershey, 1989). Sulphide may also undergo a number of reactions within the sediments, such as oxidation and metal sulphide and organic S formation (Luther and Church, this volume).

Early work on SO_4^{2-} reduction in lake sediments assumed that the products were H_2S and iron monosulphide (acid-volatile sulphides: Sorokin, 1975; Berner, 1984). Howarth and colleagues (Howarth, 1979; Howarth and Teal, 1979; Howarth and Merkel, 1984) showed that pyrite (FeS_2) could also be formed rapidly as a result of SO_4^{2-} reduction in salt marsh sediments, and the same process was later shown to occur in freshwater sediments (Rudd, Kelly and Furutani, 1986). In addition, ^{35}S from $^{35}SO_4^{2-}$ was also shown to be incorporated into organic matter in lake sediments (Rudd, Kelly and Furutani, 1986), although this apparently does not occur to any significant extent in salt-marsh sediments (Howarth and Merkel, 1984). Because other phases besides acid-volatile sulphide may be formed, use of the production of only the acid-volatile sulphide to estimate bacterial SO_4^{2-} reduction rates will yield underestimates (Howarth, 1979; Giblin and Wieder, this volume). The conditions determining which reaction is followed (oxidation *vs* metal S formation *vs* organic S formation) are poorly understood, but their occurrence in sediments has been studied by several groups, as discussed below.

7.2.3.1 Formation of Metal Sulphides

In addition to H_2S, reduced forms of metals are produced in anoxic zones of sediments. Use of oxidized forms of metals as electron acceptors by bacteria releases sufficient Fe^{2+} and other metals (e.g. Cd^{2+}, Pb^{2+}) to cause these environments to become supersaturated with respect to the sulphides of iron and other reduced metals (Davison, 1980; Cook, 1984; Lovley, 1987). For some lakes, precipitation of iron sulphides (amorphous iron monosulphide, hydrotroilite, mackinawite, pyrrhotite, greigite, and pyrite; Berner, 1984; Davison, 1980) may be an important mechanism for incorporating S into lacustrine sediments and removing S from the water column (Figure 7.3; Cook and Schindler, 1983; Cook, 1984; White *et al.*, 1989; Giblin *et al.*, 1990).

The factors controlling the formation of the various reduced S minerals are poorly known. Pyrite and acid-volatile metal sulphides (hydrotroilite, mackinawite, and greigite) have been identified in lake sediments (Cook, 1981; Kelly and Rudd, 1984; Nriagu and Soon, 1985; Giblin *et al.*, 1990), but determining the exact chemical form of reduced S present and conditions that lead to the formation of these species requires further research.

The flux of iron to the zone of highest sulphide concentrations in the sediments may limit the ability of sediments to sequester S (Schindler, 1985; Carignan and Tessier, 1988; Rudd, Kelly and Furutani, 1986; Herlihy *et al.*, 1987; Giblin *et al.*, 1990). Where measured, rates of iron accumulation in lakes are either comparable in magnitude to S accumulation or greater by five-fold to ten-fold (Table 7.3). Rudd, Kelly and Furutani (1986) found that the proportion of extractable Fe associated with S in several acidified Adirondack lakes was a small fraction of the total Fe. Much higher proportions of S-bound Fe were measured in the sediments of acidic lakes in Quebec (Carignan and Tessier, 1988), indicating that Fe supply could become limiting to further formation of iron sulphide compounds in those lakes. Giblin *et al.*, 1990) found similar results for several New England lakes with organic-rich sediments and low concentrations of non-S-bound Fe. It should be noted that some of the total Fe measured may be unavailable for microbial reduction or may be bound within mineral lattices and released slowly (Lovley and Phillips, 1986). Fe^{2+} available for reaction with sulphide may also be supplied by diffusion from sediment horizons beneath the zone of SO_4^{2-} reduction. Concentration gradients of Fe^{2+} in the porewater are typically high and may,

Table 7.3. Rates of sulphur and iron accumulation in lake sediments

Site	S accumulation (mmol m^{-2} a^{-1})	Fe accumulation (mmol m^{-2} a^{-1})	Reference
Little Rock Lake	12	12	Baker *et al.* (1989)
McNearney Lake	12	16	Cook *et al.* (1990)
Lake 223	15	20	Cook (1984)
Lake 227	13	25	Cook (1981)
Cone Pond	21–34	40–140	Giblin *et al.* (1990)
Lake 96760	37	47	Carignan and Tessier (1988)
Spectacle Lake	27–41	48–71	Giblin *et al.* (1990)
Mares Lake	22–27	58–244	Giblin *et al.* (1990)
Lakes of the Clouds	5	125–250	Giblin *et al.* (1990)
Miles Lake	72–80	146–205	Giblin *et al.* (1990)
Mirror Lake	18–25	200–250	Giblin *et al.* (1990)
Big Moose Lake	47	500	White *et al.* (1989)

over long periods of time, cause the transport of Fe from the entire sediment column to the zone where sulphide is produced and iron sulphide is formed (Cook, 1984; Cook *et al.*, 1987). Once the inventory of reactive iron is exhausted (Anderson and Schiff, 1987), the capacity for S retention by reaction with iron is limited by the rate of supply of Fe. If precipitation of metal sulphides or incorporation into organic forms does not occur, H_2S may build up to high concentrations or diffuse to oxic zones where it will eventually be oxidized back to SO_4^{2-}. A decrease in the capacity of the sediments to retain S would lead to greater cycling of S from the sediments and higher SO_4^{2-}concentrations in the water (Giblin *et al.*, 1990).

7.2.3.2 Reaction with Organic Matter

In addition to reacting with metals to form sulphide compounds, H_2S and S^0 may react with organic matter to form organic-S compounds (e.g. thiols, mercaptans, and thioethers), as has been shown in both marine (Casagrande *et al.*, 1979; Casagrande and Ng, 1979; Francois, 1987; Vairavamurthy and Mopper, 1987, 1989; Luther and Church, this volume) and freshwater sediments (Nriagu and Soon, 1985; Rudd, Kelly and Furutani, 1986). Organic-S compounds are produced either by the replacement of a carbon-bonded oxygen atom with an S(II) atom or by addition of H_2S to a carbon–carbon double bond (Anderson and Schiff, 1987; Luther and Church, this volume). The exact conditions under which organic S is formed from H_2S and the factors controlling the relative amounts of metal sulphide and organic S formed from H_2S are not known (Nriagu and Soon, 1985; Rudd, Kelly and Furutani, 1986; Carignan and Tessier, 1988). The quality and quantity of organic carbon in lake sediments may possibly limit the synthesis of organic-S compounds and may be responsible for the unexplained variability in organic S content in lake sediments.

Other factors affecting the relative reactivity of H_2S for iron or organic C may be the concentrations of Fe^{2+} and total sulphide in the porewater, and pH, which affects the portion of total sulphide that is S^{2-}. Baker *et al.* (1989) showed that adding Fe^{2+} or lowering the pH decreased the proportion of $^{35}SO_4^{2-}$ reacting to form organic ^{35}S in sediment incubations. However, examination of $^{35}SO_4^{2-}$ incubation data from a variety of lakes studied by Rudd, Kelly and Furutani (1986) reveals no clear pattern. Determination of the factors controlling organic S formation from sulphide requires additional research.

Rudd, Kelly and Furitani (1986) added $^{35}SO_4^{2-}$ to epilimnetic sediments in situ and observed that the tracer initially was reduced and incorporated into both inorganic and organic S in the sediments. After an eight-month period, some of the tracer was lost from both the organic-S and inorganic-S pools through oxidation or volatilization (Rudd, Kelly and Furutani, 1986). More tracer was lost from the inorganic-S pools, leaving the organic form as the predominant phase, particularly in the upper sediment horizons. Other studies have not observed large increases in organic-S compounds in sediments following S amendments. Studies in which SO_4^{2-} has been added to whole lakes (e.g. Lake 223; Figure 7.3a) or to intact sediment cores (Giblin *et al.*, 1990) revealed that the inorganic S phase exhibited increases in hypolimnetic sediments. In an examination of S sediment accumulation in the hypolimnetic sediments of a lake in the Adirondacks (New York, USA), White *et al.* (1989) showed that both inorganic-S and organic-S concentrations have increased, but inorganic S has increased more during the past 100 years. In an experimentally acidified lake, Kelly and Rudd (unpub. data) found that in epilimnetic cores, the sulphur increase near the surface was due to an increase in both inorganic and organic S. In hypolimnetic cores, the increase was due mostly to inorganic S. Thus, diverse patterns can be found within the same lake.

The study of the factors affecting organic-S accumulation has been impeded by the lack of a direct analytical method. Organic S is typically estimated by subtracting inorganic S (usually measured by using the chromium reduction method of Howarth and Merkel, 1984; see Giblin and Wieder, this volume) from total S. Total S may be measured on wet samples with an alkaline oxidation (Tabatabai and Bremner, 1970) followed by reduction to H_2S (Johnson and Nishita, 1952; Johnson and Ulrich, 1959). Because this method is labour intensive, total S is more frequently analysed by drying samples and combusting the S to SO_2, which can then be measured by fluorescence or infrared spectrophotometry. However, drying sediments can lead to significant and variable losses of total S (Amaral *et al.*, 1989), especially when lyophilization is used as the drying method. Drying also leads to a change in stable S isotope content (Amaral *et al.*, 1989), a result which shows that the loss is selective rather than equally distributed among the various S species. Other investigators have shown that drying does not lead to loss of inorganic-S content (Canfield *et al.*, 1986; Carignan and Tessier, 1988; Morse and Cornwell, 1987). The stability of inorganic S with respect to drying suggests that the losses from the total S pool are from the organic fraction and that organic S estimated by difference may be too low. Organic-S volatilization upon drying is not as much of a probelm in studies using ^{35}S. In these studies, the organic ^{35}S is measured by aqua regia digestion (Howarth and Teal, 1979), following removal of inorganic ^{35}S, with all analyses usually performed quickly on wet sediments. A standard technique is needed for the assay of organic S, as well as for all sedimentary S species, to allow intercomparison of results and determination of the factors affecting organic-S accumulation in lake sediments.

7.2.3.3 Oxidation to Elemental Sulphur and Higher Oxidation States

Dissolved and solid sulphide compounds are readily oxidized both chemically (Avrahami and Golding, 1968; Chen and Morris, 1972; Pankow and Morgan, 1980) and biologically (Schiff and Hodson, 1973; Roy and Trudinger, 1970). In both oxidation mechanisms, polysulphides, S^0, $S_2O_3^-$, SO_4^{2-}, and SO_4^{2-} may be formed (Luther and Church, this volume). Although SO_4^{2-} is the most stable phase, the others are metastable forms that may persist for long periods of time.

Biological oxidation of sulphide and other reduced S species (thiosulphate, thiocyanate, elemental sulphur, and tetrathionate) is carried out by both the chemotrophic sulphur oxidizers, which may use reduced sulphur as an energy source and fix CO_2, and the phototrophic green and purple bacteria, which use reduced sulphur as an electron donor in the fixation of CO_2 (Table 7.1). In some lakes, these activities can be significant sources of organic matter productivity (Parkin and Brock, 1981), particularly meromictic salt lakes (Gorlenko, 1978). The relative potential of these reduced S species to be oxidized and the environmental factors that regulate the extent of oxidation have not been studied nearly as much as SO_4^{2-} reduction and are much less understood.

Oxidation of reduced S occurs at interfaces: the oxic–anoxic interface for chemical and chemotrophic oxidation and, in the anoxic zone, the light–dark interface for phototrophic oxidation (Table 7.1). Such interfaces may occur within or on sediments or in the water column during stratification. Depending on the depth of light penetration, phototrophic bacteria may grow below the chemoautotrophic bacteria (Gorlenko, 1978). The phototrophic bacteria may migrate up and down in the water column, following the diurnal changes in depth of maximum concentration of reduced sulphur (Sorokin, 1970), or the bacteria may remain at fixed depth (Parkin and Brock, 1981). During the winter period, the oxic–anoxic interface in sediments moves downward, due to lower rates of microbial oxygen consumption, and reduced S that accumulated during the antecedent anoxic period is subject to partial oxidation (Rudd, Kelly and Furutani, 1986; Kling *et al.*, 1991), with attendant effects on the partitioning of S species.

7.2.4 FORMATION OF VOLATILE SULPHUR SPECIES

The release of volatile S species is a potential pathway for export of S from lakes. Volatile S species include H_2S and a variety of organic-S compounds, including carbonyl sulphide (COS), dimethyl sulphide (DMS), methane thiol

(MSH), dimethyl disulphide (DMDS), and carbon disulphide (CS_2). In general, these organic-S gases are more soluble than methane and even carbon dioxide, but they are volatile at lake temperatures and may contribute more to water–air fluxes than H_2S because of their greater stability in the presence of oxygen.

Most research on the water–air exchange of S gases has been concerned with DMS in marine systems (Andreae and Barnard, 1984; Andreae and Jaeschke, this volume) and with emissions from salt marshes, soils, and vegetation (e.g. Steudler and Peterson, 1984; Goldan *et al.*, 1987; Giblin and Wieder, this volume; Andreae and Jaeschke, this volume). In marine systems, DMS is present in ocean surface waters at concentrations ranging from 0.031 to 34.3 nM (1 to 1100 ng (S) l^{-1}), with a mean of 2.6 nM (83 ng (S) l^{-1}) (Andreae and Barnard, 1984; Turner *et al.*, 1988).

Few measurements have been made on volatile S production and release from freshwater systems. Holdway and Nriagu (1988) measured DMS concentrations in Lake Ontario surface waters ranging from 0.6 to 1.4 nM (20 to 45 ng (S) l^{-1}). Concentrations of DMS measured in Lake Mendota, a eutrophic Wisconsin lake, were much lower (0.017 to 0.16 nM; 0.5 to 5.2 ng (S) l^{-1}; Zinder and Brock, 1978), whereas two Antarctic lakes had higher concentrations (3.2 to 4.8 nM; 100 to 150 ng (S) l^{-1}; Deprez, Franzmann and Burton, 1986). A recent survey of a variety of volatile S species in the epilimnia of Canadian Shield lakes during summer stratification showed average summer DMS values ranging from 0.48 to 1.6 nM (15 to 51 ng (S) l^{-1}; Richards, Kelly and Rudd, 1991; Table 7.4). This same study also showed low concentrations of COS (0.091 to 0.38 nM; 2.9 to 12 ng (S) l^{-1}) as well as MSH (0.018 to 1.1 nM; 0.58 to 35 ng (S) l^{-1}) and DMDS (0.02 to 0.097 nM; 0.64 to 3.1 ng (S) l^{-1}). Dimethyl disulphide may be produced by MSH oxidation after sample collection, so that measured concentrations of both gases may not reflect in situ concentrations. Much higher values of S gases were found in one shallow unstratified lake, with DMS averaging 6.6 nM (210 ng (S) l^{-1}), COS 0.61 nM (20 ng (S) l^{-1}), DMDS 1.4 nM (45 ng (S) l^{-1}), and MSH 8.6 nM (275 ng (S) l^{-1}) (Lake 114; Table 7.4). These concentrations were as high as those measured in a nearby bog pool, and in other Ontario bogs (Nriagu, Holdway and Coker, 1987), demonstrating a high potential flux for at least some freshwater systems.

The source of volatile S species in marine waters is primarily algal (Andreae and Jaeschke, this volume), but in freshwaters a mixture of water column and sediment sources may exist, depending on algal physiology and species and lake bathymetry. Some freshwater algal species have been shown to produce DMS under axenic growth conditions (Bechard and Rayburn, 1979). The absence of DMS production by many freshwater algae might be because dimethylsulphonium propionate (DMSP), the algal-produced precursor of DMS, is thought to be an osmolyte in high-salinity water. In

Table 7.4. Estimates of gaseous sulphur flux and its significance in lake sulphate budgets. From Richards, Kelly and Rudd (1991)

Lake (mean depth)	k^a (cm hr^{-1})	Species	Mean Surface concentration (nmol l^{-1})	Sulfur flux[b] (nmol m^{-2} d^{-1})	Sulfur flux[b] (mol a^{-1})	Flux as percentage of total S lost
302S (5.1 m)	0.95	DMS	1.3	290	7.8	0.08
		DOS	0.4	82	2.2	0.02
		MSH	not detected	—	—	—
		DMDS	0.09	21	0.6	0.006
		Total	1.7	400	10.0	0.100
Loss at fall overturn[d] (hypolimnion)					7.4	0.07
239 (10.5 m)	0.95	DMS	1.1	250	35	0.331
		COS	0.2	60	8.1	0.077
		MSH	Not detected	—	—	—
		DMDS	0.020	4	0.6	0.006
		Total	1.4	320	43	0.414
114 (1.7 m)	0.9515	DMS	6.6	1500	44	—
		COS	0.6	140	4	—
		MSH	8.6	1965	58	—
		DMDS	1.4	325	10	—
		Total	17.2	3900	116	—

[a]k, mass transfer coefficient.
[b]Assuming a negligible concentration of S gases in the atmosphere, and an ice-free period of 8 months.
[c]The percentage of the total S lost from the water column (sediment accumulation and gas loss) that is volatile sulphur gas flux (Rudd *et al.*, 1990).
L302S; total lost is 99.61 keq (total over 5 years from 1982–1986; Rudd *et al.*, 1990).
L239; total lost is 20.9 keq (annual mean for 1981–1983; Schindler *et al.*, 1986).
[d]Loss at overturn *assumes* that the volatile species were lost to the atmosphere, and chemical oxidations and conversions were negligible. Thus, it represents a *maximum* loss.

support of this function of DMSP, Iversen, Nearhoof and Andreae (1989) observed that DMSP and DMS concentrations increased with increasing salinity along a brackish–marine salinity gradient. Howarth and Stewart (this volume) have suggested that the low rate of DMS production by algae in most freshwaters may also reflect a relatively high abundance of molybdenum (indicated by a low sulphate : molybdenum ratio; Howarth, Marino and Cole, 1988) in these waters.

Production of volatile S, primarily DMS, is also known to occur in decomposing freshwater algal cultures (Bechard and Rayburn, 1979) and in decomposing algal mats (Zinder, Deomel and Brock, 1977). Production of volatile S species during decomposition in other systems (e.g. bacterial cultures, soils, and manures) is well known (Bremner and Steele, 1978). Decomposition of methionine has been studied by adding it to anoxic sediments from a lake (Zinder and Brock, 1978) and a salt marsh (Kiene and Visscher, 1987). In these studies the major volatile S product was MSH, along with small amounts of DMS. Both MSH and DMS were present in concentrations of 10 to 120 nM (320 to 3840 ng (S) l^{-1}) in the anoxic porewater of several lakes on the Precambrian Shield (Richards, Kelly and Rudd, 1991), at levels similar to marine porewater DMS concentrations (Andreae, 1985) and 2 to 14 times higher than in the overlying water. Bubbles produced from the sediments of an Arctic lake contained COS, DMS, and CS_2 (Hines and Morrison, 1989). Thus, in small shallow lakes such as Lake 114 (Table 7.4), volatile S release from sediments may be the primary source of S gases in surface water. However, the similarity of volatile S gas concentrations in most lakes with mean depths greater than a few metres (e.g. Lake 302S and Lake 239, Table 7.4; Lake Ontario, Holdway and Nriagu, 1988), suggests that either algal production or some low rate of organic matter decomposition in the water column is supporting the concentrations observed in deeper lakes.

Richards, Kelly and Rudd (1991) examined the effect of concentration of SO_4^{2-} on surface water volatile S production by comparing two lakes with similar bathymetries but contrasting chemistries: one lake with low SO_4^{2-} and circumneutral pH and one with high SO_4^{2-} and low pH. Volatile S concentrations over the summer were virtually indistinguishable in surface waters of the two lakes (Figure 7.6), suggesting no effect of either SO_4^{2-} concentration or pH. However, the hypolimnetic production rate of DMS was much greater in the lake with greater SO_4^{2-} concentrations. If the absence of change in surface water volatile S compounds over the range of freshwater SO_4^{2} concentrations is borne out, present-day measurement of volatile S compounds in the atmosphere may be used to estimate natural (non-anthropogenic) sources of atmospheric S from lakes.

Volatile S compounds are catabolized, as well as synthesized, in sediments. Zinder and Brock (1978) were only able to detect the production of volatile S compounds in sediments by suppressing breakdown of these compounds with

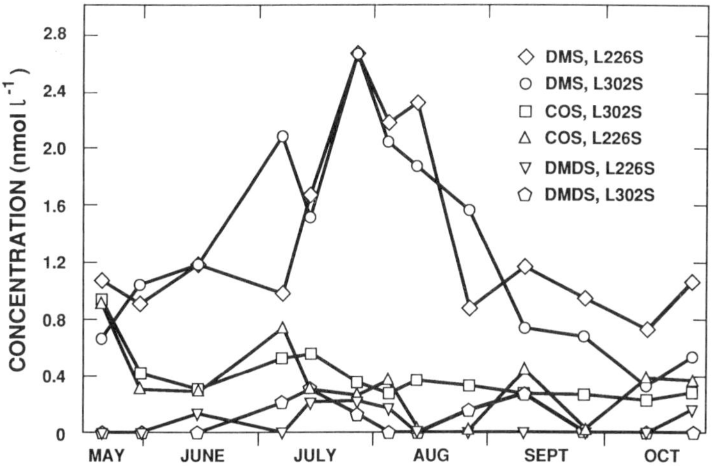

Figure 7.6. Surface water concentrations of inorganic and organic volatile S species in Lake 302 South and in Lake 226 Southwest during 1988. The SO_4^{2-} concentration in Lake 302 South was 106 to 135 μmol L^{-1} and the pH was 4.5 to 4.6 during 1988. In Lake 226 Southwest, SO_4^{2-} was 26 to 28 μmol L^{-1} and the pH was 7.0 during 1988. From Richards, Kelly and Rudd (1991)

chloroform, an inhibitor of C-1 metabolism. Both MSH and DMS are broken down to CO_2, CH_4, and H_2S (Zinder and Brock, 1978; Kiene *et al.*, 1986). Oremland *et al.* (1989) isolated a methanogenic bacterium that was able to grow on DMS. Also, phototrophic purple bacteria have been shown to oxidize DMS to DMSO (Zeyer *et al.*, 1987). Undoubtedly, other transformations of volatile S species are carried out by bacteria. Zinder and Brock (1978) suggested that sediments may be a sink for volatile S species, rather than a source. However, the porewater measurements mentioned above suggest that sediments are a net source of volatile organic-S species in lakes.

The volatile organic-S species are present in porewater in nanomolar concentrations. In contrast, H_2S is often present in anoxic zones in micromolar concentrations (Cook, 1984; Nurnberg, 1984; Carignan, 1988). In oxic zones, H_2S is unstable and is oxidized with a half-life of 50 hours (Millero *et al.*, 1987; Millero and Hershey, 1989). The organic-S species are stable in cold oxic water for up to 24 hours (Richards, Kelly and Rudd, 1991). Because of its low concentration and chemical instability, there are probably few reliable measurements of H_2S concentration in oxic surface waters. Therefore, the importance of fluxes of H_2S in relation to the volatile organic-S species is difficult to assess.

The relative role of gaseous S flux in whole-lake budgets has recently been assessed in one lake and appears to be a minor term. As discussed below, up to 50% of incoming SO_4^{2-} may be removed from the water column by all mechanisms, and it is not known whether the sediments are the sole sink for this S. Some S lost from the water column could be converted into volatile compounds and lost to the atmosphere. However, in a whole-lake $^{35}SO_4^{2-}$ addition experiment, Stuiver (1967) was able to account for all of the ^{35}S added to the lake either in the sediments or the water column, indicating little if any volatilization. This agrees with Richards, Kelly and Rudd (1991) who compared estimates of volatile organic-S fluxes with SO_4^{2-} budgets for three Canadian Shield lakes (Table 7.4). These fluxes were insignificant compared to SO_4^{2-} reduction rates (Table 7.2) or to whole-lake S retention estimated from mass balances (Table 7.4).

An additional issue is the role of volatile S fluxes from lakes as a source of S to the atmosphere. The lake–air fluxes of organic-S gases for a shallow unstratified lake (Lake 114; Table 7.4) are similar on an areal basis to fluxes from tropical forests (Andreae and Andreae, 1988) and are about one-third of the mean areal flux from the ocean (Andreae and Raemdonck, 1983). The range of S fluxes from lakes (Table 7.4) is similar to the range measured from midcontinent soils in the United States (Goldan *et al.*, 1987; Lamb *et al.*, 1987). Because Lake 114 is located in a low SO_4^{2-} deposition area (13 to 43 μmol m^{-2} day^{-1}; 0.42 to 1.4 mg (S) m^{-2} day^{-1}; Linsey, Schindler and Stainton, 1987), as much as 10 to 30% of atmospheric S deposition falling directly on the lake surface may be returned to the atmosphere by gas exchange of organic-S compounds. Some areas of North America (e.g. Hudson Bay lowlands), northern Europe, and northern Asia contain many shallow unstratified lakes that may have rates of volatilization comparable to those for Lake 114. Also, lakes in which SO_4^{2-} concentrations are unusually high due to evaporation (> 20 g l^{-1}; 0.6 M) have recently been shown to have DMS fluxes to the atmosphere that are greater than those in the ocean (Richards, Rudd and Kelly, in review). The importance of the lake-to-air flux of S gases for the S budgets of these lakes and as an input to the atmosphere deserves further attention. However, because lakes cover a small percentage of the surface area of the continents, lakes are probably a very small source of sulphur to the atmosphere.

7.3 SULPHUR CYCLING IN WHOLE LAKES

The previous section describes the various reactions that S undergoes in small lakes. In this section, these reactions will be put into a whole-lake perspective by examining S budgets and a simple model relating inputs, outputs, and retention of S in small lakes.

7.3.1 WHOLE-LAKE SULPHUR BUDGETS

Construction of whole-lake S budgets has emphasized inputs and outputs of SO_4^{2-}, primarily because SO_4^{2-} is the most abundant and most commonly measured S species. David and Mitchell (1985) were the only workers to examine other forms of S in addition to SO_4^{2-}; they found that organic S was a minor constituent in the inputs and outflows of South Lake (New York).

From a whole-lake perspective, SO_4^{2-} is removed from the water column and retained in the lake sediments by two processes: bacterial SO_4^{2-} reduction, followed by iron sulphide or organic sulphide formation, and sedimentation of organic S from the water column. The net sedimentation of organic S is difficult to quantify because rates of mineralization are not known and are indirectly estimated, as discussed above. Net retention of sulphides derived from bacterial SO_4^{2-} reduction is also difficult to quantify. In systems in which bacterial SO_4^{2-} reduction has been measured directly, oxidation of reduced S species after formation may be significant and is rarely measured (Rudd, Kelly and Furutani, 1986). Usually net loss of S from the water column is calculated from a mass balance by difference, and includes both organic S sedimentation and accumulation of S formed from bacterial SO_4^{2-} reduction.

Whole-lake S budgets have been calculated for six lakes that have been investigated for the mechanisms of S removal (Table 7.5). For three lakes in midcontinental North America (Lake 223, Lake 302N, and Lake 302S, all of which are located at ELA), SO_4^{2-} is removed from the water column, primarily by bacterial SO_4^{2-} reduction, at large areal rates. Part of the S removal is due to the enhancement of bacterial SO_4^{2-} reduction from experimental additions of SO_4^{2-}. Rates of SO_4^{2-} reduction for these three lakes are higher than that for Little Rock Lake (Wisconsin) and South Lake (New York) (Table 7.5). Interestingly, although seasonal rates of bacterial SO_4^{2-} reduction in Lake 227 were quite high because of stimulation by eutrophication (73 mmol m^{-2} a^{-1}; 2.3 g (S) m^{-2} a^{-1}; Table 7.2), the reduced S was not sequestered efficiently by the hypolimnetic sediments, because the annual rate of SO_4^{2-} reduction and retention was estimated to be only 2 mmol m^{-2} a^{-1} (0.06 g (S) m^{-2} a^{-1}. Apparently, a fraction of the S derived from bacterial SO_4^{2-} reduction remained in the hypolimnion as H_2S, to be oxidized to SO_4^{2-} at fall and spring overturn (Cook, 1981).

The rates of S removal due to bacterial SO_4^{2-} reduction were variable among the lakes, yet rates of organic-S sedimentation were similar among the lakes, ranging from 3 to 11 mmol m^{-2} a^{-1} (0.1 to 0.35 g (S) m^{-2} a^{-1}) (Table 7.5). The rates of organic-S sedimentation for Lake 302N and 302S are gross sedimentation, because estimates of recycling at the sediment–water interface were not made.

Except for South Lake, these lakes retain S at relatively high rates (Table 7.5), so that a large proportion of the SO_4^{2-} input to the lake is removed from the water column. Differences among lakes in the proportion of input SO_4^{2-} retained are related to differences in morphometry and hydrology of the lakes and are discussed below.

Table 7.5. Whole-lake annual sulphur budgets for six lakes, showing the processes by which sulphur is removed from the water column and retained in the sediments on a net basis. Units are mmol m^{-2} a^{-1}

						In-lake retention process	
	Natural inputs	Experimental SO_4^{2-} inputs	Total inputs	Outflow	Change in storage	Bacterial SO_4^{2-} reduction	Organic-S sedimentation
Lake 223[a]	76	227	303	101	80	119–122	<3
Lake 227[b]	82	—	82	74	—	2–8	<6
Lake 302S[c]	83	151	234	64	65	94–105	<11
Lake 302N[c]	116	—	116	58	12	35–46	<11
Little Rock Lake[d]	21	—	21	11	—	6	9
South Lake[e]	421	—	421	403	—	9	9

[a]Average for the period 1976 through 1983, from Cook *et al.* (1986). The organic sulphur accumulation in the sediments was estimated by Cook (1981) by using net nitrogen accumulation and the N : S ratios of seston from the deepest point in the lake. Because organic sedimentation at the deepest point in the lake is a maximum value, this estimate is likely an upper limit on a whole-lake basis. Retention of S associated with bacterial SO_4^{2-} reduction is the residual in the budget.

[b]Average for the period 1974–1975 and 1978–1979, from Cook (1981). See footnote '*a*' for method of calculating organic S accumulation and bacterial SO_4^{2-} retention.

[c]For the period 1981 through 1985. Organic sulphur sedimentation is an estimate of gross sedimentation; no estimates are available for decomposition at the sediment–water interface. Retention of S associated with bacterial SO_4^{2-} reduction is the residual in the budget. From Rudd *et al.* (1990).

[d]Estimated budget for preacidification conditions. Organic sulphur sedimentation is estimated by using carbon to sulphur ratios. From Baker *et al.* (1989).

[e]For the period 1981 through 1982 [from Mitchell *et al.* (1985) and Landers and Mitchell (1988)], except for SO_4^{2-} reduction rates, which were estimated for 1986 (Owen and Mitchell, 1989).

7.3.2 MODELLING SULPHATE REMOVAL IN LAKES

An apparent dichotomy that arose during research on SO_4^{2-} removal in acidified lakes was the great difference among lakes in the importance of in-lake SO_4^{2-} retention in whole-lake S budgets. As mentioned above, in the first ELA acidification experiment (Lake 223), 40% of the SO_4^{2-} input to the lake was biologically removed from the water column and retained in the lake sediments (Cook *et al.*, 1986). This finding was in marked contrast to some atmospherically acidified lakes where little or none of the SO_4^{2-} input was removed within the lake (Wright, 1983; Galloway *et al.*, 1983). This difference between the experimentally and atmospherically acidified lakes was not solely due to the presence of an anoxic hypolimnion in Lake 223, because the anoxic hypolimnion is less than 2% of the area of Lake 223. Furthermore, SO_4^{2-} reduction was also occurring in littoral sediments of Lake 223 at rates similar to those in the profundal sediments (Kelly and Rudd, 1984; Cook *et al.*, 1986). When SO_4^{2-} reduction rates were measured directly in a variety of acidified lakes in North America and northern Europe, areal rates were found to be fairly similar from lake to lake (Rudd *et al.*, 1986); the SO_4^{2-} reduction rates collected in Table 7.2 support this broad similarity among many lakes. The relatively small range in SO_4^{2-} reduction rates led to the hypothesis that the most important factors controlling whole-lake SO_4^{2-} retention in acidified lakes were water residence time and mean depth (i.e. the extent of contact between the water and the sediments) (Kelly *et al.*, 1987; Baker *et al.*, 1986).

The models developed for SO_4^{2-} removal in lakes with different residence times and mean depths are mathematically similar to those for phosphorus removal (e.g. Dillon, 1974). Conceptually, however, the models are different in that the phosphorus removal is essentially a water column (algal) process, whereas usually SO_4^{2-} removal via bacterial reduction is a sediment process (Cook *et al.*, 1986; Rudd *et al.*, 1990). The key parameters for the S models are water residence time, mean depth, and a mass transfer coefficient that quantifies the areal rate of SO_4^{2-} reduction by the sediments (Kelly *et al.*, 1987; Baker, Brezonik and Pollman, 1986).

The model formulation assumes that SO_4^{2-} reduction rate is first order with respect to SO_4^{2-} concentration. Sulphate reduction in sediments has been shown to be first order within a lake as SO_4^{2-} concentration increases. In comparing many lakes, Rudd *et al.* (1986) found that the concentration–rate relationship was weaker among lakes than that within one lake; however, the concentration range of the measurements was relatively small (30 to 120 μM; 0.1 to 3.8 mg (S) l^{-1}) and thus was not suitable for observing differences among lakes. The average mass transfer coefficient [S_s = annual rate of S reduction (mmol m^{-2} a^{-1}) divided by the SO_4^{2-} concentration (mmol m^{-3})] calculated from these direct measurements was 0.36 to 0.17 m a^{-1} (Kelly *et al.*, 1987). This mass transfer is analogous to a piston velocity for SO_4^{2-} flux from the water column to the sediments. The proportion of input SO_4^{2-} removed within the lake (R_s) is then calculated from

$$R_s = 1 - [q_a/(q_a + S_s)] \tag{7.1}$$

where q_a is the areal outflow (m a^{-1}) and is equal to mean depth (m) divided by the outflow-based water residence time (year). S_s can also be calculated for lakes where SO_4^{2-} removal has been measured by mass balance budgets, by rearranging Equation (7.1). When calculated this way, good agreement was reached by two independent groups: 0.54 to 0.13 m a^{-1} ($n = 8$; Kelly *et al.*, 1987) and 0.52 to 0.34 m a^{-1} ($n = 14$; Baker and Brezonik, 1988). Use of these values predicts SO_4^{2-} removal very well (Figure 7.7; Kelly *et al.*, 1987; Baker and Brezonik, 1988).

The cause of the slightly higher mean mass transfer coefficients calculated from the whole-lake budgets relative to the direct measurements of SO_4^{2-} reduction is expected because of the contribution of algal uptake and sedimentation. In two experimental lake basins at ELA, where one had elevated SO_4^{2-} concentrations and one did not, the contribution of gross S sedimentation to SO_4^{2-} removal was 14% in the low SO_4^{2-} basin and 6.5% in the high SO_4^{2-} basin (Rudd *et al.*, 1990).

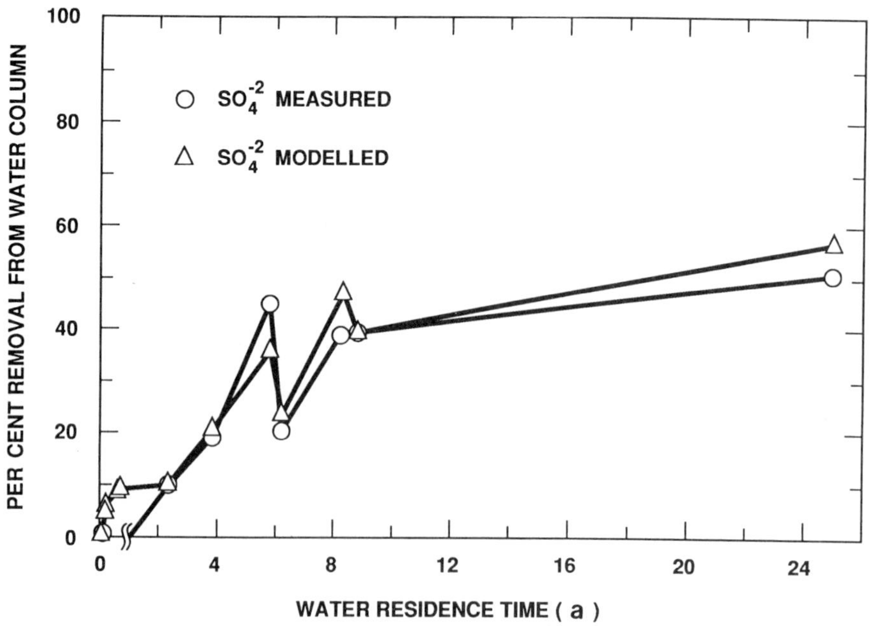

Figure 7.7. In-lake retention of SO_4^{2-} as a function of lakewater residence time. The model-predicted values are derived from Kelly *et al.* (1987) by using a mass transfer coefficient of 0.5 m a^{-1}

This model, and these mass transfer coefficients, are only suitable for relatively oligotrophic lakes without extensive anoxic hypolimnia (Kelly *et al.*, 1987). One might expect lakes with large anoxic hypolimnia to have much larger mass transfer coefficients [e.g. Lake Malawi, Africa (Kelly *et al.*, 1987)], because of the more rapid SO_4^{2-} transport from eddy diffusion across the oxic–anoxic interface in the sediments. However, Lake 227, which has a seasonally anoxic hypolimnion, has a relatively high mass transfer coefficient on a seasonal basis (5.5 m a^{-1}; Cook, 1981), yet has a low mass transfer coefficient derived from an annual whole-lake budget (0.14 m a^{-1}; Table 7.5). This annual mass transfer coefficient is smaller than that for oligotrophic lakes. In Lake 227, SO_4^{2-} was being reduced at high rates yet the efficiency of retention of reduced S was low. The H_2S produced from bacterial SO_4^{2-} reduction remained in the anoxic water column where it was oxidized to SO_4^{2-} at spring and fall overturn.

Lake Anna also does not fit the model because SO_4^{2-} is conveyed to the sediments by sedimentation of particles and not by diffusion, giving this lake an unusually high mass transfer coefficient (Mills, Bell and Herlihy, 1989). Lakes with low SO_4^{2-} reduction rates [e.g. South Lake (David and Mitchell, 1985) and McNearney Lake, discussed above], are also exceptions to the model. In these lakes, algal uptake and sedimentation may be the dominant mechanism of S retention. No predictive model has been developed for these lake systems and would be difficult to construct because the removal is probably related to algal biomass accumulation, which would be best predicted by phosphorus loading. Also, no *a priori* method exists to identify lakes that have low SO_4^{2-} reduction rates.

Another approach to determining the coefficients for SO_4^{2-} mass transfer is to use the net rate of accumulation of total S (from all mechanisms, including algal sedimentation and bacterial reduction) in the sediments as the measure of net SO_4^{2-} removal

$$S_s = [\text{Accum. rate}]/[SO_4^{2-}] \tag{7.2}$$

where the sediment accumulation rate is in mmol m^{-2} a^{-1} and SO_4^{2-} is in mmol m^{-3}. The use of this technique in lakes experiencing changes in atmospheric deposition introduces uncertainty because the accumulation rate must be averaged for several decades, a period during which the lakewater SO_4^{2-} concentration has likely changed. Norton *et al.* (1988) measured S accumulation rates and SO_4^{2-} concentrations in a number of lakes in eastern North America. From the data of Norton *et al.*, which include 15 lakes not used in the whole-lake budget approaches discussed above, we calculated a mass transfer coefficient of 0.8 to 0.5 m^{-2} a^{-1}, using Equation (7.2). This is a slightly higher mean value than coefficients calculated from whole-lake budgets (0.5 to 0.3 m^{-2} a^{-1}; Kelly *et al.*, 1987; Baker and Brezonik, 1988), perhaps because the sediment accumulation rates were measured at the deepest points in the lakes, where S accumulation is likely to be greater than the average lake value. However, the overlap of the sediment accumulation and whole-lake budget data is strong support for the SO_4^{2-} removal model.

Extrapolation of in-lake SO_4^{2-} retention from a particular lake or group of lakes requires knowledge of mean depth and residence time. In-lake SO_4^{2-} retention will be larger for lakes with longer residence times and smaller for shorter residence time lakes (Equation (7.1); Figure 7.7). Norton *et al.* (1988) examined S accumulation rates and concluded that, in general, these rates represent a small component of the whole-lake S budget that would remove less than 10 to 20 μM (0.32 to 0.64 mg (S) l^{-1}) of SO_4^{2-}. However, their data set included many lakes with short residence times, and their conclusions probably would have been different had more lakes with long residence times been included.

Some workers (Norton *et al.*, 1988; Shaffer *et al.*, 1988) have tried to simplify further the models of SO_4^{2-} removal by substituting the ratio of the watershed area to lake area and some generalized estimate of runoff for measurements of areal water loading (mean depth divided by water residence time). These approaches can lead to serious errors unless the value used for runoff is close to the true net value. For example, the estimated q_a values using 1 m a^{-1} precipitation (Norton *et al.*, 1988) are an order of magnitude larger than the literature values of q_a for two of the lakes examined by Norton *et al.* [McNearney Lake (Cook *et al.*, 1990) and Hustler Lake (Kanciruk *et al.*, 1986)]. The use of too high a value for net precipitation leads to an underestimate of the in-lake SO_4^{2-} retention, as was done by Norton *et al.* (1988). However, even though the most accurate calculation requires good hydrologic data, a rough estimate of water residence time will at least indicate whether SO_4^{2-} retention should be an important or a negligible process in a given lake.

7.4 INTERACTIONS OF THE C, N, P, AND S BIOGEOCHEMICAL CYCLES IN LAKES

The biogeochemistry of S is closely linked to biogeochemistry of C, N, and P (Howarth and Stewart, this volume). Without the link to the C cycle, the behaviour of S in small lakes would be conservative. In turn, sulphur's chemical forms, concentrations, and reactions affect the cycles of the other elements. Because of these interactions, the biogeochemistry of C, N, P, and S should not be examined in isolation.

A number of interactions of the S and C biogeochemical cycles have been treated above in Section 7.2. For example, methane fluxes in lakes may be affected by changes in the S cycle. Increasing SO_4^{2-} concentrations may cause increases in catabolism by bacterial SO_4^{2-} reduction and decreases in that by methanogenic bacteria (Figure 7.4; Kelly *et al.*, 1982; Kelly, Furutani and Schindler, 1984). In addition to the effect of SO_4^{2-} on the rate of methane formation, SO_4^{2-} may increase the rate of anoxic methane oxidation, potentially causing a reduction in the flux of methane from the sediments to the water column and to the atmosphere.

SO_4^{2-}-reducing microorganisms may be able to metabolize organic molecules (e.g. alkanes and low-oxygen content compounds) that are not suitable for mixed populations of methanogenic bacteria. Thus, increasing the relative importance of bacterial SO_4^{2-} reducers may result in more decomposition and recycling of organic matter and less burial of organic matter in lake sediments, although this has only been shown in peatland soils (Yavitt, Lang and Wieder, 1987).

Whereas addition of SO_4^{2-} to lakes probably only affects the pathway of decomposition (SO_4^{2-} reduction *vs* methanogenesis) rather than the total amount, additions of sulphuric acid may affect total decomposition if the pH becomes low enough. For example, acidification with sulphuric acid to pHs below 5 may inhibit decomposition (Kelly, Furutani and Schindler, 1984).

Another interaction of the S and C cycles in the production of HCO_3^- during bacterial SO_4^{2-} reduction. This production of HCO_3^- can affect the alkalinity of lakes and can counteract the effects of atmospheric deposition (Cook *et al.*, 1986; Schindler *et al.*, 1986; Howarth and Stewart, this volume). Increases in SO_4^{2-} deposition will cause increases in bacterial SO_4^{2-} reduction and increases in production of HCO_3^-, thereby providing a homeostatic mechanism for maintaining alkalinity concentrations and slowing the rate of acidification (Cook *et al.*, 1986). The efficiency with which this homeostatis maintains alkalinity depends on the amount of SO_4^{2-} retained by the lake, which is a function of SO_4^{2-} concentration and hydraulic residence time, and the ability of the sediments to sequester reduced S [S(II)] in an inorganic or organic S form. In the Lake 223 experiment, the sediments retained nearly all of the S produced by bacterial SO_4^{2-} reduction, much of it as iron sulphides (Figure 7.3(a); Cook *et al.*, 1986). These SO_4^{2-} losses counteracted the acidification by neutralizing between 66 and 81% of the added acid. The efficiency of S retention in the lake sediments also appears to be high in the Lake 302S experiment (Rudd *et al.*, 1990). The rate of supply of iron to the sediments may eventually limit the ability of sediments to sequester S and produce alkalinity (Schindler, 1985; Carignan and Tessier, 1988; Giblin *et al.*, 1990).

Alkalinity is also produced during ester sulphate formation, which does not involve reduction of SO_4^{2-}. Theoretical considerations show that ester sulphate formation may produce either one equivalent of alkalinity per SO_4^{2-} equivalent incorporated (Kelly and Rudd, 1989) or two equivalents when the esters contain base cations (Baker *et al.*, 1989; Urban and Baker, 1989). However, the chemical identities and characteristics of these esters are unknown, and further research is needed on the relationship between ester sulphate formation and alkalinity production.

Changes in the S cycle caused either by increased inputs of nutrients (eutrophication) or by increased atmospheric SO_4^{2-} deposition may result in an increased rate of sulphide production. Accumulation of produced H_2S, a toxic substance, in the water column may have severe effects on a variety of organisms. In some lakes, sulphide is used by chemosynthetic and photosynthetic anaerobic bacteria in the production of organic matter (Howarth and Stewart, this volume). Primary production by these organisms should increase with increased rates of bacterial SO_4^{2-} reduction and H_2S accumulation.

The formation of organic-S compounds from reduced S species, which has been demonstrated for marine sediments (Francois, 1987; Vairavamurthy and Mopper, 1987, 1989; Luther and Church, this volume), is another interaction of the carbon and sulphur cycles. Although the formation of organic-S compounds in lacustrine sediments has not received adequate study, the synthesis of organic-S compounds may be limited by the amount and types of organic compounds present, as well as by the amount of reduced S, particularly polysulphides (Vairavamurthy and Mopper, 1987, 1989). Thus the trophic state of a lake may affect not only the rate of SO_4^{2-} reduction, as discussed above, but also the distribution of S into the various end-products.

There is at present no evidence that primary productivity is ever limited by availability of SO_4^{2-}. In most lakes, we can assume that productivity is determined by the supply of other nutrients. However, S may play an indirect role in controlling the supply of nutrients by affecting the mobility of P. Under conditions of increased SO_4^{2-} concentrations, enhanced iron sulphide formation may remove iron from the iron–phosphorus cycle and make it unavailable to bind phosphate (Wetzel, 1975; Baccini, 1985). Increased fluxes of phosphorus to the water column may result in increases in primary productivity. However, in a set of enclosure experiments in a softwater lake, Curtis (1989) found that P release was not related to sulphide accumulation of Fe(III) accumulation. In these experiments, low H^+ concentrations alone or in combination with high sulphide concentrations were the factors controlling P release from sediments. In contrast, Caraco, Cole and Likens (1989) presented evidence for enhanced P release in lake sediments from New England with relatively low Fe content and relatively high S content. Caraco, Cole and Likens examined the relationship between P release and total sediment respiration, in a large set of lakes with a wide range of SO_4^{2-} concentrations. Relative P release was significantly correlated to SO_4^{2-} concentration under both oxic and anoxic conditions, suggesting that iron sulphide formation may be a controlling factor. However, sediment Fe alone did not explain the variation in relative P release. After SO_4^{2-}, conductivity was the next best predictor of relative P release. The importance of conductivity suggests that the observation could also be related to the findings of Mayer and Kramer (1986), who showed that softwater lake sediments adsorb P more efficiently than sediments from hardwater lakes. As Caraco, Cole and Likens (1989) state, the mechanism behind the empirical relationship between P release and SO_4^{2-} has yet to be sorted out.

Increasing SO_4^{2-} concentrations may affect the rates of nitrogen fixation, an important source of nitrogen for many lakes. Due to a similarity in stereochemistry between SO_4^{2-} and molybdate, SO_4^{2-} blocks the uptake of molybdenum, a necessary micronutrient for nitrogen-fixing organisms (Howarth, Marino and Cole, 1988; Howarth and Stewart, this volume). Inhibition of nitrogen fixation in freshwaters may alter the supply of nitrogen and cause a shift in nutrient limitation from phosphorus to nitrogen if the sulphate : molybdenum ratio is sufficiently high. Such an effect has been demonstrated in sulphate-rich salt lakes (Marino *et al.*, 1990).

Another effect of the N cycle occurs when SO_4^{2-} enters freshwater as sulphuric acid and the pH of lake is lowered below about 5.4. When this occurs, nitrification is inhibited, causing accumulation of NH_4^+ (Rudd *et al.*, 1988). In addition, because the N cycle is blocked and does not proceed from NH_4^+ to NO_3^-, denitrification decreases, leading to a build-up of fixed nitrogen.

7.5 SUMMARY

The key features of the S cycle in dimictic freshwater lakes are assimilatory reduction and uptake of SO_4^{2-} in the water column by algae and bacteria; dissimilatory SO_4^{2-} reduction by bacteria in anoxic zones; reaction of reduced S species with organic matter and reduced metals (e.g. Fe^{2+}); and production of S gases (COS, DMS, DMDS, and MSH), probably accompanying decomposition of S-bearing organic matter.

For many lakes, water column concentrations of SO_4^{2-} are great enough to provide all of the S required by algae and bacteria. Uptake of SO_4^{2-} by algae appears to be independent of SO_4^{2-} concentrations from 0.03 to 43 mM (1. to 1400 mg (S) l^{-1}). Seston S undergoes decomposition in the sediments, providing a supply of S and volatile S compounds to the overlying water. The production of S gases in lakes is independent of SO_4^{2-} concentrations and pH. DMS, COS, DMDS, and MSH are sparingly soluble and chemically reactive in water and thus are present in lake water in low (< 10 nM; < 320 ng (S) l^{-1}) concentrations. Water–air gas exchange of S gases may contribute significantly to atmospheric S concentrations, particularly in areas remote from industrial sectors and marine S sources.

Rates of bacterial SO_4^{2-} reduction depend on SO_4^{2-} concentrations and organic matter supply. Experimental additions of SO_4^{2-} to mesocosms and whole lakes stimulated the rate of SO_4^{2-} reduction, indicating that other factors (e.g. organic matter) were not limiting bacterial activity for these oligotrophic systems. However, the trophic status of lakes may also influence the extent of SO_4^{2-} reduction, with ultraoligotrophic systems not producing enough organic matter to support bacterial SO_4^{2-} reduction. Experimental eutrophication with P and N stimulated bacterial SO_4^{2-} reduction in Lake 227, probably because of the increased supply of labile organic matter.

The capacity of sediments to form metal sulphide compounds depends on the supply of reducible metals, especially Fe. Lakes studied in north central and eastern North America have sufficient supply of reducible Fe from sedimentation, so that iron is not limiting the formation of Fe-sulphides. For lakes studied in Quebec, rates of Fe accumulation in sediments are comparable in magnitude to rates of S accumulation, which suggests that Fe may be limiting the formation of Fe-sulphide compounds. Hydrogen sulphide may also react to form organic-S compounds. The factors controlling the relative amounts of metal sulphides and organic S formed H_2S are not known.

Sulphur budgets for five lakes show that rates of organic-S formation were similar, but retention of S produced from bacterial SO_4^{2-} reduction was highly variable, with 4 to 60% of the input S retained by this mechanism. Differences in the percentage of input S retained among these five lakes are primarily related to differences among lakes in water residence time and bathymetry. A simple steady-state model, based on inputs, outputs, and first-order SO_4^{2-} reduction, successfully explains differences in the retention of input SO_4^{2-}.

REFERENCES

Amaral, J. A., Hesslein, R. H., Rudd, J. W. M. and Fox, D. E. (1989). Loss of total sulfur and changes in sulfur isotopic ratios due to drying of lacustrine sediments. *Limnol. Oceanogr.*, **34**, 1351–8.

Anderson, R. F. and Schiff, S. L. (1987). Alkalinity generation and the fate of sulfur in lake sediments. *Can. J. Fish. Aquat. Sci.*, **44** (Suppl. 1), 188–93.

Anderson, R. F., Schiff, S. L. and Hesslein, R. H. (1987). Determining sediment accumulation and mixing rates using ^{210}Pb, ^{137}Cs, and other tracers: problems due to postdepositional mobility or coring artifacts. *Can. J. Fish. Aquat. Sci.*, **44**, (Suppl. 1), 231–50.

Andersson, B. I. (1985). Chemical composition of water across the sediment–water interface in the acidified Lake Gårdsjön, SW Sweden. *Ecol. Bull. (Stockholm)*, **37**, 263–75.

Andreae, M. O. (1985). Dimethylsulfide in the water column and the sediment porewaters of the Peru upwelling area. *Limnol. Oceanogr.*, **30**, 1208–18.

Andreae, M. O. and Andreae, T. W. (1988). The cycle of biogenic sulfur compounds over the Amazon Basin. 1. Dry season. *J. Geophys. Res.*, **93**, 1487–97.

Andreae, M. O. and Barnard, W. R. (1984). The marine chemistry of DMS. *Mar. Chem.*, **14**, 267–79.

Andreae, M. O. and Jaeschke, W. A. This volume.
Andreae, M. O. and Raemdonck, H. (1983). Dimethyl sulfide in the surface ocean and the marine atmosphere: a global view. *Science*, **221**, 744–7.
Avrahami, M. and Golding, R. M. (1968). The oxidation of sulfide ion at very low concentrations in aqueous solutions. *J. Am. Chem. Soc.*, 1968, 647–51.
Baccini, P. (1985). Phosphate interactions at the sediment–water interface. In: Stumm, W. (Ed.). *Chemical Processes in Lakes*. Wiley Interscience, New York, pp. 189–205.
Baker, L. A. and Brezonik, P. L. (1988). Dynamic model of in-lake alkalinity generation. *Water Resources Res.*, **24**, 65–74.
Baker, L. A., Brezonik, P. L. and Edgerton, E. S. (1986). Sources and sinks of ions in a softwater acidic lake in Florida. *Water Resources Res.*, **22**, 715–22.
Baker, L. A., Brezonik, P. L., and Pollman, C. D. (1986). Model of internal alkalinity generation: sulfate retention component. *Water Air Soil Pollut.*, **31**, 89–94.
Baker, L. A., Urban, N. R., Brezonik, P. L. and Sherman, L. A. (1989). Sulfur cycling in a seepage lake. In: Saltzman, E. and Cooper, W. (Eds). *Biogenic Sulfur in the Environment*. American Chemical Society, Washington DC, pp. 79–100.
Bechard, M. J. and Rayburn, W. R. (1979). Volatile organic sulfides from freshwater algae. *J. Phycol.*, **15**, 379–83.
Berner, R. A. (1984). Sedimentary pyrite formation: an update. *Geochim. Cosmochim. Acta*, **48**, 605–15.
Boudreau, B. P. and Westrich, J. T. (1984). The dependence of bacterial sulfate reduction on sulfate concentration in marine sediments. *Geochim. Cosmochim. Acta*, **48**, 2503–16.
Brakke, D. F., Landers, D. H. and Eilers, J. M. (1988). Chemical and physical characteristics of lakes in the northeastern United States. *Environ. Sci. Technol.*, **22**, 155–63.
Bremner, J. M. and Steele, C. G. (1978). Role of microorganisms in the atmospheric sulfur cycle. In: Alexander, M. (Ed.). *Adv. Microb. Ecol.*, Vol. 2, pp. 155–201.
Canfield, D. E., Raiswell, R., Westrich, J. T., Reaves, D. M. and Berner, R. A. (1986). The use of chromium reduction in the analysis of reduced inorganic sulfur in sediments and shales. *Chem. Geol.*, **54**, 149–55.
Capone, D. G. and Kiene, R. P. (1988). Comparison of microbial dynamics in marine and freshwater sediments: Contrasts in anaerobic carbon catabolism. *Limnol. Oceanogr.*, **4**, 725–49.
Caraco, N. F., Cole, J. J. and Likens, G. E. (1989). Evidence for sulphate-controlled phosphorus release from sediments of aquatic systems. *Nature*, **341**, 316–18.
Carignan, R. (1985). Quantitative importance of alkalinity flux from the sediments of acidic lakes. *Nature*, **317**, 158–60.
Carignan, R. (1988). Seasonal dynamics of sulfate and hydrogen sulfide near the sediment–water interface of an oligotrophic acid lake. *Verh. Int. Ver. Theor. Angew. Limnol.*, **23**, 106–15.
Carignan, R. and Tessier, A. (1988). The co-diagenesis of sulfur and iron in acid lake sediments of southwestern Quebec. *Geochim. Cosmochim. Acta*, **52**, 1179–88.
Casagrande, D. J. and Ng, L. (1979). Incorporation of elemental sulphur in coal as organic sulphur. *Nature*, **282**, 598–9.
Casagrande, D. J., Idowu, G., Friedman, A., Rickert, P., Siefert, K. and Schlenz, D. (1979). H_2S incorporation in coal precursors: origins of organic sulphur in coal. *Nature*, **282**, 599–600.
Chen, K. Y. and Morris, J. C. (1972). Kinetics of oxidation of aqueous sulfide by O_2. *Environ. Sci. Technol.*, **6**, 529–37.

Cook, R. B. (1981). The biogeochemistry of sulfur in two small lakes. *Ph.D. Thesis*, Columbia University, New York, 248pp.
Cook, R. B. (1984). Distributions of ferrous iron and sulfide in an anoxic hypolimnion. *Can. J. Fish. Aquat. Sci.*, **41**, 286–93.
Cook, R. B. and Schindler, D. W. (1983). The biogeochemistry of sulfur in an experimentally acidified lake. *Ecol. Bull. (Stockholm)*, **35**, 115–27.
Cook, R. B., Kelley, C. A., Schindler, D. W. and Turner, M. A. (1986). Mechanisms of hydrogen ion neutralization in an experimentally acidified lake. *Limnol. Oceanogr.*, **31**, 134–48.
Cook, R. B., Kelley, C. A., Kingston, J. C. and Kreis, R. G. (1987). Chemical limnology of soft water lakes in the Upper Midwest. *Biogeochemistry*, **4**, 97–117.
Cook, R. B., Kreis Jr, R. G., Kingston, J. C., Camburn, K. E., Norton, S. A., Mitchell, M. J., Fry, B. and Shane, L. C. K. (1990). Paleolimnology of McNearney Lake: an acidic lake in northern Michigan. *Paleolimnology*, **3**, 13–34.
Curtis, P. J. (1989). Effects of hydrogen ion and sulphate on the phosphorus cycle of a Precambrian Shield lake. *Nature*, **337**, 156–8.
Datko, A. H., Mudd, S. H., Giovanelli, J. and Macnicol, P. K. (1978). Sulfur-containing compound in *Lemna perpusilla* 6746 grown at a range of sulfate concentrations. *Plant Physiol.*, **62**, 629–35.
David, M. B. and Mitchell, M. J. (1985). Sulfur constituents and cycling in water, seston, and sediments of an oligotrophic lake. *Limnol. Oceanogr.*, **30**, 1196–207.
Davison, W. (1980). A critical comparison of the measured solubilities of ferrous sulphide in natural waters. *Geochim. Cosmochim. Acta*, **44**, 803–8.
Deprez, P. P., Franzmann, P. D. and Burton, H. R. (1986). Determination of reduced sulfur gases in Antarctic lakes and seawater by gas chromatography after solid adsorbent preconcentration. *J. Chromatogr.*, **362**, 9–21.
Dillon, P. J. (1974). A critical review of Vollenweider's nutrient budget model and other related models. *Water Resources Bull.*, **10**, 969–89.
Dunnette, D. A. (1973). Chemical ecology of hydrogen sulfide production in freshwater lake sediment. *Ph.D. Dissertation*, University of Michigan, 123pp.
Dunnette, D. A. (1989). Origin of hydrogen sulfide in freshwater sediments. In: Saltzman, E. and Cooper, W. (Eds). *Biogenic Sulfur in the Environment*. American Chemical Society, Washington DC, pp. 72–8.
Eilers, J. M., Brakke, D. F., Landers, D. H. and Overton, W. S. (1989). Chemistry of lakes in designated wilderness areas in the western United States. *Environ. Monit. Assess.*, **12**, 3–21.
Fee, E. J. (1979). A relation between lake morphometry and primary productivity and its use in interpreting whole-lake eutrophication experiments. *Limnol. Oceanogr.*, **24**, 401–16.
Francois, R. (1987). A study of sulfur enrichment in the humic acid fraction of marine sediments during early diagenesis. *Geochim. Cosmochim. Acta*, **51**, 17–27.
Fry, B. (1986). Stable sulfur isotopic distributions and sulfate reduction in lake sediments of the Adirondack Mountains, New York. *Biogeochemistry*, **2**, 329–43.
Fry, B. (1990). Sulfur stable isotope studies of lake cores from the Paleoecological Investigation of Recent Lake Acidification project. *J. Paleolimnol.*, **4**, 23–29.
Galloway, J. N., Schofield, C. L., Peters, N. E., Hendrey, G. R. and Altwicker, E. R. (1983). Effect of atmospheric sulfur on the composition of three Adirondack lakes. *Can. J. Fish. Aquat. Sci.*, **10**, 799–806.
Giblin, A., Peterson, B. J. and Fry, B. (1989). Controls on sulfur storage in lake sediments: the importance of iron and organic matter. Abstracts of the Annual Meeting, American Society of Limnol. and Oceanogr., Fairbanks, Alaska, June 1989.

Giblin, A. E. and Wieder, R. K. This volume.
Giblin, A. E., Likens, G. E., White, D. and Howarth, R. W. (1990). Sulfur storage and alkalinity generation in New England lake sediments. *Limnol. Oceanogr.*, **35**, 852–69.
Goldan, P. D., Kuster, W. C., Albritton, D. L. and Fehsenfield, F. C. (1987). The measurement of natural sulfur emissions from soils and vegetation: three sites in the eastern United States revisited. *J. Atmos. Chem.*, **5**, 439–67.
Gorham, E., Lund, J. W. G., Sanger, J. E. and Dean, W. E. (1974). Some relationships between algal standing crop, water chemistry, and sediment chemistry in the English Lakes. *Limnol. Oceanogr,*, **19**, 601–17.
Gorlenko, V. M. (1978). Phototrophic sulfur bacteria in salt meromictic lakes and their role in the sulfur cycle. In: Krumbein, W. E. (Ed.). *Environmental Biogeochemistry and Geomicrobiology*. Vol. 1: *The Aquatic Environment*. Ann Arbor Science, Ann Arbor, MI, pp. 109–19.
Herlihy, A. T., Mills, A. L., Hornberger, G. M. and Bruckner, A. E. (1987). The importance of sediment sulfate reduction to the sulfate budget of an impoundment receiving acid mine drainage. *Water Resources Res.*, **23**, 287–92.
Hines, M. E. and Morrison, M. C. (1989). Emissions of biogenic sulfur compounds from Alaskan tundra. *Eos Trans. Am. Geophys. Union*, **70**, 284.
Holdren, G. R., Brunelle, T. M., Matisoff, G. and Whalen, M. (1984). Timing the increase in atmospheric sulphur deposition in the Adirondack Mountains. *Nature*, **311**, 245–8.
Holdway, D. A. and Nriagu, J. O. (1988). A purge and trap gas chromatographic method for dimethyl sulfide in freshwater. *Intern. J. Environ. Anal. Chem.*, **32**, 177–86.
Howarth, R. W. (1979). Pyrite: Its rapid formation in a salt marsh and its importance in ecosystem metabolism. *Science*, **203**, 49–51.
Howarth, R. W., Marino, R. and Cole, J. J. (1988). Nitrogen fixation in freshwater, estuarine, and marine ecosystems. 2. Biogeochemical controls. *Limnol. Oceanogr.* **33**, 688–701.
Howarth, R. W. and Merkel, S. (1984). Pyrite formation and the measurement of sulfate reduction in salt marsh sediments. *Limnol. Oceanogr.*, **9**, 598–608.
Howarth, R. W. and Stewart, J. W. B. This volume.
Howarth, R. W. and Teal, J. M. (1979). Sulfate reduction in a New England salt marsh. *Limnol. Oceanogr.*, **28**, 70–82.
Ingvorsen, K. and Brock, T. D. (1982). Electron flow via sulfate reduction and methanogenesis in the anaerobic hypolimnion of Lake Mendota. *Limnol. Oceanogr.*, **27**, 559–64.
Ingvorsen, K. and Jørgensen, B. B. (1984). Kinetics of sulfate uptake by freshwater and marine species of *Desulfovibrio. Arch. Microbiol.*, **139**, 61–6.
Ivanov, M. V. (1968). *Microbiological Processes in the Formation of Sulfur Deposits*. Israel Programme for Scientific Translations, Ltd. 298pp.
Iversen, R. L., Nearhoof, S. L. and Andreae, M. O. (1989). Production of dimethylsulfonium propionate and dimethylsulfide in estuarine and coastal waters. *Limnol. Oceanogr.*, **34**, 53–67.
Jensen, H. S. and Andersen, F. O. (1987). Sulfate reduction in a coastal lake during summer stratification. *Arch. Hydrobiol.*, **108**, 465–81.
Johnson, C. M. and Nishita, H. (1952). Microestimation of sulfur in plant materials, soils, and irrigation waters. *Anal. Chem.*, **24**, 736–42.
Johnson, C. M. and Ulrich, A. (1959). Analytical methods for use in plant analysis. *Calif. Agric. Exp. Stn. Bull.*, **766**, 55.

Jones, J. G., Simon, B. M. and Roscoe, J. V. (1982). Microbiological sources of sulphide in freshwater lake sediments. *J. Gen. Microbiol.*, **128**, 2833–9.

Kanciruk, P., Eilers, J. M., McCord, R. A. Landers, D. H., Brakke, D. F. and Linthurst, R. A. (1986). *Characteristics of Lakes in the Eastern United States*. Vol. III: *Data Compendium of Site Characteristics and Chemical Variables*. EPA/600/4-86/007c. US Environmental Protection Agency, Washington DC.

Kelly, C. A., Furutani, A. and Schindler, D. W. (1984). Effects of lake acidification on rates of organic matter decomposition in sediments. *Limnol. Oceanogr.*, **29**, 687–94.

Kelly, C. A. and Rudd, J. W. M. (1984). Epilimnetic sulfate reduction and its relationship to lake acidification. *Biogeochemistry*, **1**, 63–77.

Kelly, C. A. and Rudd, J. W. M. (1989). Reply to comment by Urban and Baker, *Limnol. Oceanogr.*, **34**, 1146–8.

Kelly, C. A., Rudd, J. W. M. and Schindler, D. W. (1988). Carbon and electron flow via methanogenesis, SO_4^{2-}, NO^{3-}, Fe^{3+}, and Mn^{4+} reduction in the anoxic hypolimnia of three lakes. *Arch. Hydrobiol. Beih. Ergebn. Limnol.*, **31**, 333–44.

Kelly, C. A., Rudd, J. W. M., Cook, R. B. and Schindler, D. W. (1982). The potential importance of bacterial processes in regulating the rate of lake acidification. *Limnol. Oceanogr.*, **27**, 868–82.

Kelly, C. A., Rudd, J. W. M., Hesslein, R. H., Schindler, D. W., Dillon, P. J., Driscoll, C. T., Gherini, S. A. and Hecky, R. E. (1987). Prediction of biological acid neutralization in acid-sensitive lakes. *Biogeochemistry*, **3**, 129–40.

Kiene, R. P. and Visscher, P. T. (1987). Production and fate of methylated sulfur compounds from methionine and dimethylsulfonium propionate in anoxic salt marsh sediments. *Appl. Environ. Microbiol.*, **53**, 2426–34.

Kiene, R. P., Oremland, R. S., Catena, A., Miller, L. G. and Capone, D. G. (1986). Metabolism of reduced methylated sulfur compounds in anaerobic sediments and by a pure culture of an estuarine methanogen. *Appl. Environ. Microbiol.*, **52**, 1037–45.

King, G. M. and Klug, M. J. (1980). Sulfhydrolase activity in sediments of Wintergreen Lake, Kalamazoo County, Michigan. *Appl. Environ. Microbiol.*, **39**, 950–6.

King, G. M. and Klug, M. J. (1982). Comparative aspects of sulfur mineralization in sediments of a eutrophic lake basin. *Appl. Environ. Microbiol.*, **43**, 1406–12.

Kipphut, G. W. (1978). An investigation of sedimentary process in lakes. *Ph.D. Thesis*, Columbia University, New York, 180pp.

Kling, G. W., Giblin, A. E., Fry, B. and Peterson, B. J. (1991). The role of seasonal turnover in lake alkalinity dynamics. *Limnol. Oceanogr.*, **36**, 106–22.

Knauss, H. J. and Porter, J. W. (1954). The absorption of inorganic ions by *Chlorella pyrenoidosa*. *Plant Physiol.*, **29**, 229–34.

Kuenen, J. G. and Tuovinen, O. H. (1981). The genera *Thiobacillus* and *Thiomicrospira*. In: *The Prokaryotes*. Springer-Verlag, Berlin, pp. 1023–36.

Kuivila, K. M., Murray, J. W., Devol, A. H. and Novelli, P. C. (1989). Methane production, sulfate reduction, and competition for substrates in the sediments of Lake Washington. *Geochim. Cosmochim. Acta*, **53**, 409–16.

Lamb, B., Westberg, H., Allivine, G., Bamesberger, L. and Guenther, A. (1987). Measurement of biogenic sulfur emissions from soils and vegetation: application of dynamic enclosure methods with Natusch filter and GC/FPD analysis. *J. Atmos. Chem.*, **5**, 469–91.

Landers, D. H. and Mitchell, M. J. (1988). Incorporation of $^{35}SO_4^{2-}$ into sediments of three New York lakes. *Hydrobiol.*, **160**, 85–95.

Landers, D. H., Overton, W. S., Linthurst, R. A. *et al.* (1988). Eastern Lake Survey. *Environ. Sci. Technol.*, **22**, 128–35.

Landers, D. H., Ford, J., Eilers, J., Mitch, M., Morrison, M., Newell, A. and Hurley, K. (1989). Regional evaluation of lakes on the Kenai Peninsula of Alaska. Abstracts of the Annual Meeting, American Society of Limnology and Oceanography. Fairbanks, Alaska, June 1989.

La Riviere, J. W. M. and Schmidt, K. (1981). Morphologically conspicuous sulfur-oxidizing eubacteria. In: *The Prokaryotes*. Springer-Verlag, Berlin.

Linsey, G. A., Schindler, D. W. and Stainton, M. P. (1987). Atmospheric desposition of nutrients and major ions at the Experimental Lakes Area in northwestern Ontario, 1970–1982. *Can. J. Fish. Aquat. Sci.*, **44**, (Suppl. 1), 206–14.

Lovley, D. R. (1987). Organic matter mineralization with the reduction of ferric iron: a review. *Geomicrobiol. J.*, **5**, 375–99.

Lovley, D. R. and Klug, M. J. (1986). Model for the distribution of sulfate reduction and methanogenesis in freshwater sediments. *Geochim. Cosmochim. Acta*, **50**, 11–18.

Lovley, D. R. and Phillips, E. J. P. (1986). Availability of ferric iron for microbial reduction in bottom sediments of the freshwater tidal Potomac River. *Appl. Environ. Microbiol.*, **52**, 751–7.

Luther, G. W. and Church, T. M. This volume.

Marino, R., Howarth, R. W., Shamess, J. and Prepas, E. (1990). Molybdenum and sulfate as controls on the abundance of nitrogen-fixing cyanobacteria in saline lakes in Alberta. *Limnol. Oceanogr.*, **35**, 245–59.

Martens, C. S. and Jannasch, H. W. (1983). Cycling of metabolizable C, N, P, and S in organic-rich marine sediments. In: Bolin, B. and Cook, R. B. (Eds). *The Major Biogeochemical Cycles and their Interactions*. Wiley, Chichestesr, pp. 509–15.

Mayer, T. and Kramer, J. R. (1986). Effect of lake acidification on the adsorption of phosphorus by sediments. *Water Air Soil Pollut.*, **31**, 949–58.

Millero, F. J. and Hershey, J. P. (1989). Thermodynamics and kinetics of hydrogen sulfide in natural waters. In: Saltzman, E. and Cooper, W. (Eds). *Biogenic Sulfur in the Environment*. American Chemical Society, Washington DC, pp. 282–313.

Millero, F. J., Hubinger, S., Fernandez, M. and Garnett, E. S. (1987). Oxidation of H_2S in seawater as a function of temperature, pH, and ionic strength. *Environ. Sci. Technol.*, **21**, 439–43.

Mills, A. L., Bell, P. E. and Herlihy, A. T. (1989). Microbes, sediments, and acidified water: the importance of biological buffering. In: Rao, S. S. (Ed.). *Acid Stress and Aquatic Microbial Interactions*. CRC Press, Boca Raton, Florida, pp. 1–19.

Mitchell, M. J., David, M. B. and Harrison, R. B. Chapter 9 in this volume.

Mitchell, M. J., David, M. B. and Uutala, A. J. (1985). Sulfur distribution in lake sediment profiles as an index of historical depositional patterns. *Hydrobiol.*, **121**, 121–7.

Mitchell, M. J., Owen, J. S. and Schindler, S. C. (1990). Factors affecting sulfur incorporation into lake sediments: paleoecological implications. *J. Paleolimnol.*, **4**, 1–22

Mitchell, M. J., Schindler, S. C., Owen, J. S. and Norton, S. A. (1988). Comparison of sulfur concentrations within lake sediment profiles. *Hydrobiol.*, **157**, 219–29.

Morse, J. W. and Cornwell, J. C. (1987). Analysis and distribution of iron sulfide minerals in recent anoxic marine sediments. *Mar. Chem.*, **22**, 55–69.

Norton, S. A., Mitchell, M. J., Kahl, J. S. and Brewer, G. R. (1988). In-lake alkalinity generation by sulfate reduction: a paleolimnological assessment. *Water Air Soil Pollut.*, **39**, 33–45.

Nriagu, J. O., Holdway, D. A. and Coker, R. D. (1987). Biogenic sulfur and the acidity of rainfall in remote areas of Canada. *Science*, **237**, 1189–92.

Nriagu, J. O. and Soon, Y. K. (1985). Distribution and isotopic composition of sulfur in lake sediments of northern Ontario. *Geochim. Cosmochim. Acta*, **49**, 823–34.

Nurnberg, G. K. (1984). Iron and hydrogen sulfide interference in the analysis of soluble reactive phosphorus in anoxic waters. *Water Res.*, **18**, 369–77.

Oremland, R. S., Kiene, R. P., Mathrani, I., Whiticar, M. J. and Boone, D. R. (1989). Description of an estuarine methylotrophic methanogen which grows on dimethyl sulfide. *Appl. Environ. Microbiol.*, **55**, 994–1002.

Owen, J. S. and Mitchell, M. J. (1989). Sulfur biogeochemistry of an acidic lake in the Adirondack Region of New York. In: Rao, S. (Ed.). *Acid Stress and Aquatic Microbial Interactions*. CRC, Boca Raton, Florida, pp. 59–68.

Pankow, J. F. and Morgan, J. J. (1980). Dissolution of tetragonal ferrous sulfide (mackinawite) in anoxic aqueous systems. 2. Implications for the cycling of iron, sulfur, and trace metals. *Environ. Sci. Technol.*, **14**, 183–6.

Parkin, T. B. and Brock, T. D. (1981). The role of phototrophic bacteria in the sulfur cycle of a meromictic lake. *Limnol. Oceanogr.*, **26**, 880–90.

Perry, T. E., Baker, L. A. and Brezonik, P. L. (1985). Comparison of sulfate reduction rates in laboratory microcosms, field mesocosms, and in situ at Little Rock Lake, Wisconsin. *Lake Reservoir Manage.*, **2**, 309–12.

Pfennig, N. and Biebl, H. (1981). The dissimilatory sulfate-reducing bacteria. In: *The Prokaryotes*. Springer-Verlag, Berlin, pp. 941–7.

Pfenning, N., Widdel, F. and Truper, H. G. (1981). The dissimilatory sulfur-reducing bacteria. In: *The Prokaryotes*. Springer-Verlag, Berlin, pp. 726–940.

Ramm, A. E. and Bella, D. A. (1974). Sulfide production in anaerobic microcosms. *Limnol. Oceanogr.*, **19**, 110–18.

Richards, S. R., Kelly, C. A. and Rudd, J. W. M. (1991). Organic volatile sulfur in lakes of the Canadian Shield and its loss to the atmosphere. *Limnol. Oceanogr.* **36**, 468–482.

Richards, S. R., Rudd, J. W. M. and Kelly, C. A. (in review). Organic volatile sulfur in systems ranging in sulfate and dissolved salt concentrations over five orders of magnitude. *Limnol. Oceanogr.,* in press.

Roy, A. B. and Trudinger, P. A. (1970). *The Biochemistry of Inorganic Compounds of Sulfur*. Cambridge University Press, 400pp.

Rudd, J. W. M., Kelly, C. A. and Furutani, A. (1986). The role of sulfate reduction in the long term accumulation of organic and inorganic sulfur in lake sediments. *Limnol. Oceanogr.*, **31**, 1281–91.

Rudd, J. W. M., Kelly, C. A., St Louis, V., Hesslein, R. H. and Holoka, M. H. (1986). Microbial consumption of nitric and sulfuric acids in acidified north temperate lakes. *Limnol. Oceanogr.*, **31**, 1267–80.

Rudd, J. W. M., Kelly, C. A., Schindler, D. W. and Turner, M. A. (1988). Disruption of the nitrogen cycle in acidified lakes. *Science*, **240**, 1515–17.

Rudd, J. W. M., Kelly, C. A., Schindler, D. W. and Turner, M. A. (1990). A comparison of the acidification efficiencies of nitric and sulfuric acids by two whole-lake addition experiments. *Limnol. Oceanogr.* **35**, 663–679.

Schiff, S. and Anderson, R. F. (1987). Limnocorral studies of chemical and biological acid neutralization in two freshwater lakes. *Can. J. Fish. Aquat. Sci.*, **44**, (Suppl. 1), 173–87.

Schiff, J. A. and Hodson, R. C. (1973). The metabolism of sulfate. *Annu. Rev. Plant Physiol.*, **24**, 381–414.

Schindler, D. W. (1985). The coupling of elemental cycles by organisms: evidence from whole-lake chemical perturbations. In: Stumm, W. (Ed.). *Chemical Processes in Lakes*. Wiley-Interscience, New York, pp. 255–50.

Schindler, D. W., Turner, M. A., Stainton, M. P. and Linsay, G. A. (1986). Natural sources of acid neutralizing capacity in low-alkalinity lakes of the Precambrian Shield. *Science*, **232**, 844–7.

Shaffer, P. W., Hooper, R. P., Eshleman, R. N. and Church, M. R. (1988). Watershed *vs* in-lake alkalinity generation: a comparison of rates using input–output studies. *Water Air Soil Pollut.*, **39**, 263–73.

Smith, R. L. and Klug, M. J. (1982). Reduction of sulfur compounds in the sediments of a eutrophic lake basin. *Appl. Environ. Microbiol.*, **41**, 1230–7.

Sorokin, Y. I. (1970). Interrelations between sulphur and carbon turnover in meromictic lakes. *Arch. Hydrobiol.*, **66**, 391–446.

Sorokin, Y. I. (1975). Sulphide formation and chemical composition of bottom sediments of some Italian lakes. *Hydrobiol.*, **47**, 231–40.

Steudler, P. A. and Peterson, B. J. (1984). Contribution of gaseous sulphur from salt marshes to the global sulphur cycle. *Nature*, **311**, 455–7.

Stuiver, M. (1967). The sulfur cycle in lake waters during thermal stratification. *Geochim. Cosmochim. Acta*, **31**, 2151–67.

Sullivan, T. J., Eilers, J. M., Church, M. R., Blick, D. J., Eshleman, K. N., Landers, D. H. and DeHaan, M. S. (1988). Atmospheric wet sulphate deposition and lakewater chemistry. *Nature*, **331**, 607–9.

Sweerts, J. P., Rudd, J. W. M. and Kelly, C. A. (1986). Metabolic activities in flocculent surface sediments and underlying sandy littoral sediments. *Limnol. Oceanogr.*, **31**, 330–8.

Tabatabai, M. A. and Bremner, J. M. (1970). An alkaline oxidation method for the determination of total sulfur in soils. *Soil Sci. Soc. Am. Proc.*, **34**, 62–5.

Turner, S. M., Malin, G., Liss, P. S., Harbour, D. S. and Holligan, P. M. (1988). The seasonal variation of dimethyl sulfide and dimethylsulfonium propionate concentrations in nearshore waters. *Limnol. Oceanogr.*, **33**, 364–75.

Urban, N. R. and Baker, L. A. (1989). Alkalinity generation associated with the formation of ester sulfates. *Limnol. Oceanogr.*, **34**, 1144–6.

Utainier, R. Y., Pfenning, N. and Truper, H. G. (1981). Introduction to the phototrophic prokaryotes. In: *The Prokaryotes*. Springer-Verlag, Berlin, pp. 197–211.

Vairavamurthy, A. and Mopper, K. (1987). Geochemical formation of organosulphur compounds (thiols) by addition of H_2S to sedimentary organic matter. *Nature*, **329**, 623–5.

Vairavamurthy, A. and Mopper, K. (1989). Mechanistic studies of organosulfur (thiol) formation in coastal marine sediments. In: Saltzman, E. and Cooper, W. (Eds). *Biogenic Sulfur in the Environment*. American Chemical Society, Washington DC, pp. 231–42.

Weissner, W. (1981). The Family Beggiatoaceae. In: *The Prokaryotes*. Springer-Verlag, Berlin, pp. 380–9.

Wetzel, R. G. (1975). *Limnology*. Saunders Co., Philadelphia, 743pp.

White, J. R., Gubala, C. P., Fry, B., Owen, J. and Mitchell, M. J. (1989). Sediment biogeochemistry of iron and sulfur in an acidic lake. *Geochim. Cosmochim. Acta*, **53**, 2547–59.

Wright, R. F. (1983). Input–output budgets at Langtjern, a small acidified lake in southern Norway. *Hydrobiol.*, **101**, 1–12.

Yavitt, J. B., Lang, G. E. and Wieder, R. K. (1987). Control of carbon mineralization to CH_4 and CO_2 in anaerobic, *Sphagnum*-derived peat from Big Run Bog, West Virginia. *Biogeochemistry*, **4**, 141–57.

Zeyer, J., Eicher, P., Wakeham, S. G. and Schwarzenbach, R. P. (1987). Oxidation of dimethyl sulfide to dimethyl sulfoxide by phototrophic purple bacteria. *Appl. Environ. Microbiol.*, **53**, 2026–32.

Zinder, S. H. and Brock, T. D. (1978). Methane, carbon dioxide, and hydrogen sulfide production from the terminal methiol group of methionine by anaerobic lake sediments. *Appl. Environ. Microbiol.*, **35**, 344–52.

Zinder, S. H., Deomel, W. N. and Brock, T. D. (1977). Production of volatile sulfur compounds during the decomposition of algal mats. *Appl. Environ. Microbiol.*, **34**, 859–60.

8 Interaction of Carbon, Sulphur, and Oxygen Cycles in Continental and Marginal Seas

A. YU. LEIN

and

M. V. IVANOV
USSR Academy of Sciences, Moscow, USSR

8.1 INTRODUCTION

Most of the requirements of contemporary society are satisfied at the expense of the exploitation of terrestrial ecosystems and minerals mined from the continental part of our planet. The exploitation of biological and mineral resources of the oceans is considerably less. Consequently, the potential impact of anthropogenic pollution in the oceans often draws less attention than on land. However, the oceans are more and more being turned into a pit for garbage and the cesspool of our planet. Industrial, agricultural, and household wastes are being released in the oceans in increasing amounts. The near-continental regions of the ocean such as coastal or marginal seas, and especially the continental seas—such as the Persian Gulf, Black, Baltic, Mediterranean, and Caspian Seas—are the greatest sufferers from anthropogenic impact.

The recent literature contains alarming data indicating the drastic changes in hydrochemical and biological conditions in continental and marginal seas. Hydrochemists more and more often observe uncommonly high concentrations of nitrogen, phosphorus, and heavy metals in surface horizons of littoral sediments and deficiency of dissolved oxygen in bottom waters. These oxygen deficiencies cause the death of benthic animals over enormous shelf areas. A worrisome change, at least in the deeper waters of some continental seas, has been the rise of the border of oxygen-free waters. In deep-water depressions of the Baltic Sea, H_2S has become a constant component of the bottom water.

Sulphur Cycling on the Continents
Edited by R.W. Howarth, J.W.B. Stewart and M.V. Ivanov

Biologists have observed considerable changes in species composition of fauna and flora of these seas: the number of species has decreased, commercially valuable species of fish and shellfish have been displaced by medusae and other gelatinous animals. Blue–green algae have become dominant in some seas such as the Baltic, and 'red tide' blooms have become more common in many seas, including the Black Sea coast of Bulgaria and the eastern coast of the Adriatic Sea.

The above circumstances induced us to study in detail biogeochemical processes of the sulphur, carbon and oxygen cycles in the continental and marginal seas of the largest continent—Euroasia. A considerable part of the data used in this review is already published, e.g. the materials on the Baltic Sea (Lein *et al.*, 1981; Lein and Ivanov, 1983; Lein, 1983), Black Sea (Ivanov *et al.*, 1984; Lein *et al.*, 1986; Lein, Ivanov and Vainshtein, 1990; Lein and Ivanov, 1989, 1990; Vainstein *et al.*, 1985) and Caspian Sea (Rivkina, Tokarev and Gorlatov, 1985). The main part of the materials on the Far East seas and Norwegian Sea is presented for the first time.

8.2 GENERAL PROCESSES OF ELEMENT CYCLES IN THE SEAS

The sulphur cycle in seas is tied to other element cycles through organic matter cycling. In situ photosynthesis is the dominant source of organic matter, with allochthonous inputs from land being much smaller. Such allochthonous inputs of organic matter typically account for 10% of the total organic matter inputs to continental seas, somewhat less for marginal seas, and still less for open oceans (Romankevich, 1977). It should be stressed that the amounts of organic matter and nutrients flowing into the sea from land increase with the development of industry and agriculture on the land. These anthropogenic fluxes, particularly of nutrients and of toxics, disturb the natural element cycles of the marine ecosystems.

The rates of mineralization of organic matter are usually comparable to the rates of photosynthesis, although in some areas with considerable inputs of allochthonous matter, mineralization rates may exceed photosynthesis. In the oxic zone, which as a rule covers most of the water column and the upper millimetres or so of silt sediments, mineralization is carried out by aerobic organotrophic microorganisms. When constructing biogeochemical models, one usually assumes that in oxic conditions carbon-, nitrogen-, phosphorus-, and sulphur-containing organic compounds are oxidized to form inorganic oxides. However, microbial biomass containing these elements is also synthesized. This new organic matter, differing considerably in its composition from phytoplankton biomass, is also involved in subsequent element cycles. It is important to emphasize that in the course of organic matter biosynthesis, organotrophic bacteria incorporate not only organic molecules but also carbon dioxide, which is fixed through the pathway of heterotrophic CO_2 assimilation. The portion of mineral carbon assimilated by organotrophs can reach 12–18% of the total carbon of new bacteria biomass (Saralov, 1990).

Anaerobic mineralization of organic matter occurs in anoxic zones characteristic of most of the silt sediments of continental and marginal seas. Under some conditions, anoxia is also observed in part of the water column. Anoxic water columns often occur in narrow and deep gulfs or fiords and are now found in the deep-water basins of the Baltic Sea. Anoxia characterizes the entire deep-water portion of the Black Sea, which makes up more than two-thirds of the total area of the sea.

The processes of anaerobic mineralization result in production of such reduced inorganic compounds as H_2, H_2S, and NH_4^+ as well as various low-molecular organic compounds, such as alcohols and fatty acids, which are the products of carbohydrate fermentation and amino-acid deamination. Some of the anaerobic mineralization products are used by the so-called secondary anaerobes. The following three groups of these microorganisms are of utmost geochemical significance: acetogens, synthesizing acetate from CO_2 and H_2; sulphate reducers, reducing sulphate to sulphide through oxidation of hydrogen and low-molecular organic acids and alcohols; and methanogenes forming methane either from methyl groups of organic compounds or via CO_2 reduction by hydrogen.

A portion of the final products of anaerobic metabolism (such as sulphides and methane) remains buried in the sediments. However, the bulk of reduced products of anaerobic metabolism migrates according to concentration gradients towards the interface of the anoxic and oxic zones. Above this border the processes of H_2, H_2S, NH_4^+, and CH_4 oxidation are carried out by specific groups of microorganisms, such as hydrogen-oxidizing, thionic, sulphur, nitrifying, and methane-oxidizing bacteria. Most representatives of these groups do not require complex organic compounds to synthesize their biomass. As the main carbon source, they use either only carbon dioxide (chemoautotrophs and photoautotrophs) or carbon dioxide and intermediate products of methane oxidation (methanotrophs). Due to the intensive activity of these autotrophic and methanotrophic microorganisms, a significant formation of organic matter takes place at the interface of the oxic and anoxic zones. The role of this biomass in the nutrition of the sea organisms draws increasing attention of biogeochemists.

The above general scheme of biogeochemical processes qualitatively characterizes interactions of element cycles and the role of the main groups of living organisms. Unfortunately, except for photosynthesis, almost all of the biogeochemical processes involving sulphur, carbon, and oxygen interactions have not received sufficient quantitative study. Therefore, the main objective of our long-term research work was to quantify these processes occurring in anoxic areas and at the oxic–anoxic interfaces. The main task of this chapter is to summarize the data obtained for continental and marginal seas of Eurasia. The details of methods used are described elsewhere (Ivanov, Lein and Kashparova, 1976; Laurinavichus and Belyaev, 1978; Lein, 1983; Galchenko, Gorlatov and Tokarev, 1986).

8.3 THE NORWEGIAN SEA

Sediments from the eastern part of the Norwegian Sea near Medvezhiy Island (the region of the wreck of the Soviet submarine *Komsomolets*) were studied in May 1989. Five cores were taken at stations in water depths ranging from 1670 to 1685 metres. At stations 2072 (73° 42.78 N; 13° 18.31 E), 2073 (73° 43.40 N; 13° 14.70 E), and 2074 (73° 42.70 N; 13° 16.10 E), only the top 15 to 20 cm were sampled. At station 2082 (73° 43.02 N; 13° 17 O E), sediments were sampled between 30 and 85 cm depth. At station 2083 (73° 43.06 N; 13° 15.08 E), sediments between 90 and 310 cm were sampled. The combined sediments from cores 2074, 2082 and 2083 can be considered representative of the entire upper holocene sediments, whose thickness in this region is approximately 150 cm, assuming a sedimentation rate of 0.2 mm a^{-1}.

Comparatively weak reducing conditions are observed in the sediments, as indicated by relatively high oxidation–reduction potentials and by insignificant decreases of sulphate with depth (Table 8.1). None the less, reduced sulphur is present even in the surface sediments; at depth reduced sulphur makes up more than 1% of the weight of the sediment (0.3 mol S g^{-1}; Table 8.1). The organic carbon content varies from 0.41 to 1.00% (Table 8.1).

Rates of sulphate reduction as measured with radiolabelled sulphate varied from 22 to 117 ng (S) g^{-1} day^{-1} (0.7 to 3.7 nmol (S) g^{-1} day^{-1}) in the upper 20 cm of the sediment (Table 8.1). The highest rates were observed in the interval from 30 to 50 cm (up to 190 ng (S) g^{-1} day^{-1}; 5.9 nmol (S) g^{-1} day^{-1}). Rates were low below 80 cm. Integrating these measured rates of sulphate reduction over depth (specific gravity = 1.3), and assuming the rates do not vary seasonally, we estimate an annual rate of 45 g (S) m^{-2} a^{-1} (1.4 mol (S) m^{-2} a^{-1}). From the reduced sulphur data in Table 8.1, we estimate a rate of sulphur burial of 9.4 g (S) m^{-2} a^{-1} (0.3 mol (S) m^{-2} a^{-1}), or 21% of the sulphate that is reduced. The rest of the reduced sulphur formed during sulphate reduction must be re-oxidized, perhaps supporting production by chemoautotrophic bacteria.

Rates of anaerobic methane oxidation (measured using radiolabelled ^{14}C) in these sediments are fairly high (Table 8.1). Methane would appear to be a major substrate fuelling sulphate reduction in the surface sediments, if indeed sulphate-reducing bacteria are responsible for this methane oxidation. Assuming two moles of methane consumed for every mole of sulphate

Table 8.1. Chemical characteristics and bacterial processes of Norwegian Sea sediments

Station	(cm)	E_h (mV)	Water content (%)	Methane content ($\mu g(C) l^{-1}$)	SO_4^{2-} reduction (ng (S) g^{-1} day^{-1})	CH_4 formation from CO_2 ($\mu g(C) l^{-1}$ day^{-1})	CH_4 Oxidation ($\mu g(C) l^{-1}$ day^{-1})	Total reduced S (%)	SO_4^{2-} (g/kg)	Organic C (%)
2072	0–4	+360	58	19	—	0.083	1	0.008	2.7	0.84
	8–10	+250	52	21	58	0.027	15	0.005	2.5	0.41
	15–18	+200	47	34	92	—	14	0.006	2.6	0.41
2073	0–4	+280	59	18	24	—	19	0.007	2.5	0.72
	10–14	+190	49	22	15	—	8	0.006	2.6	0.49
2074	0–3	+220	60	18	45	—	11	0.2	2.7	2.68
	7–10	+130	48	23	22	—	8.8	0.2	2.6	0.46
	15–19	+140	49	33	117	—	10	0.007	2.6	0.51
2082	30–36	+245	46	35	190	0.045	11	0.06	2.7	—
	40–45	+73	47	35	143	0.005	14	0.06	2.8	—
	47–53	−50	47	480	151	0.087	211	0.09	2.7	0.61
	62–68	−132	49	430	120	0.044	212	0.08	2.6	0.79
	78–85	−143	49	520	6.4	0.013	274	0.17	2.7	1
2083	90–110	−140	31	960	4.8	0.014	204	1.5	2.3	0.75
	190–210	−185	29	2000	—	0.009	409	1.1	2.7	0.38
	290–310	−215	27	3400	—	0.001	1015	0.89	2.6	0.35

reduced, measured rates of methane oxidation above 45 cm at all stations are sufficient to account for an average of almost 75% of the rate of sulphate reduction at each depth (range = 17 to 178%). Below 45 cm, rates of sulphate reduction are too low to account for the measured rate of of methane oxidation unless some oxidant other than sulphate is being used (Table 8.1).

It is also interesting to compare the rate of methane oxidation with the rate of methane formation (also measured using radiolabelled ^{14}C). The rates of methane oxidation are at least two to three orders of magnitude higher than the rates of methane formation (Table 8.1). These data, together with rather high methane concentrations in the sediments (Table 8.1), suggest an important source of methane coming from deeper horizons, creating an upward flux of methane which is partially intercepted by methane-oxidizing bacteria. The origin of this deep methane remains unknown and requires further investigation.

8.4 THE SEA OF OKHOTSK

Most of the sediments of the Sea of Okhotsk are represented by pale-green diatomic silts, and are characterized by anoxia and possess high concentrations of reduced sulphur (Volkov and Rozanov, 1983). Geological information on the area is presented in Zonenshayn *et al.* (1987), Morozov (1987), and Chertkova, Bilichenko and Stunzas (1987). Our biogeochemical investigations were carried out on the east continental slope southwest of Alaid Island in June–July 1986 (Lein *et al.*, 1989). The rate of sedimentation in the study region is about 0.2 mm a^{-1}.

A major gas jet emitting bubbles 1 cm in diameter occurs in this region. The rate of the gas flow is up to 0.2 m s^{-1} (Zonenshayn *et al.*, 1987). In the zone of the gas-gusher outlet, the pale-green sediments common elsewhere in the Sea of Okhotsk are replaced by black sulphidic sediments, covered unevenly by a white coating of native sulphur or whitish bacterial mats. Carbonate nodules and plates of irregular form, ranging from 1–2 to 20 cm in diameter, also occur around the gas outlet.

We sampled cores from stations underlying water of depths of 786 to 796 metres (station 1391—50° 30 91 N; 155° 18.00 E; station 1395—50° 30.90 N; 155° 18.16 E; and station 1404—50° 30.80 N; 155° 18.15 E). Grains and nodules of calcium carbonate and abundant carbonate cementation of sediments were observed in the sediments of all stations. Snow-like accumulations of gas hydrate were discovered at a depth of 185 to 225 cm at station 1395. Methane composed almost 98% of the volume of the hydrate. Gas hydrates were also observed at station 1404.

Sulphate reduction rates at stations 1391 and 1404 are highest near the surface and then decrease with depth over the top metre, a pattern which is quite typical for sediments of the continental slope (Table 8.2). Porewater characteristics are also typical, with sulphate and calcium ions decreasing over depth, accompanied by a gradual decrease in alkalinity. Oxidation–reduction potentials are positive in the upper horizons and moderately reduced by a depth of 1 m. The rates of methanogenesis considerably exceed the rates of anaerobic methane oxidation in the top metre at these two stations, resulting in an accumulation of methane in the upper horizons (Table 8.2).

A quite different situation is observed in the sediments of station 1395 and in the lower portions of the sediments at station 1404. In both cases the sediments contain methane gas hydrates, and dissolved methane increases sharply in the porewaters just above the hydrates, probably because of partial decomposition of the gas hydrate under the thermodynamically unstable conditions in the subsurface sediments. Anaerobic methane oxidation also sharply increases, and its intensity in sediments at a depth of 130–140 cm at station 1395 is 25–100 times higher than at depths above 60 cm (Table 8.2). The oxidation of the methane results in the porewaters being supersaturated with carbonates, leading to the calcium carbonate concretions and nodules. That the carbon of the carbonate concretions originates from methane is confirmed by stable isotopic data (Lein *et al.*, 1989). $\delta^{13}C$ value of gas hydrate methane are −54.6‰, and the values of $\delta^{13}C$ of six samples of diagentic carbonates from the sediments of stations 1396 and 1404 varied from −47.5 to −49.2‰, whereas $\delta^{13}C$ of shell carbonate in the same silts ranged from 1 to −5‰.

Unlike the deeper sediments of the Norwegian Sea, rates of sulphate reduction are amply sufficient to account for the anaerobic methane oxidation observed in all of the depths at all three stations in the Okhotsk Sea (Table 8.2).

8.5 THE BALTIC SEA

The Baltic Sea is a body with a high bioproductivity and great amount of organic matter in the sediments, making it interesting for the study of biogeochemical processes. Like the Black and Azov Seas, the Baltic Sea belongs to small continental seas (Vigliere, 1974). However, unlike these two reservoirs, the microflora of the bottom sediment and its geochemical activity in the Baltic Sea have received only occasional study. Schneider (1977) determined the number of sulphate-reducing bacteria at three stations in the Kiel fiord, and Bagander (1977) used benthic chambers to characterize the turnover of phosphorus, oxygen, and sulphur compounds in the surface horizon of sediments in the Stockholm archipelago. Our investigations, conducted during the summer of 1978, included the study of number, distribution, and geochemical activity of different groups of microorganisms in the

Table 8.2. Chemical characteristics and bacterial processes in sediments of the Okhotsk Sea

Station	(cm)	Water content (%)	Methane content (ng (C) l^{-1})	SO_4^{2-} reduction (ng (S) g^{-1} day^{-1})	CH_4 formation from CO_2 (ng (C) l^{-1} day^{-1})	CH_4 Oxidation (ng (C) l^{-1} day^{-1})	$^{14}CO_2$ assimilation (mg (C) l^{-1} day^{-1})	Production of total reduced sulphur (g (S) m^{-2} a^{-1})
1391	0–15	45.2	17	158	174.6	3.4	22.7	11.24
	15–30	45.2	62	121	150.2	15.1	17.1	17.22
	60–70	35.8	23	68	179.5	6.2	12.2	—
	120–130	43.9	3.3	60	69.5	0.8	12.1	—
	185–195	46.1	6300	146	100.1	147.7	14.5	—
								38.46
1395	0–15	44.4	100	355	38.5	23.3	14.9	25.27
	15–30	35.3	181	353	27.9	36.7	9.2	25.12
	30–40	39.9	276	84	41.6	61.6	38.8	5.98
	50–60	78.2	65	230	13.1	15.2	18.3	32.74
	90–100	36.3	4300	12	139.7	758.4	19.5	—
	130–140	46.8	8160	9	183.3	1439	49.1	—
								89.11

1404	0–15	48	318	339	324.4	740	89.9	24.13
	15–25	55.8	98	254	559.2	43.2	30.3	12.05
	25–40	53.1	80	269	98.2	19.8	14.1	19.1
	40–50	49.2	71	192	59	8.8	53.7	9.11
	50–65	43.5	116	200	94.1	24.7	13.4	14.23
	65–80	49	61	165	31.3	8.8	12	11.74
	80–95	36.8	49	101	54	8.5	10.2	7.19
	95–105	32.9	100	57	16.9	13.8	20	—
	105–120	39.1	306	112	56	36.8	18.3	—
	120–130	43	520	—	7.8	29.2	11.8	—
	130–145	38.4	416	—	7.8	29.2	11.8	—
	145–160	33.4	305	322	64.7	58.4	17.9	—
	160–172	42.6	350	—	45.5	115.4	11	—
	172–182	40.4	318	229	25.4	73.3	11.5	—
	182–195	38.2	1070	153	16.3	177.9	9.4	—
	195–205	52.8	9445	301	31.6	1684.3	12.3	—
	205–220	48	3812	366	2.8	636.2	9.1	—
	220–230	39.5	4388	680	8.2	761.7	7.4	—
	230–240	34.3	5575	165	4.8	798.7	14.8	—
	240–250	40.9	1190	181	9.1	4	8.8	—
								97.55

sediments, determination of organic carbon assimilation via bacterial processes under conditions of anaerobic diagenesis, and calculation of the carbon isotopic and material balances. At the same time we studied the biogeochemical processes of the sulphur cycle in sediments of the Baltic Sea and calculated the sulphur material–isotope balance (Lein *et al.*, 1981; Lein and Ivanov, 1983). In this paper we characterize rates of sulphate reduction and the role of sulphate reduction in the carbon cycle in sediments of the Upper Holocene, the thickness of which varied in different areas of the sea from the first centimetres to 13 m in the region of Gdansk depression.

We sampled shallow-water sediments from the Gulf of Riga (station 2589—38 m, 57° 09.6; 21° 01.9; and station 2601—53 m, 57° 43.7; 23° 38.3), sediments of the open Baltic Sea (station 2611—130 m, 57° 45.8; 20° 37.3; station 2631—120 m; 57° 23.5; 19° 26.3; station 2636—160 m, 56° 43.5; 20° 00.0; and station 2656—43 m, 54° 50.4; 13° 40.6) and sediments of the deep-water Gotland depression (station 2618—215 m, 57° 24.3; 20° 00.0; and station 2622—240 m, 57° 25.2; 20° 18.1). All the sediments are rich in organic matter and have measured oxidation–reduction potentials which vary from −90 to −340 mV. Reducing conditions are observed either just under a thin layer of oxidized sediment or at the sediment–water interface. Free hydrogen sulphide in near-bottom water was detected only in the Gotland depression at the depth of 226 m in concentration of 0.04 mg l^{-1}.

The lowest concentrations of total reduced sulphur are observed in the surface layers of coarse-grained sediments at the coastal stations, which contain about 2% organic carbon. In sediments of most of the other stations, organic carbon ranges from 2.1 to 5.9% (average = 4.8%), and the total content of reduced sulphur compounds reaches 1% (0.3 mmol g^{-1}) even in the upper horizons. Methane concentrations range from 160 to over 500 μg (C) g^{-1} (wet weight; Geodekyan, Trotsyuk and Avilov, 1978). Highest concentrations are observed in the sediments of the Gulf of Riga and Gdansk depression (Geodekyan, Trotsyuk and Avilov, 1978).

The organic matter in sediments of the Baltic Sea comes both from inflow of terrestrially derived carbon with river runoff and from sedimentation of dead phytoplankton. If we assume that organic carbon in river runoff of the Baltic region has a $\delta^{13}C$ value of −27.1% (Wiekman, 1952) and that the $\delta^{13}C$ value of phytoplankton is −21.5% (Galimov, 1968), we can calculate the share of phytoplankton-derived and terrestrially derived carbon in the sediments of our stations (also assuming no fractionation in the carbon isotopic ratio during diagenesis and no other sources of organic carbon inputs). The $\delta^{13}C$ values for the sediments at most of our stations range from −21.7 to −22.8‰, suggesting that the main bulk of organic matter carbon is formed from phytoplankton residues (82–100%). Values for some coastal stations are as low as −24.9‰, suggesting that some 60% of the carbon may be terrestrially derived.

Rates of sulphate reduction are shown in Table 8.3. Rates are moderately high at most stations, and the maximum activities are generally found in the upper 20 cm of sediment. Non-acid-volatile products (pyrite and organic sulphur) are a major product of the sulphate-reduction measurement at many of the stations (Table 8.3). As with oceanic sediments and the sediments in other seas, sulphate reduction is the major process consuming organic carbon under anoxic conditions in the Baltic Sea (Lein and Ivanov, 1980; Belyaev, Lein and Ivanov, 1980; Lein, 1984, 1989).

Our data on the rates of biogeochemical processes together with literature data allow us to calculate a material–isotope balance of the carbon cycle processes in the Baltic Sea and describe quantitatively the interaction of the carbon and sulphur cycles (Table 8.4, after Lein, 1983). In situ primary production by phytoplankton contributes some 50×10^6 t (C) a^{-1} to the Baltic Sea, while rivers deliver some 10×10^6 t (C) a^{-1} of terrestrially derived organic matter. In order to calculate the amount of organic carbon permanently buried in sediments (1.55×10^6 t a^{-1}), we used the average value of organic carbon in the sediments at our stations (4.8%) and data on the rate of sediment accumulation. The rate of organic carbon consumed through sulphate reduction (16.9×10^6 t a^{-1}) is calculated from our average rate data (Table 8.3), assuming two moles of carbon consumed for every mole of sulphate reduced. The flux of organic carbon to the sediments from the water column (18.5×10^6 t a^{-1}) is simply the sum of the rate of permanent burial and the rate of carbon consumption in sulphate reduction (Table 8.4).

The total flux of organic carbon to the sediments accounts for less than 30% of the inputs of organic carbon to the Baltic Sea from phytoplankton production and terrestrial inputs from the continent. Hence, most of the organic carbon inputs (more than 70%) are mineralized in the water column of the Baltic Sea.

It should be noted that the average carbon isotope composition of organic matter in the upper horizons of the Baltic Sea sediments from our measurements ($\delta^{13}C = -22.8‰$), is quite close to the estimated average value for all organic matter inputs to the Baltic Sea given in Table 8.4 ($\delta^{13}C = -22.6‰$).

8.6 THE BLACK SEA

The existence of a huge zone of hydrogen-sulphide-containing water in the deep part of the Black Sea was first noted just over a 100 years ago by Academician N. J. Andrusov. Recently, the problem of sulphide in the waters of the Black Sea has again attracted the attention not only of scientists but of the public at large, because of reports that the upper border of the anoxic waters has moved up rather quickly during the last 15–20 years. There is a resulting worry that anoxic waters may reach all of the way to the sea surface.

Table 8.3. Chemical characteristics and rates of sulphate reduction in Baltic Sea sediments

Station	(cm)	SO_4^{2-} (mg kg^{-1})	Rate of sulphate reduction (ng (S) g^{-1} day^{-1}): Acid-volatile S	S^0	FeS_2 + organic S	Total	Organic C consumption during sulphate reduction (ng (C) g^{-1} day^{-1})	Production of reduced sulphur (g (S) m^{-2} a^{-1})
2589	0–5	125	1434	235	38	1707	1279	40.5
	5–28	82	—	—	—	360	—	47.8
	38–47	18	—	—	—	1	—	0.7
								89
2601	0.5–15	4.5	117	30	1	147	109.6	32.8
	80–100	4.3	1	1	15	17	11	5.3
	125–135	0	1	1	1	1	—	—
								38
2611	0–20	179	623	51	378	1052	789	164.7
	46–54	142	22	3	39	64	46.6	14
	10–115	63.5	4	1	88	92	68.5	—
								179
2622	0–5	90.6	3	10	33	46	35.6	1.53
	10–15	89	87	170	24	281	210.9	17.3
	25–50	88	91	12	1145	1248	937	473.7
	150–165	56	36	1	219	255	191.7	
								492.5

2618	2–20	211	317	24	1042	1383	1027	150.9
	26–30	204	71	10	1233	1314	986.3	168.3
	70–85	155	38	1	231	269	202.7	85.5
	150–160	113	7	—	382	389	361.6	—
	175–191	94	37	1	889	926	693.1	—
								404.7
2631	0–5	151.5	16	23	1943	1982	1487	47
	5–10	144	9	22	580	611	466	14.5
	10–15	136.4	12	3	419	434	326	48.5
	150–160	4.5	6	—	38	44	33	—
								110
2636	0–10	162.8	333	—	694	1027	770	—
	10–15	29.3	3	1	47	50	38	—
								—
2656	0–36	241	273	6	1526	1805	1350	296.9
	46–56	169.4	55	1	924	979	734	153.4
	96–106	103.6	14	1	389	403	301	49.7
	150–160	20	3	1	67	70	52	—
								500
2682	0–5	285.7	1283	131	414	1828	1370	190.8
	40–50	128.8	14	1	116	130	98.6	29.6
	90–100	106.7	2	2	319	323	241	49.6
	170–180	86.7	2	49	33	84	63	—
	270–280	71.7	20	140	23	183	137	—
								270

Table 8.4. Mass-isotopic balance of organic carbon in Baltic Sea sediments (after Lein, 1983)

	Quantity	δ^{13} C%
Primary production (10^6 t (C) a^{-1})	50	−21.5
Organic C from continent (10^6 t (C) a^{-1})	10	−27.0
Total flux of organic C (10^6 t (C) a^{-1})	60	−22.4
Average content organic C in sediments (% dry weight)	4.8	
Rate of sedimentation (mm a^{-1})	25	
Area of sediments (km^2)	297.8	
Rate of sediment accumulation (mg cm^{-1} a^{-1})	10.9	
Total accumulation of sedimentary material (10^6 t a^{-1} dry wt)	32.4	
Accumulation of organic carbon in sediments (10^6 t a^{-1})	1.55	−22.8
Organic C consumption by sulphate reduction (10^6 t a^{-1})	16.9	
Flux of organic C from water column to sediments (10^6 t a^{-1})	18.5	

Most investigators think the Black Sea sulphide has been produced by sulphate-reducing bacteria. Since rates of sulphate reduction in marine systems are generally limited by organic matter supply, an increase in hydrogen sulphide in the water may suggest that greater inputs of organic matter are stimulating increased rates of sulphate reduction. That is, eutrophication may be leading to the increased volume of anoxic waters.

Early work on sulphate reduction in the Black Sea suggested rates of 25 g (S) m^{-2} a^{-1} (0.8 mol (S) m^{-2} a^{-1}; Sorokin, 1962, 1982). About half of this sulphate reduction was thought to occur in the sediments and half in the anoxic water column. More recent measurements made by Soviet micro-biologists during a series of expeditions between 1980 and 1988 (Ivanov *et al.*, 1984; Vainshtein *et al.*, 1985; Lein and Ivanov, 1990; Lein *et al.*, 1986, 1989; Lein, Ivanov and Vainshtein, 1990) have found higher rates. Data collected in December 1984 found rates of 84 g (S) m^{-2} a^{-1} (2.6 mol (S) m^{-2} a^{-1}), with the preponderance of sulphate reduction occurring in the water column (66 g (S) m^{-2} a^{-1}; 2.1 mol (S) m^{-2} a^{-1}) and only 18 g (S) m^{-2} a^{-1} (0.5 mol (S) m^{-2} a^{-1}) in the sediments. These sediment rates are 50% higher than those originally reported by Sorokin (1962). American investigators (Albert, Taylor and Martens, 1988) working on the R. V. *Knorr* in March of 1988 found rates of sulphate reduction in the sediments to be virtually identical to those found by the recent Soviet studies, but their rates of reduction in the water column were much lower, only 13 to 26% of the 1984 Soviet measurements, or virtually equivalent to the water-column rates reported by Sorokin (1962).

These differences in rates of sulphate reduction in the water column are of much interest. However, the measurements are all made on the brink of the sensitivity of the radioisotopic method, and so differences may be exaggerated, and data should be extrapolated with caution. On the other hand, the rates of all microbiological processes in the upper horizons of the hydrogen-sulphide-containing water probably undergo substantial seasonal fluctuations. Data obtained by B. B. Namsaraev and his colleagues (pers. comm.) are an indication of such a possibility. Sulphate reduction rates in samples from the upper zone of the anoxic waters from the western basin, taken in March–April 1988, are an order of magnitude higher even than our data obtained in December 1984.

The comparatively light stable sulphur isotopic composition of H_2S in the water column of the Black Sea tends to confirm that the majority of the sulphide is indeed formed from sulphate reduction in the water column (Lein, Ivanov and Vainshtein, 1990). This gives us some confidence to use our rates of sulphate reduction from 1984 in the sediments and water column (84 g (S) m^{-2} a^{-1}; 2.6 mol (S) m^{-2} a^{-1}) to analyse carbon consumption by this pathway. Assuming two moles of the organic carbon consumed per mole of sulphate reduced, we estimate carbon consumption by sulphate reduction in the Black Sea as 63 g (C) m^{-2} a^{-1} almost 80% of this occurring in the anoxic water column. This rate of carbon oxidation in the anoxic zone is significantly higher than the mass-balance calculations of Deuser (1971) and Scopincev (1975), who obtained estimates of 17 and 9 g (C) m^{-2} a^{-1} respectively. Both of these authors assumed anaerobic carbon consumption in the water column was less than or at most equal to the rate in the sediments.

Since the 1950s the potential for chemoautotrophic production to contribute a significant input of organic carbon to the Black Sea has been appreciated. The contact of hydrogen-sulphide-containing waters (enriched also with ammonium, methane, and other reducing compounds) with the overlying oxygenated waters creates a rather favourable environment for chemoautotrophic bacteria. Sorokin (1971) reported rates of dark CO_2 fixation, due mainly to the activity of sulphur-oxidizing bacteria, as high as 4.0 to 8.0 μg (C) l^{-1} day^{-1} in the zone of overlap between oxygen and sulphide, or up to 66 g (C) m^{-2} a^{-1}. However, Scopincev (1975) noted that such high rates of chemoautotrophic production, if based on sulphur oxidation, would require a supply of dissolved sulphide of at least 200 g (S) m^{-2} a^{-1} (6.3 mol (S) m^{-2} a^{-1}). This calculation, based on an assumption of three moles sulphur oxidized to fix one mole of carbon, yields an estimated supply of sulphide which is quite high compared to measured rates of sulphate reduction. Sorokin (1971) himself had measured sulphide oxidation rates averaging 250 g (S) m^{-2} a^{-1} (7.8 mol (S) m^{-2} a^{-1}), but this now seems almost certainly too high.

More recent data by both American and Soviet workers on dark CO_2 assimilation in the Black Sea have suggested rates about half the value of the original Sorokin data, or 2 to 5 μg (C) l^{-1} day^{-1} (Jannasch, 1988; Nesterov, Namsaraev and Borzenkov, 1989, 1990). These data suggest an annual rate of dark fixation of 33 g (C) m^{-2} a^{-1}. However, Nesterov, Namsaraev and Borzenkov (1989) used different inhibitors of bacterial activity to estimate that on average only 37% of the dark CO_2 assimilation is due to production by chemoautotrophic sulphur bacteria (Table 8.5). Half was thought to be due to production by nitrifying bacteria, and the rest is fixation by heterotrophic organisms. This rate of production by sulphur-oxidizing bacteria requires a supply of sulphide of only 37 g (S) m^{-2} a^{-1} (1.2 mol (S) m^{-2} a^{-1}), or 44% of our estimated rate of sulphate reduction.

The work of Nesterov, Namsaraev and Borzenkov (1989, 1990) is the first showing the importance of chemoautotrophic nitrifying bacteria in the Black Sea. It is obviously of interest to estimate independently the potential for such production. Using our estimate of 63 g (C) m^{-2} a^{-1} for anaerobic carbon mineralization, and assuming from the Redfield ratio that the ratio of C : N in the organic matter being decomposed is 6.6 : 1 (molar), annual nitrogen mineralization in the anoxic zone is estimated as 11 g (N) m^{-2} a^{-1}. Based on a ratio of nitrogen oxidized to carbon fixed by nitrifiers of 30 : 1, this nitrogen flux would support a chemoautotrophic production of only 0.37 g (C) m^{-2} a^{-1}, some 40-fold less than the estimate of Nesterov and colleagues. This discrepancy is probably the result either of an incomplete inhibition of other chemoautotrophic bacteria, or of unusually favourable hydrochemical conditions for nitrifying bacteria during the course of the study by Nesterov, Namsaraev and Borzenkov (1989) yielding rates which cannot be extrapolated to annual rates.

We suspect that the actual rate of chemoautotrophic production by sulphur-oxidizing bacteria in the Black Sea is in the range of 16 to 28 g (C) m^{-2} a^{-1}. The maximum rate which could be supported if all of the sulphide

Table 8.5. Production of organic matter by chemoautrophy in the water column of the Black Sea (after Nesterov, Namsaraev and Borzenkow, 1990)

Station	Date	Production of organic carbon (mg (C) m^{-2} day^{-1})				
		heterotrophs	nitrifiers	sulphur oxidizers	nitrifiers + sulphur oxidizers	Total
2114	March 12	16	49	82	131	147
2156	April 3	0	26	24	50	50
2193	April 15	37	116	35	151	188
Average		21	64	47	111	129

produced in our estimate of sulphate reduction is re-oxidized, rather than stored in the sediments, is 28 g (C) m^{-2} a^{-1}. Karl and Knauer (1988), however, estimate a value of at least 32 g (C) m^{-2} a^{-1}. Deuser (1971) used a value of 15 g (C) m^{-2} a^{-1} in his calculations.

The isotopic composition of organic carbon in the surface sediments of the Black Sea ranges from −25.5 to −26.6‰. This is anomalously light compared to most marine sediments, which usually reflect the isotopic composition of phytoplankton and are isotopically more light by 1 to 2‰. The average value of $\delta^{13}C$ of the phytoplankton in the Black Sea is −21.2‰ and of zooplankton is −23‰ (Deuser, 1971). Significant inputs of terrigenous organic matter ($\delta^{13}C$ = −25%) in the deep-water sediments of the Black Sea is quite unlikely. The most likely explanation for the light isotopic composition of organic carbon in the sediments of the Black Sea is an input of very isotopically light carbon produced during chemoautotrophic production.

A final statement on the carbon budget for the anoxic zone of the Black Sea will have to await further study, since it now appears that the amplitude of seasonal changes may be rather high, and there are no complete seasonal data. None the less, preparation of a carbon budget, even with the currently available data, leads to some interesting conclusions. The use of our sulphate reduction data requires a larger input of carbon to the anoxic zone than was assumed in the early budgets of Deuser (1971) and Scopincev (1975). If our sulphate reduction data are correct, the organic carbon input to the anoxic zone of the Black Sea must be approximately 72 g (C) m^{-2} a^{-1}. Of this amount, 63 g (C) m^{-2} a^{-1} are mineralized in the process of sulphate reduction in water column and sediments, 5 g (C) m^{-2} a^{-1} are transferred in soluble form, and 4 g (C) m^{-2} a^{-1} are buried in bottom sediments. Anaerobic production by chemoautotrophs may be supposed to supply some 22 g (C) m^{-2} a^{-1} of this carbon input. The rest, 50 g (C) m^{-2} a^{-1}, must come from photosynthesis in the waters above the anoxic zone. In order to maintain such a downward flux of organic matter, net primary production in the Black Sea must be of the order of 200 to 250 g (C) m^{-2} a^{-1}, since only 20–25% of the organic matter produced in primary production reaches a depth as great as 100 to 150 m as it sediments (Karl and Knauer, 1988). Our estimates of net primary production and of chemoautotrophic production agree well with the estimates of Karl and Knauer (1988), who give values of 212 and 33 g (C) m^{-2} a^{-1} for these two processes.

8.7 THE CASPIAN SEA

The Caspian Sea is a continental sea which lost the link with the ocean, resulting in considerable changes in the composition of water chemistry; this sea has changed considerably under the influence of river discharge and precipitation of chlorides and sulphates in Kara-Bogas-gol Bay. Compared to average ocean water, the water of the Caspian Sea is highly desalinated, depleted in chloride, and enriched with sulphate and calcium. A considerable admixture of terrigenic sulphate from river discharge is indicated by stable sulphur isotopic composition; the $\delta^{34}S$ value ranges from +8.7 to +13.8% (Mekhtieva and Rabinovich, 1975), quite unlike the typical seawater value of +21‰. At the present time, all the water in the Caspian is saturated with oxygen, even in the deepest waters. In earlier work in the Caspian, an intensive smell of sulphide was noted in some near-bottom waters. However, at present it appears that hydrogen sulphide is produced only in the sediments.

Rates of sulphate reduction and other sedimentary biogeochemical processes were studied using ^{35}S and ^{14}C tracers in the shallow sediments of the southeastern part of the Caspian Sea in 1977 (Ivanov *et al.*, 1980) and in sediments located along the western shoreline near the estuary of the Kura river in October 1982 (Rivkina *et al.*, 1985). Here we just summarize the sulphate reduction data. Most of the hydrogen sulphide formed in these sediments comes from sulphate reduction (Ivanov *et al.*, 1980; Rivkina *et al.*, 1985), as is the case for the other seas discussed in this chapter. Rates of sulphate reduction were the highest in the sediments of stations 62 and 57 (Table 8.6), where high rates were observed down to 100 and 160 cm. At station 60, in contrast, high rates of sulphate reduction were found only in the upper 10 cm of sediment. At most stations and depths, acid-volatile sulphides were the dominant short-term product of $^{35}SO_4^{2-}$ reduction, but non-acid-volatile products (pyrite and elemental sulphur) dominated in some cases (Table 8.6).

Integrating the data on rates of sulphate reduction over depth yields an average annual estimate of 50.6 g (S) m^{-2} a^{-1} (1.6 mol (S) m^{-2} a^{-1}). Rates

Table 8.6. Characteristics of sediments and rates of sulphate reduction in the south part of the Caspian Sea (after Mechtieva and Rabinovich, 1975; Rivkina *et al.*, 1985; Ivanov *et al.*, 1980, 1984)

Station		E_h	SO_4^{2-}	Organic C	Pyrite S and S^0	Acid-volatile sulphur	Total reduced S	SO_4^{2-} reduction	
								Pyrite S and S^0	Total reduced S
	(cm)	(mV)	(mg/l)	(%)	(%)	(%)	(%)	(ng g^{-1} day^{-1})	(ng g^{-1} day^{-1})
57	0–3	+90	890	0.8	0.064	0.027	0.09	41.9	827.13
	3–25	+110	700	0.6	0.053	0.016	0.07	19.03	677.13
	25–50	−125	400	0.4	0.046	0.064	0.11	—	—
	58–69	—	420	0.4	0.045	0.105	0.15	—	—
	69–107	—	420	0.4	0.027	0.010	0.04	18.32	462.82
58	0–5	−200	890	0.4	0.093	0.005	0.10	52.66	418.96
	5–15	−180	870	0.2	0.163	0.002	0.16	—	—
	60–70	−120	600	0.2	0.189	0.003	0.19	2.2	66.87
59	0–2	+55	980	0.7	0.060	—	0.1	1	18.5
	5–45	−100	380	0.8	—	—	0.2	2.5	48.84
	90–100	−180	—	0.8	—	—	0.1	2.4	62.45
60	0–9	—	—	0.4	—	—	—	100	131
	9.6–9.9	—	—	—	0.098	—	0.10	—	—
	9.9–32	—	800	1.0	0.232	0.017	0.25	0.4	2.4
	40–45	—	—	—	—	—	—	—	0.1
52	0–50	−165	821	—	—	—	—	232.5	541.8
	50–100	−175	801	—	—	—	—	1562	1829.1

may be higher elsewhere in the Caspian Sea since it may be assumed by analogy with other seas and marine systems that sulphate reduction is highest in fine-grained sediments with high organic carbon content. In the coarse and non-uniform-grained sediments which we studied, the organic carbon content ranged from 0.2 to 1% (Table 8.6), whereas elsewhere in the Caspian Sea the organic carbon content ranges up to 3% (Bordovskiy, 1974).

The concentration of total reduced sulphur in the sediments varies from 0.03 to 0.24% (0.9 to 7.5 mmol (S) g^{-1}) in the sediments underlying shallow water we report on here (Table 8.6). In most of the sediment samples, pyrite is the dominant form, but in some cases, acid-soluble sulphides prevail (see station 57, horizon 25–69 cm, in Table 8.6). The rate of sedimentation on the eastern shelf of the southern Caspian is approximately 0.3 mm a^{-1}; in the west, sedimentation averages 0.6 mm a^{-1} (Bordovskiy, 1974). Using these data, we estimate that the permanent burial of total reduced sulphur is approximately 60% of the rate of sulphate reduction. About 40% of the reduced sulphur formed during sulphate reduction is re-oxidized to sulphate in the upper layers of sediments and at the sediment–water interface.

Distribution of reduced sulphur in the deeper sediments of the Caspian Sea and its stable isotopic composition were studied by Mekhtieva and Rabinovich (1975). Using a mass-balance approach for sulphur isotopes, they estimated a permanent burial of sulphur in all of the sediments of the Caspian Sea of 2.5×10^6 t (S) a^{-1}. Assuming that for the Caspian as a whole 60% of sulphate reduced during sulphate reduction remains buried, as in our stations, then a total of 6×10^6 t (S) a^{-1} of sulphate are reduced in the Caspian Sea.

8.8 CONCLUSIONS

Table 8.7 summarizes the rates of sulphate reduction and several related biogeochemical processes in anoxic zones (sediments generally, but including anoxic waters in the Black Sea) in the various seas reviewed in this chapter. Organic carbon consumption during sulphate reduction is estimated assuming that two moles of carbon are oxidized for every mole of sulphate reduced. Mineralization of ammonium is estimated from the rate of carbon oxidation, assuming stoichiometry defined the Redfield ratio. From the amount of reduced sulphur and ammonium which come from the anoxic zone into the oxic zone, it is possible to calculate the amount of oxygen consumed in the oxidation of these reduced substances, and to estimate the maximal value of secondary organic matter produced by chemoautotrophic production by sulphur- and nitrogen-oxidizing bacteria. Although the calculated values of biogeochemical processes given in Table 8.7 are only crude estimates in some cases, they give a representation of the scale of these processes in different regions of marginal and continental seas.

Table 8.7. Rates of different biogeochemical processes in surface sediments of some marginal and continental seas

Process	Sea of Okhotsk depth 780–800 m	Norwegian Sea depth 1670–1690 m	Baltic Sea depth 50–270 m	Black Sea depth 20–200 m	Black Sea depth 200–2000 m	Caspian Sea depth 10–15 m
Sulphate reduction (g (S) m^{-2} a^{-1})	77.1	44.8	260	18	68	50.6
Organic C consumption during sulphate reduction (g (C) m^{-2} a^{-1})	56.5	33.6	195	13.5	51	37.9
NH_4 production in anaerobic condition (g (N) m^{-2} a^{-1})	8.5	5	29.3	2	7.7	5.4
O_2 consumption for H_2S oxidation (g (O) m^{-2} a^{-1})	193.5	112	650	45	170	126.5
O_2 consumption for NH_4 oxidation (g (O) m^{-2} a^{-1})	39	23	174	63	35.2	24.7
Possible chemosynthesis by sulphur bacteria (g (C) m^{-2} a^{-1})	19.5	11.2	65	12	17	12.7
Possible chemosynthesis by nitrifyers (g (C) m^{-2} a^{-1})	0.24	0.16	0.94	0.06	0.25	0.17

The rate of sulphate reduction in the studied continental and marginal seas varies from 18 to 260 g (S) m^{-2} a^{-1} (0.56 to 8.1 mol (S) m^{-2} a^{-1}; Table 8.7). With the exception of the shallow depths of the Black Sea, these rates are all higher than average sulphate reduction rates in the sediments of the precontinental areas of the ocean, which range from 18 to 26 g (S) m^{-2} a^{-1} (0.56 to 0.8 mol (S) m^{-2} a^{-1}; Lein, 1989). The highest rates of sulphate reduction are found in the sediments of the shallow continental Baltic Sea (Table 8.7).

The intensity of oxygen consumption used up in oxidation of reduced compounds (H_2S, NH_4^-) formed in anaerobic organic matter decomposition in sediments varies from 205 to 824 g (O_2) m^{-2} a^{-1} (Table 8.7).

Sulphate reduction is responsible for most carbon oxidation in the anoxic portions of marginal and continental seas, as in most coastal marine systems. The rates of this carbon consumption can be quite significant, ranging from 14 to 195 g (C) m^{-2} a^{-1} in the various seas we reviewed (Table 8.7). Likewise, significant quantities of ammonium are released.

Much of the reduced sulphur formed during sulphate reduction is re-oxidized at the oxic–anoxic interface. Also, most of the ammonium released during the decomposition accompanying sulphate reduction is later oxidized. These oxidation processes—particularly the oxidation of reduced sulphur to sulphate—consume large quantities of oxygen (Table 8.7). Such consumption undoubtedly plays a major role in the potential depletion of oxygen in the bottom waters of these seas, and it is the seas with the largest rates of oxygen consumption by sulphur oxidation which in fact have anoxic waters at depth: the Baltic and Black Seas.

The potential rates of synthesis of new organic carbon by sulphur-oxidizing chemoautotrophic bacteria appear large, again particularly in the Baltic Sea. Production by nitrogen-oxidizing bacteria must certainly be smaller (Table 8.7). However, more direct measurements of the rates of these processes are required in most of the seas we reviewed.

REFERENCES

Albert, D. B., Taylor, C. D. and Martens, C. S. (1988). The rates of sulfate reduction in the sediments and anaerobic water column of the Black Sea. *Eos Trans. Am. Geophys. Union*, **69**(44), 1241.

Bagander, L. E. (1977). Sulphur fluxes of the sediment and water interface. In situ study of closed systems, Eh, pH. *Ph.D. Thesis*, University of Stockholm, Stockholm, 136pp.

Belyaev, S. S., Lein, A. Yu. and Ivanov, M. V. (1980). The role of methane producing and sulfate-reducing bacteria in the destruction of organic matter. In: *Biogeochemistry of Ancient and Modern Environments*. Springer-Verlag, pp. 235–42.

Bordovsky, O. K. (1974). *The Organic Matter of Marine and Oceanic Sediments in Early Diagenesis*. Nauka, Moscow.

Chertkova, L. V., Bilichenko, A. A. and Stunzas, P. A. (1987). Revealing CH_4 gas hydrates of Okchotsk Sea. Third Annual Meeting of Soviet Oceanographers. Gidrometeoizdat, Leningrad, p. 130.

Deuser, W. G. (1971). Organic carbon budget of the Black Sea. In: *Deep Sea Research*, Vol. 18, p. 995.

Fairbridge, R. W. (Ed.). (1974). *Encyclopedia of Oceanography*. Gidrometeoizdat, Leningrad p. 631.

Galchenko, V. F., Gorlatov, S. N. and Tokarev, V. G. (1986). Microbial CH_4 oxidation in the Bering Sea sediments. *Microbiologiya*, **55**(4), 669–73.

Galimov, E. M. (1968). *Geochemistry of Carbon Stable Isotopes*. Nedra, Moscow, p. 224.

Geodekyan, A. A., Trotsyuk, V. Ya. and Avilov, V. J. (1978). Gaseous hydrocarbons in water and bottom sediments as potential indicators of oil production sources in the sedimentary entrails of World Oceans. In: *Genetic Prerequisites of Oil and Gas Content of the World Ocean*. Nauka, Moscow.

Ivanov, M. V., Lein, A. Yu, and Kashparova, E. V. (1976). The intensity of formation and diagenetic transformation of reduced sulfur compounds in the Pacific Ocean sediments. in: *Biogeochemistry of Diagenesis of Ocean Sediments*. Nauka, Moscow, pp. 171–8.

Ivanov, M. V., Belyaev, S. S., Laurinavichus, K. S. and Obrazcova, A. Ya. (1980). Microbial formation H_2S and CH_4 in recent sediment of the Caspian Sea. *Geochimiya*, **3**, 416.

Ivanov, M. V., Vainshtein, M. B., Galchenko, V. F., Gorlatov, S. N. and Lein, A. Yu. (1984). Distribution and geochemical activity of the bacteria in the sediments. In: *Oil A. Gas-genetic Studies of the Bulgarian Sector of the Black Sea*. Bulgarian Academy of Science, Sofia, pp. 150–81.

Jannasch, H. W. (1988). Microbial processes in the oxic–anoxic interface of the Black Sea water column. *Eos Trans. Am. Geophys. Union*, **69**(44), 1230.

Karl, D. and Knauer, G. (1988). Microbial production and particle flux in the epipelagic zone of the Black Sea. *Eos Trans. Am. Geophys. Union*, **69**(4), 1241.

Laurinavichus, K. S. and Belyaev, S. S. (1978). Study of the microbial CH_4 formation rate using radioisotopes. *Microbiologiya*, **47**(6), 1115.

Lein, A. Yu. (1983). Biogeochemistry of the anaerobic diagenesis of recent Baltic Sea sediments. In: *Environmental Biogeochemistry. Ecol. Bull. Stockholm*, **35**, 444–61.

Lein, A. Yu. (1984). Anaerobic consumption of organic matter in modern marine sediments. *Nature*, **312**(8), 148–50.

Lein, A. Yu. (1989). Interaction of the sulphur and carbon cycles in recent marine sediments. In: *Evolution of the Global Biogeochemical Sulphur Cycle*. SCOPE 39. Wiley, Chichester, pp. 143–9.

Lein, A. Yu. and Ivanov, M. V. (1980). C_{org} consumption during reduction processes in Pacific Ocean sediments (^{35}S, ^{14}C). *DAN SSSR*, **252**(1), 217–19.

Lein, A. Yu. and Ivanov, M. V. (1983). Reduced sulphur accumulation in sediments of marine basins with high rates of sulphate reduction. In: *The Global Biogeochemical Sulphur Cycle*. SCOPE 19. Wiley, Chichester, pp. 413–20.

Lein, A. Yu. and Ivanov, M. V. (1989). Balance of sulfur and carbon in the Black Sea. In: *9th International Symposium on Environment Biogeochemistry*. Abstract. Moscow, p. 107.

Lein, A. Yu. and Ivanov, M. V. (1990). Hydrogen sulfide production in shelf sediments and its balance in the Black Sea. *Microbiologiya*, **59**(5), 921–8.

Lein, A. Yu., Ivanov, M. V. and Vainshtein, M. B. (1990). Hydrogen sulfide balance in the deep-sea zone of the Black Sea. *Microbiologiya*, **59**(4), 656–65.

Lein, A. Yu., Namsarayev, B. B., Trotsyuk, V. Ya. and Ivanov, M. V. (1981).

Bacterial methanogenesis in Holocene sediments of the Baltic Sea. *J. Geomicrobial.*, **2**(4), 299–317.

Lein, A. Yu., Vainshtein, M. B. and Namsaraev, B. B. *et al.* (1982). Biogeochemistry of anaerobic diagenesis of recent sediments of the Baltic Sea. *Geochimiya*, **3**, 428–40.

Lein, A. Yu., Vainshtein, M. B., Shakola, V. A. and Romankevich, E. A. (1986). A rate of biogeochemical processes in the shallow water sediments of the Bulgarian shelf of the Black Sea. *Geochimiya*, **10**, 1477–86.

Lein, A. Yu., Galchenko, V. F., Pokrowsky, B. G. *et al.* (1989). Marine carbonate nodules as a result of processes of microbial oxidizing of gashydrate methane of the Ochotsk Sea. *Geochimiya*, **10**, 1396–1406.

Malceva, V. J., Esikov, A. D. and Pavlova, G. A. (1987). Chemical and isotopic composition of pore waters during gas hydrate formation in the sediments of Okhotsk Sea. Third Annual Meeting of Soviet Oceanographers. Gidrometeoizdat, Leningrad, p. 125.

Mechtieva, V. L. and Rabinovich, A. L. (1975). Sulphur isotopic balance of the Caspian Sea. *Oceanologiya*, **15**(1), 66–73.

Morozov, A. A. (1987). Speciality of early diagenesis in sediments with gas seepage (island Paramushir). Third Annual Meeting of Soviet Oceanographers. Gidrometeoizdat, Leningrad, p. 130.

Murray, J. W., Jannasch, H. W. and Honjo, S. *et al.* (1989). Unexpected changes in the oxic/anoxic interface in the Black Sea. *Nature*, **338**(6214), 411–13.

Namsaraev, B. B., Nesterov, A. I., Borzenkov, I. A. and Belyaev, S. S. (1989). Microorganisms participation in carbon cycling in the Black Sea water column. 9th ISEB, abstract, Moscow, p. 112.

Nesterov, A. I., Namsaraev, B. B. and Borzenkov, I. A. (1989). Contribution of litoautotrophic nitrification in organic matter production in the Black Sea water column. 9th ISEB, abstract, Moscow, p. 113.

Nesterov, A. I., Namsaraev, B. B. and Borzenkov, I. A. (1990). Processes of bacterial chemosynthesis in western cyclonic part of the Black Sea. In: *Changing of Ecosystems of the Black Sea*. Nauka, Moscow.

Practical Ecology of Marine Regions (The Black Sea). (1990). Naukova Dumka, Kiev, p. 251.

Rivkina, E. M., Tokarev, V. G. and Gorlatov, S. N. *et al.* (1985). Biogeochemical processes in bottom sediments of south-west shelf of Caspian Sea. *Geochimiya*, **10**, 1486–502.

Romankevich, E. A. (1977). *Geochemistry of Organic Matter in the Ocean*. Nauka, Moscow, p. 190.

Saralov, A. I. (1990). Microbial processes of nitrogen cycle in lakes. *Dissertation thesis*, Inst. of Microbial., Moscow.

Schneider, I. (1977). Desulfuration and sulphur oxidation of a brackish water environment. In: *Microbial Ecology of Brackish Water Environment*. Springer-Verlag, Berlin, pp. 244–8.

Scopincev, B. A. (1975). *Formation of the Modern Chemical Composition of the Black Sea Waters*. Gidrometeoizdat, Leningrad, p. 336.

Sorokin, Yu. I. (1962). Experimental investigation of bacterial sulphate reduction in the Black Sea, by ^{35}S. *Microbiologiya*, **31**(3), 402.

Sorokin, Yu. I. (1971). Experimental data on the rates of H_2S oxidation in the Black Sea. *Oceanologiya*, **11**(3), 423.

Sorokin, Yu. I. (1982). *The Black Sea*. Nauka, Moscow, p. 216.

Tambiev, S. B. and Zhabina, N. N. (1988). Formation of pyrite in sulfide containing waters of the Black Sea: the scale and its influence on recent sediments. *Dokl. Akad. Nauk SSSR*, **299**(5), 1216–21.

Vainshtein, M. B., Tokarev, V. G., Shakola, V. A., Lein, A. Yu. and Ivanov, M. V. (1985). Geochemical activity of the sulfate reducing bacteria in sediments of western part of the Black Sea. *Geochimiya*, **7**, 1032–44.

Volkov, I. I. and Rozanov, A. G. (1983). The sulphur cycle in oceans. In: Ivanov, M. V. and J. Freney (Eds). *The Global Biogeochemical Sulphur Cycle*. SCOPE 19. Wiley, Chichester, pp. 357–76.

Wiekman, F. E. (1952). Variation in the relative abundance of the carbon isotopes in plants. *Geochim. Cosmochim. Acta*, **2**(3).

Zonenshain, L. P., Murdmaa, I. O. and Baranov, B. V. *et al.* (1987). Underwater gas seepage in the Sea of Okhotsk to west from Paramushir Island. *Oceanologiya*, **27**(5),

9 Sulphur Dynamics of Forest Ecosystems

M. J. MITCHELL
Environmental and Forest Biology, State University of New York, Syracuse NY, USA

M. B. DAVID
Department of Forestry, University of Illinois, Urbana IL, USA

and

R. B. HARRISON
College of Forest Resources, University of Washington, Seattle WA, USA

9.1 INTRODUCTION

Sulphur is an important macronutrient within forest ecosystems throughout the world. In general, S availability does not limit production. However, S deficiencies have been found for forests in some regions, especially where atmospheric inputs are low or where there is excess N. Examples of such forests include *Pinus radiata* in Australia and *Pseudotsuga menziesii* (Douglas fir) fertilized with N along the Pacific Coast of North America (Lambert, 1986; Lambert and Turner, 1977; Lambert, Turner and Edwards, 1976; Turner, Johnson and Lambert, 1980). Sulphur deficiencies in trees may not only decrease productivity, but also influence their susceptibility to certain fungal pathogens such as *Dilodia pinea* and *Dothistroma septospora* (Lambert, 1986).

Sulphur also plays a major role in elemental cycling by its interactions with other elements. The movement of sulphate in solutions through the strata of an ecosystem affects the dynamics of other elements, especially basic and acidic cations (Johnson, Turner and Kelly, 1982; Reuss and Johnson, 1986). If H^+ is associated with the sulphate anion, this cation has strong affinity for exchange complexes in foliage as well as in soil. Sulphate is a common counter-ion which affects the flux of cations in those forest ecosystems where it is not retained by either biotic (i.e. plant uptake and microbial immobilization) or abiotic (i.e. adsorption and mineral formation) processes (Figure

Sulphur Cycling on the Continents
Edited by R.W. Howarth, J.W.B. Stewart and M.V. Ivanov

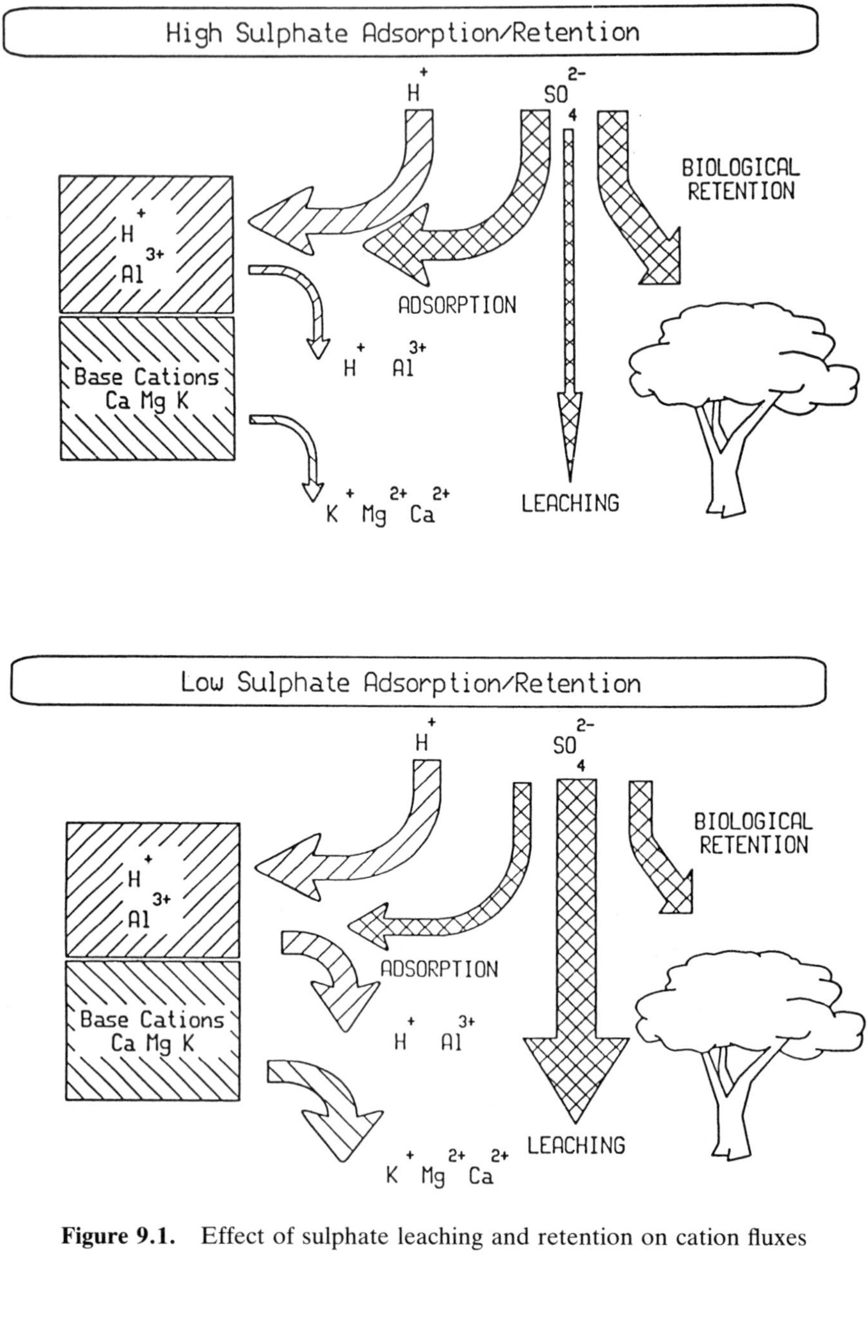

Figure 9.1. Effect of sulphate leaching and retention on cation fluxes

9.1). In forested ecosystems where there is no net retention of sulphate (e.g. S inputs are in excess of biotic demands and soils have low adsorption capacity), the flux of sulphate will generally dominate anion movement (Wiklander, 1973; Cogbill and Likens, 1974; Ulrich, Mayer and Khanna, 1980; Glass *et al.*, 1982; Mollitor and Raynal, 1983; Foster, 1985; Foster, Nicolson and Hazlett, 1989).

The cation species leached through the soil are functions of the cation exchange capacity (CEC) and base saturation. Leaching of basic cations (Ca^{2+}, Mg^{2+}, and K^+) occurs for most soils, except those with very low CEC and low concentrations of basic cations which leach H^+ and Al^{3+} (Johnson *et al.*, 1985, 1986; Foster, Morrison and Nicolson, 1986; Dewalle, Sharpe and Edwards, 1988; Stottlemyer and Hanson, 1989). The presence of Al, especially as an inorganic-monomeric form (Al^{3+}) in soil leachates and surface waters, is of major concern due to its toxicity to forest vegetation as well as aquatic biota (Cronan and Schofield, 1979; Driscoll and Schecher, 1988).

The major sulphur pools and fluxes of both inorganic and organic sulphur constituents in the atmosphere, vegetation and soil for a forest ecosystem are given in Figure 9.2. Objectives of the present chapter are to:

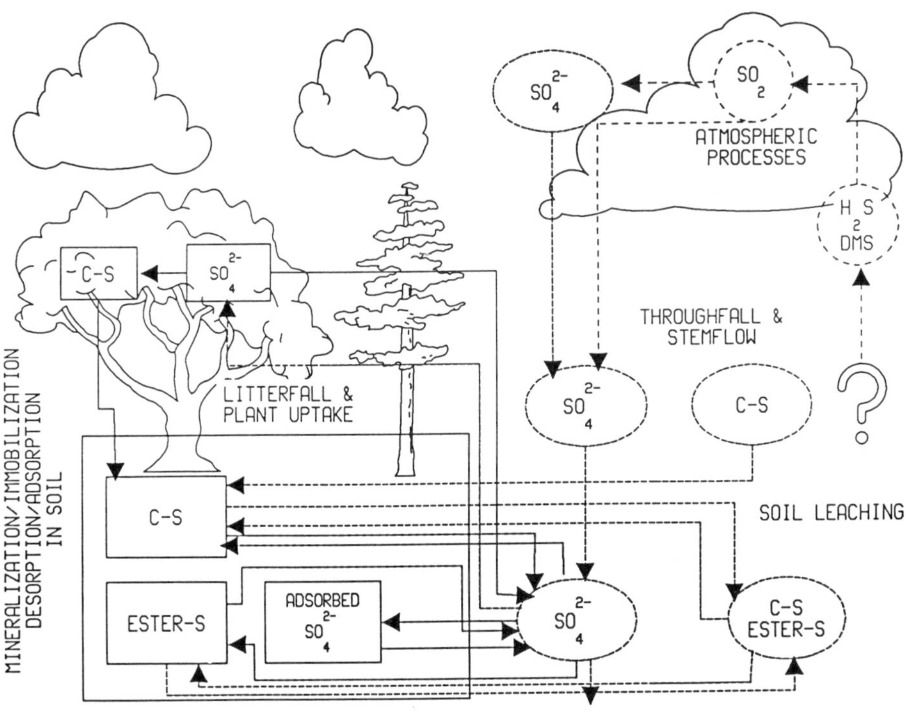

Figure 9.2. Sulphur pools and fluxes of a forest ecosystem. Solute pools and fluxes represented by short dashed lines and ovals, respectively; gases by broad dashed lines and spheres; solids by solid lines and rectangles

(1) review the various forms and quantities of sulphur found in forest ecosystems;
(2) discuss those processes which regulate sulphur dynamics and their influence on the cycling of other elements;
(3) suggest areas needing further investigation.

9.2 FORMS OF SULPHUR IN FOREST ECOSYSTEMS

9.2.1 GASEOUS FORMS

Volatile forms are important components of the sulphur cycle of forest ecosystems as either sources or losses of sulphur. The major gaseous input of sulphur to the forest occurs as the oxidant SO_2 (Lindberg *et al.*, 1986), while losses of sulphur may occur via various reduced S forms including organic H_2S as well as a range of organic S gases, including carbonyl sulphide (COS), carbon disulphide (CS_2), and dimethyl sulphide (DMS) (Lovelock, Maggs and Rasmussen, 1972; Rennenberg, 1984; Westberg, 1984).

9.2.2 SULPHUR FORMS IN FOREST VEGETATION

The major organic sulphur constituents in vegetation are C-bonded S forms, mainly as amino acid S (methionine, cystine and cysteine), but also as vitamins (biotin and thiamine), glutathione, and sulpholipids (Freney, 1967; Harwood, 1980; Harwood and Nicholls, 1979; Turner and Lambert, 1980; Turner, Lambert and Gessel, 1977). There is no evidence that ester sulphates are produced by trees (Johnson, 1984), although they may be produced by other plants as well as animals within the forest ecosystem (Fitzgerald, 1978). The metabolic pathways which synthesize various organic S constituents have been described elsewhere (Rennenberg, 1984, 1989) and will not be detailed in the present chapter. There may also be substantial concentrations of inorganic sulphate, especially in roots, due to their role in sulphate uptake (Murphy, 1980; Rennenberg, 1984).

An important biochemical linkage occurs between nitrogen and sulphur nutrition in foliage; the percentage of inorganic sulphate as fraction of the S content of foliage has been shown to reflect the availability of these nutrients (Murphy, 1980). If either S is in excess of nutritional demands or N is limiting, the percentage of sulphate in foliage increases. It has been suggested that for conifer needles the molar N : S ratio is about 33 if inorganic S is subtracted from total S concentration. This ratio reflects the relative composition of N- and S-containing amino acids (Turner, Johnson and Lambert, 1980).

9.2.3 FORMS OF SULPHUR IN SOIL

Organic S enters the soil mostly as plant litter or is formed in situ by the soil biota during the process of decomposition and humification. Soil organic S consists of two major classes: ester sulphates and C-bonded S (Fitzgerald, 1976, 1978; Nakas, 1986). These classes are the major S constituents for both agricultural and forested soils and if mineralized provide an important source of sulphate uptake by higher plants (Tabatabai and Bremner, 1972; Freney, Melville and Williams, 1975; David, Mitchell and Nakas, 1982; David, Mitchell and Schindler, 1984; David, Mitchell and Scott, 1987).

C-bonded S in soil includes amino acids, sulpholipids and sulphonic acids as well as many poorly defined compounds. Some of these compounds are formed and stabilized by humification. The chemical forms of S in humus are largely unknown, but may include both aliphatic and aromatic organic structures. Both N and S amino acids have been detected as constituents of humic substances, and these amino acids compose a substantial portion of soil S (Freney, 1967; Freney, Stevenson and Beavers, 1972; Kowalenko, 1978; Zucker and Zech, 1985). It seems likely that plant products such as thiols (—SH) and acyl thiols (R—C—SR) would be incorporated into organic matter during humification by polymerization processes.

Compounds included in the operationally defined ester sulphate class are those which contain the C—O—S linkage (ester sulphate) and the C—N—S linkage (sulphamate) along with some organic sulphites (Freney, 1961). The C—O—S linkage is found in greatest quantity, with the other forms rarely being important. Ester sulphate can originate from animals (e.g. chondroitin sulphate), microorganisms (e.g. choline sulphate) and plants (e.g. sulphated thioglycosides) (Fitzgerald, 1976, 1978).

Sulphate is the predominant form of inorganic S. It is derived from wet and dry deposition, mineral weathering, and catabolism of organic S. Most of the stable sulphate minerals in soil contain either Al, Fe or Ca. These include jurbanite ($AlOHSO_4$), basaluminite ($Al_4(OH)_{10}SO_4$), alunite ($KAl_3(SO_4)_2(OH)_6$), and gypsum ($CaSO_4xH_2O$). Under xeric conditions, some forms such as gypsum may accumulate (Freney and Williams, 1983), while in acid sulphate soils Al-sulphate forms such as jurbanite and basaluminite may occur (Van Breemen, 1983; Adams and Hajek, 1978).

Generally, the inorganic non-sulphate S fraction, which includes sulphites and thiosulphates, has been overlooked in forest soils (Aspiras, Kenny and Chesters, 1972; Landers, David and Mitchell, 1983). For northern hardwood sites in the Adirondack Mountains of New York (David, Mitchell and Nakas, 1982). White Mountains of New Hampshire (Schindler *et al.*, 1986), and coastal Maine (David, unpub. data), reduced sulphur was a major constituent of total inorganic sulphur. The role of these non-sulphate inorganic sulphur constituents as storage pools and as intermediates of biogeochemical reactions, especially under low oxygen conditions, needs further elucidation.

9.3 QUANTIFICATION OF SULPHUR IN FOREST ECOSYSTEMS

9.3.1 DISTRIBUTION OF TOTAL SULPHUR

The contribution of vegetation, forest floor and mineral soil to the sulphur content of some forest ecosystems (Table 9.1) is given in Figure 9.3. As has been found previously (Johnson, 1984; Johnson *et al.*, 1986), the major sulphur pool in these forested ecosystems (except for the site in Australia) is contained in the mineral soil and constitutes from 84 to 98% of the total ecosystem sulphur content.

Relatively few studies have reported the content (or absolute amount) of sulphur constituents in forest soils. Measurements of bulk density, horizon thicknesses, and inclusion volume (rock material > 2 mm) are difficult to obtain for many soils and often affect content calculations more than concentration differences. Figure 9.4 presents content levels of sulphur constituents for various forest soil. Total S ranged from 1.1 to 15 g (S) m^{-2} (0.034 to 0.47 mol m^{-2}) in forest floor and from 31 to 307 g (S) m^{-2} (1 to 9.6 mol m^{-2}) in mineral soil. The low value of 31 g (S) m^{-2} was from the Al horizon at Coweeta, where only 10 cm of mineral soil was sampled. Other sites generally reported mineral soil content to the rooting depth. The high value is from a bottomland site in Illinois where flooding adds organic rich material to the

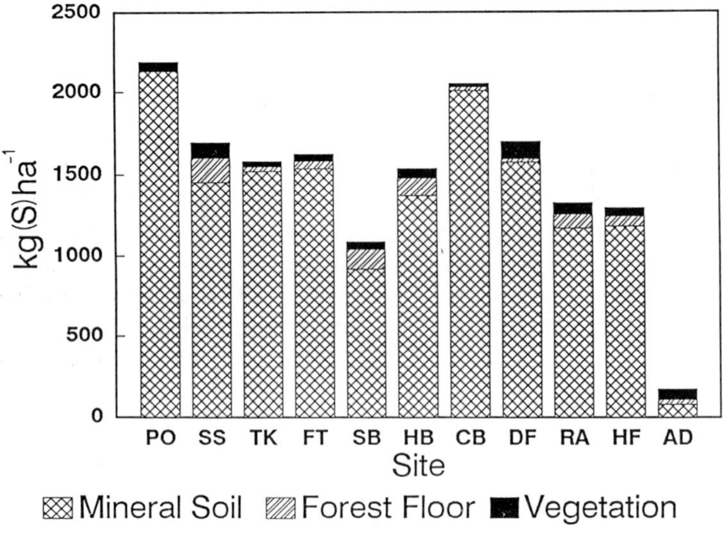

Figure 9.3. Sulphur content of selected forest ecosystems. For abbreviations see Table 9.1

Table 9.1. Selected forest sites with information on sulphur cycling

Name of site	Abbreviation	Location	Forest type	References
Ardennes Oak	FO	Ardennes, France	*Quercus* spp.	Nys and Ranger (1988)
Ardennes Spruce	FS	Ardennes, France	*Picea abies*	Nys and Ranger (1988)
Ispina Mixed Oak	PO	Ispina, Poland	*Quercus robus*	Karkanis (1976)
Walker Branch Chestnut Oak	FT[a]	Tenn., USA	*Quercus prinus* *Carya* spp.	Johnson *et al.* (1986)
Walker Branch Yellow Poplar	TK[b]	Tenn., USA	*Liriodendron tulipifera*	Johnson *et al.* (1986)
Camp Branch Mixed Oak	CB	Tenn., USA	*Quercus* spp.	Johnson *et al.* (1986)
Thompson Red Alder	RA	Washington, USA	*Alnus rubra*	Johnson *et al.* (1986)
Thompson Douglas Fir	DF	Washington, USA	*Pseudotsuga menziesii*	Johnson *et al.* (1986)
Bago, NSW Douglas Fir	AD	Australia	*Pseudotsuga menziesii*	Turner and Lambert (1980)
Solling Beech	SB	West Germany	*Fagus silvatica*	Meiwes and Khanna (1981)
Solling Spruce	SS	West Germany	*Picea abies*	Meiwes and Khanna (1981)
Huntington Forest, Adirondack Mts.	HF	New York, USA	*Fagus grandifolia* *Acer* spp.	David, Mitchell and Scott (1987) Shepard *et al.* (1989) Mitchell *et al.* (unpublished)
Hubbard Brook White Mts.	HB	New Hampshire, USA	*Acer* spp.	Likens *et al.* (1977, 1990), Mitchell *et al.* (1989)

[a]FT refers to the Fullerton soil series.
[b]TK refers to the Tarklin soil series.

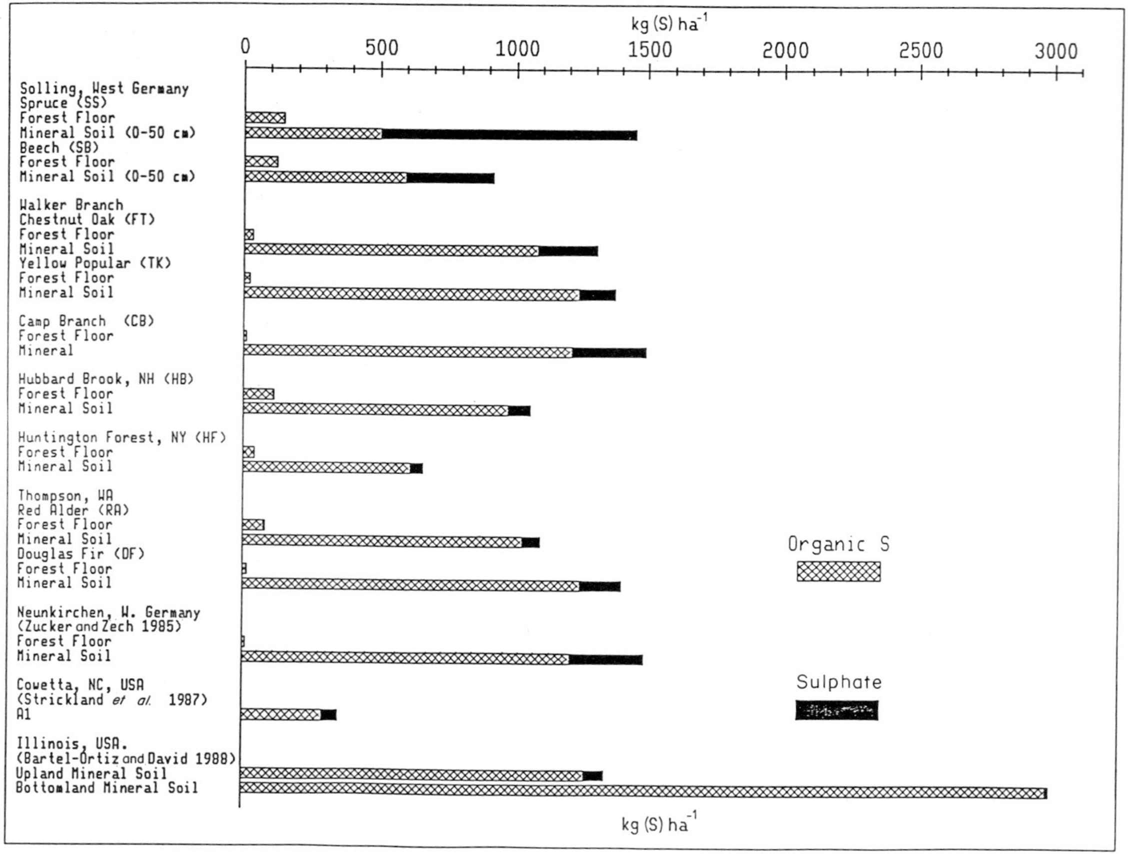

Figure 9.4. Organic sylphur and sulphate content of forest floor and mineral soils of selected forest ecosystems. For abbreviations see Table 9.1

soil. In most forest soils, the sulphur pool is primarily organic S, ranging from 35 to 98% of total S. The only sites with less than 80% organic S were the mineral soils at Solling, West Germany and Camp Branch, Tennessee; the former site received very high sulphur deposition from anthropogenic emissions.

9.3.2 THE IMPORTANCE OF ORGANIC SULPHUR

Organic constituents are the predominant form of sulphur in forest soils, even for those forest ecosystems with high concentrations. Carbon-bonded sulphur is the dominant type of sulphur in forest soils in Quebec, in the Adirondack Mountains (New York), and in the White Mountains (New Hampshire) (Lowe and DeLong, 1963; Van Loon, Hay and Goh, 1987; David, Mitchell and Nakas, 1982; David, Mitchell and Schindler, 1984; David, Mitchell and Scott, 1987; Strick *et al.*, 1982; Schindler *et al.*, 1986). Although organic S decreased with depth, it constituted 80–99% of the total S. Carbon-bonded S and ester sulphates averaged 50–80% and 20–30% of total S, respectively, in these C-rich soils. Mitchell *et al.* (1986) found similar levels in the Marmot Basin watershed of the Rocky Mountains (Alberta, Canada), with these same fractions being 67–86% and 5–32% of total S, respectively.

At a forested site in West Germany, Zucker and Zech (1985) found C-bonded sulphur and ester sulphate to dominate soil S. Bartel-Ortiz and David (1988) found organic S to be > 90% of total sulphur in upland and bottomland forests in Illinois (USA) for all soil depths. Neary, Mistry and Vanderstar (1987) reported total sulphur and sulphate levels for soils in the surrounding watershed of Plastic Lake (Ontario, Canada). Ignoring non-sulphate inorganic S (usually < 5–10% of total S; David, Mitchell and Scott, 1987), organic S in the organic (LFH) and Ae horizons would be 87% and 94% of total S, respectively.

Only a few non-glaciated sites have been examined for soil organic sulphur. At the Coweeta Hydrological Laboratory, North Carolina, Strickland *et al.* (1987) found that C-bonded sulphur and ester sulphate ranged from 39 to 87% and 13 to 30% of total sulphur in the O and A horizons, respectively. There was significant seasonal variation in all horizons examined, which was thought to result from organic transformations. Johnson *et al.* (1986) measured total sulphur and sulphate in soil at five sites in the US. Again organic sulphur can be calculated from these data if non-sulphate inorganic S is ignored. At Walker Branch, Tennessee, two sites contained 56% and 97% (*Quercus prinus*; chestnut oak) and 90% and 95% (*Liriodendron tulipifera*; yellow poplar) of total S as organic sulphur in A1 and B horizons, respectively. At the Camp Branch site (*Quercus* spp.), also in Tennessee, organic sulphur was 79% and 64% of total S in A1 and B2 horizons, respectively. Finally, at two sites in the Pacific Northwest (*Alnus rubra*; red alder, and

Table 9.2. Concentration of sulphur constituents (μg (S) g^{-1}) in forest soils with relative percentages in parentheses

Site	Total S	C-bonded S	Ester sulphate	Sulphate S	Inorganic S
Hubbard Brook, New Hampshire, USA (Schindler *et al.*, 1986) (HB)					
Oa	1563	1113(71)	442(28)	4.6(0.3)	*[a]
Bh	303	219(72)	67(22)	9.3(3.0)	*
Bs1	452	290(64)	119(26)	36.4(8.1)	*
Huntington Forest, New York, USA (David, Mitchell and Nakas, 1982) Hardwood Site (HF)					
Oa	1780	1372(77)	392(22)	3.2(0.2)	12.3(0.7)
E	124	90(73)	30(24)	1.2(1.0)	2.7(2.2)
Bh	761	633(83)	96(13)	23.7(3.1)	8.1(1.1)
Bs1	527	367(70)	115(22)	22.4(4.2)	23.2(4.4)
C	119	82(69)	23(19)	5.1(4.3)	8.4(7.1)
Conifer Site					
Oa	2003	1756(88)	227(11)	6.6(0.3)	13.0(0.6)
E	68	50(74)	15(22)	1.0(1.5)	1.6(2.4)
Bh	540	449(83)	59(11)	13.2(2.4)	18.4(3.4)
Bs1	515	293(57)	140(27)	66.8(13)	15.4(3.0)
C	126	82(65)	21(17)	18.1(14)	5.0(4.0)
Marmot Basin, Alberta, Canada (Mitchell *et al.*, 1986)					
L-F	1593	1330(83.5)	235(14.7)	11.5(0.7)	17.0(1.1)
H	962	751(78.1)	180(18.7)	19.5(2.0)	11.5(1.2)
Bf1	364	298(81.7)	30(8.3)	22.4(6.1)	14.4(4.0)
Podzol, Quebec, Canada (Van Loon, Hay and Goh, 1987)					
LFH	1003	456(45.5)	540(53.8)	7(0.7)	*
Bfh	200	170(85.0)	25(12.5)	5(2.5)	*
Coweeta, North Carolina, USA (Strickland *et al.*, 1987)					
Oie	539–1035	459–838 (79–87)	40–188 (6–17)	23–44 (3–7)	*
Oa	1038–1344	702–956 (65–78)	120–316 (10–30)	15–276 (1–22)	*
A1		(39–62)	(13–45)	(7–25)	*
West Germany (Zucker and Zech, 1985) Neunkirchen					
F-Ah	526	328(62.4)	163(31.0)	35(6.6)	*
Ah	414	243(58.7)	130(31.4)	41(9.9)	*
Ah-M	264	173(70.3)	51(19.3)	41(15.5)	*
Go-M	74	21(28.4)	32(43.2)	21(28.4)	*

Table 9.2. (*continued*)

Site	Total S	C-bonded S	Ester sulphate	Sulphate S	Inorganic S
Bottomland, Illinois, USA (Bartel-Ortiz and David, 1988)					
0–10 cm	555	345(62.2)	195(35.1)	9.4(1.7)	5.2(0.9)
10–30	399	228(57.1)	164(41.1)	4.8(1.2)	2.3(0.6)
30–60	331	179(54.0)	144(43.5)	6.6(2.2)	1.8(0.5)
60+	280	134(47.8)	138(49.2)	6.2(2.2)	1.6(0.6)
Upland, Illinois, USA					
0–10	261	168(64.4)	84(32.2)	6.0(2.3)	2.8(1.1)
10–30	120	58(48.3)	57(47.5)	4.3(3.5)	0.9(0.7)
30–60	123	50(40.6)	65(52.8)	8.0(6.5)	0.4(0.3)
60+	112	38(33.9)	63(56.3)	10.7(9.6)	0.2(0.2)

Site	Total S	Sulphate	% Organic S
Walker Branch, Tennessee, USA (Johnson *et al.*, 1986) Chestnut Oak (FT)			
A1	243	7	97
B2	129	57	56
Yellow Poplar (TK)			
A1	204	10	95
B2	144	14	90
Camp Branch, Tennessee, USA (Johnson *et al.*, 1986) Mixed Oak (CB)			
A1	204	10	79
B2	179	64	64
Thompson, Washington, USA (Johnson *et al.*, 1986) Red Alder (RA)			
A1	977	13	99
B2	295	40	90
Douglas Fir (DF)			
A1	304	13	96
B2	271	40	85
Florida, USA Spodosols (Mitchell and Blue, 1981)			
A1 or AP	104	4	96
A21	31	2	94
B21h	76	3	96
B3	51	3	94

[a]Not determined.

Pseudotsuga menziesii; Douglas fir) organic sulphur ranged from 85 to 99% of total S.

These results indicate that for all forest soils examined in the US, Canada, and Europe, organic sulphur is the major component of total S. Generally, C-bonded sulphur constitutes a greater proportion of total S than ester sulphate; however, there are soils and horizons where the reverse is true. Table 9.2 summarizes concentration data for total sulphur, organic sulphur, sulphate, and ester sulphate and C-bonded sulphur where measured. Most forest floor or organic horizons have concentrations of total S of betwen 1000 and 2000 μg (S) g^{-1} (31 to 63 $\mu mol\ g^{-1}$), whereas mineral horizons range from 50 to 800 μg (S) g^{-1} (1.6 to 35 $\mu mol\ g^{-1}$).

9.4 MAJOR FLUXES AND TRANSFORMATIONS

9.4.1 ATMOSPHERIC INPUTS

The two major sources of sulphur to forest ecosystems are atmospheric and weathering inputs. Atmospheric inputs are separated into 'wet' and 'dry' components. Wet-only deposition includes sulphur in precipitation (e.g. rain, snow and sleet). Dry deposition includes inputs via gases, aerosols, and particles. The contribution of dry deposition to total S inputs, including both particulate and gaseous forms, is greatest for those sites near sources of sulphur emission where dry deposition can contribute at least 62% of the total deposition (Lindberg *et al.*, 1986; Shepard *et al.*, 1989). For some high altitude sites, input via clouds may equal the inputs of both wet and dry inputs, but estimating this form of input is difficult (Lovett, Reiners and Olson, 1982).

Wet inputs can vary over two orders of magnitude with the largest and smallest inputs occurring for sites in Europe (e.g. PO, FS, SS, and FO) and Australia (AD), respectively (Figure 9.5). These regional and site differences are directly related to the anthropogenic inputs to the atmosphere as well as climatic effects such as wind speed and period of cloud immersion (Whelpdale, this volume). The quantification of wet inputs of sulphur can be done with high precision (Pack, 1980). The estimate of dry inputs is subject to considerable uncertainty and researchers presently are developing new techniques to rectify this problem (Hicks, 1986; Shepard *et al.*, 1989). For instance, Lindberg and Garten (1988) found excellent agreement between direct estimates of total deposition of sulphur and throughfall and stemflow fluxes for sites over a broad range of sulphur inputs and with different species compositions (Figure 9.6).

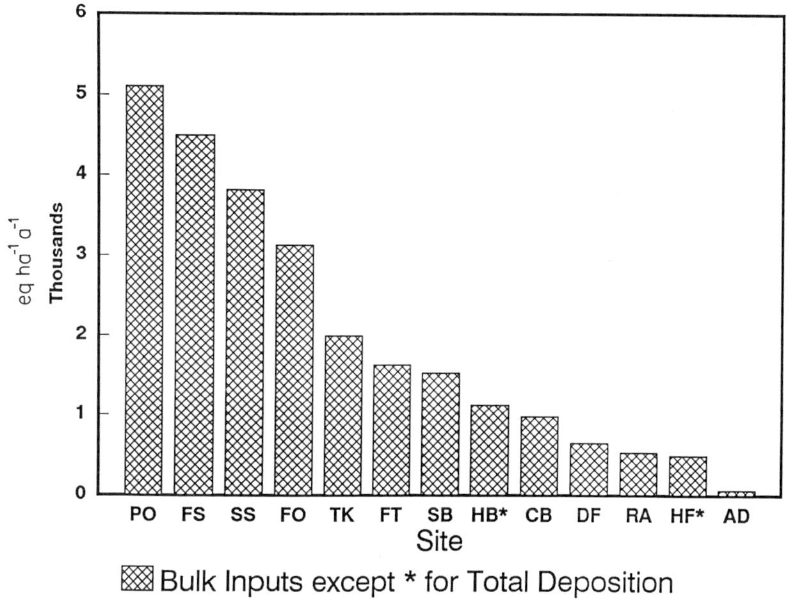

Figure 9.5. Wet inputs to sulphur to various forest ecosystems. For abbreviations see Table 9.1

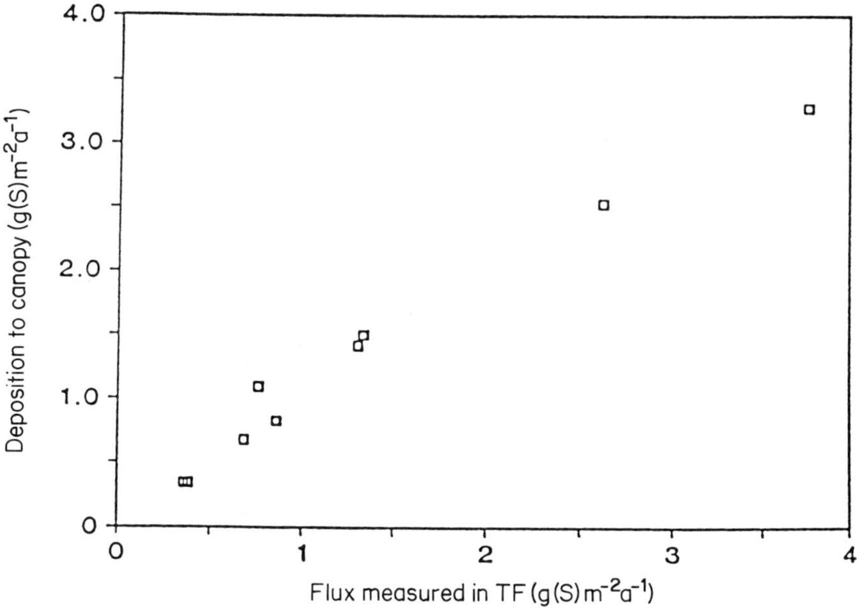

Figure 9.6. Total deposition versus throughfall and stemflow. (From Lindberg and Garten, 1988)

9.4.2 WEATHERING INPUTS

Weathering inputs are generally minor sources of sulphur for forest ecosystems, especially for those forests underlain by soils derived from glacial till and bedrock of igneous rocks; typically granite contains less than 100 μg (S) g^{-1} (3 μmol g^{-1}) (Johnson *et al.*, 1968; Murphy, 1980; David, Mitchell and Scott, 1987). For example at Hubbard Brook, it has been estimated that weathering contributes only 3% of the sulphur input (Likens *et al.*, 1977). However, for forest systems in which the soils are formed from sedimentary materials, such as shales which are sources of reduced sulphur constituents (average 2400 μg (S) g^{-1}; 75 μmol g^{-1}), the contribution to weathering to S inputs can exceed that from the atmosphere (Gibson *et al.*, 1983; Mitchell *et al.*, 1986). Even in regions where weathering inputs are considered negligible, such as the northeast US, sulphide minerals are sometimes found in bedrock and till (Everett *et al.*, 1983), but their significance as source of sulphur input is unknown.

9.4.3 GASEOUS OUTPUT FROM FOREST ECOSYSTEMS

Biogenic fluxes of sulphur gases to the atmosphere are discussed in detail elsewhere (Andreae and Jaeschke, this volume). Briefly, both the soil and forest vegetation serve as sources of biogenic S. For temperate-zone forests, fluxes directly from the vegetation are comparable to or somewhat greater than fluxes from the soil; for tropical forests, fluxes from the vegetation probably are much larger than soil fluxes (Andreae and Jaeschke, this volume). For both tropical and temperate areas, Andreae and Jaeschke (this volume) suggest total average biogenic fluxes of sulphur (from vegetation and soils) are in the range of 30 to 100 ng (S) m^{-2} min^{-1} (1 to 3 nmol m^{-2} min^{-1}). Assuming these rates hold over an entire year, they correspond to fluxes of 16 to 53 mg (S) m^{-2} a^{-1} (0.5 to 1.6 mmol m^{-2} a^{-1}). These fluxes are a minor contributor to the total sulphur in the atmosphere (Andreae and Jaeschke, this volume); the average emission rate for sulphur to the atmosphere from the surface of the world including all known natural and anthropogenic sources is some 8 to 25 times larger (Whelpdale, this volume). For most forests, biogenic fluxes also represent a small loss of sulphur from the ecosystem.

Biogenic fluxes of sulphur to the atmosphere from some tropical forests may be much greater than average, with fluxes perhaps as large as 1 to 2 g (S) m^{-2} a^{-1} (30 to 60 mmol m^{-2} a^{-1}) in extreme cases (Aneja and Cooper, 1989; Andreae and Jaeschke, this volume). Such fluxes could account for some of the discrepancies in S budgets reported for some tropical forests (Haines, 1983). However, losses (estimated from input/output budgets) from such ecosystems could also be due to other factors, including the high capacity for tropical soils to retain adsorbed sulphate (Haines, 1983; Johnson *et al.*, 1980).

Several recent studies have suggested that fluxes of sulphur from trees to the atmosphere may increase in response to more sulphur in the ecosystem (Rennenberg, 1984, 1989; Garten, 1988; Nriagu and Krouse, this volume). This clearly needs more study, but it would appear that even in forests receiving large inputs of atmospheric sulphur, biogenic fluxes back to the atmosphere are relatively small (Andreae and Jaeschke, this volume).

9.4.4 CANOPY EXCHANGE PROCESSES

The canopy of a forest affects the contribution of both wet, dry and cloud water deposition of sulphur as well as altering how sulphate interacts with other elements, especially exchangeable cations. These effects are reflected in throughfall and stemflow chemistry. In general, coniferous forests have greater net throughfall fluxes of sulphate than deciduous forests (Meiwes and Khanna, 1981; Parker, 1983; Raynal, Raleigh and Mollitor, 1985). These greater rates have been attributed to higher deposition velocities in coniferous foliage resulting from a high leaf area index and the presence of needles throughout the year (Hicks, 1986), thus increasing the amount of dry deposited sulphur (especially as SO_2). It has been suggested that sulphate exhibits little net canopy exchange (total deposition minus throughfall plus stemflow) compared to other ions such as H^+, NO_3^-, Ca^{2+} and K^+ (Lindberg and Garten, 1988). Garten, Bondietti and Lomax (1988) followed $^{35}SO_4^{2-}$ added to the boles of trees and found that little of the label was found in throughfall, thus confirming the hypothesis that there is minor leaching (< 10% of throughfall) of sulphur from the vegetation. If sulphate which passes through the canopy is associated with H^+, this acidic ion may exchange with basic cations, especially Ca^{2+}, Mg^{2+}, and K^+. Thus sites with greater deposition of sulphuric acid have higher net canopy exchange of basic cations. For example, Shepard *et al.* (1989) compared two sites with high (2.6 g (S) m^{-2} a^{-1}; 80 mmol m^{-2} a^{-1})) and low levels (800 mg (S) m^{-2} a^{-1}; 25 mmol m^{-2} a^{-1}) of sulphur input and found that the high deposition site had concomitantly higher levels of net canopy exchange of Ca^{2+} and K^+. Although it had been hypothesized that these greater loss rates of nutrient cations from foliage may be deleterious to forest health (Woodman and Cowling, 1987), this has not been substantiated by direct evidence.

9.4.5 CYCLING OF SULPHUR THROUGH VEGETATION

As shown previously (Figure 9.3), only a small proportion of the sulphur in a forest ecosystem is generally contained within the vegetation. The contribution of the vegetation to sulphur flux through the forest can be indexed by

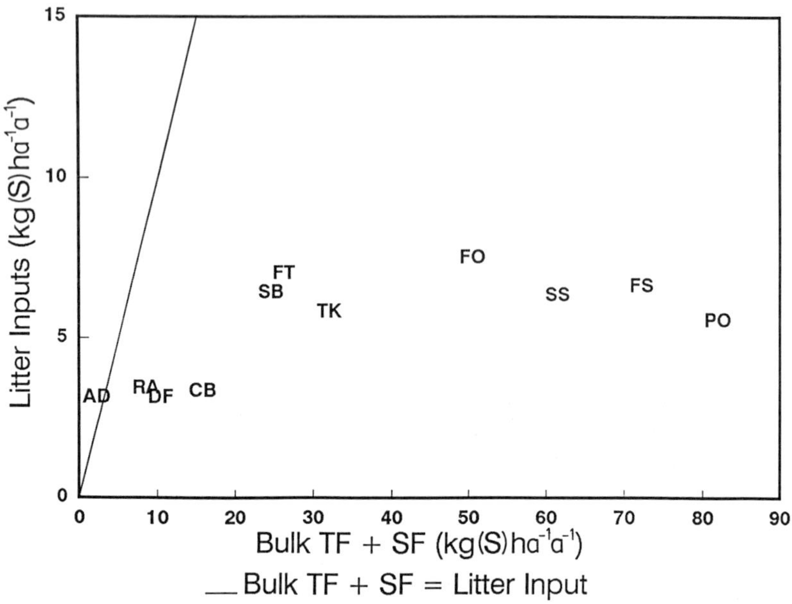

Figure 9.7. Bulk throughfall (TF) and stemflow (SF) versus litter inputs of sulphur. For abbreviations see Table 9.1

comparing bulk throughfall and stemflow to litter fluxes (Figure 9.7). The role of vegetation in cycling sulphur is most important for forest ecosystems such as Douglas fir in Australia (AD) and red alder (RA) and Douglas fir (DF) in the northwest US which have low atmospheric inputs of S and therefore S may not be in great excess of nutritional demands. Under these conditions, the uptake of sulphur by vegetation and the return of sulphur via litter may play important roles in sulphur flux (Johnson *et al.*, 1986; David, Mitchell and Scott, 1987). If the flux of sulphur to the forest floor via litter (predominantly as C-bonded sulphur) equals or even exceeds that of atmospheric inputs of inorganic sulphate, biochemical cycling may predominate over geochemical cycling of this element.

9.4.6 SULPHATE LEACHING FROM SOIL

The flux of sulphate through the soil has important implications in the transfer of basic and acidic cations from forest ecosystems to surface waters. A steady state for sulphur is sometimes assumed (Likens *et al.*, 1977; Smith and Alexander, 1986; Wright and Johannessen, 1980), and thus sulphur outputs (i.e. leaching) would equal sulphur inputs (i.e. atmospheric inputs) for forest ecosystems. A comparison of atmospheric inputs with leaching losses from

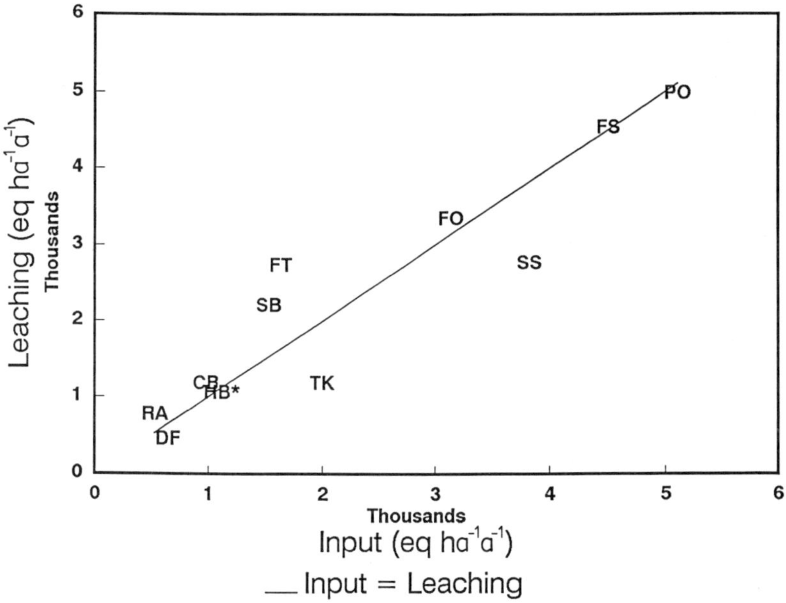

Figure 9.8. Inputs versus leaching from mineral soil. *Indicates total deposition with all other values being for bulk deposition only. For abbreviations see Table 9.1

the mineral soil for a wide range of forest ecosystems (Figure 9.8) shows that this relationship generally holds, but for some forests there is apparent divergence from steady-state behaviour. This is not surprising, and rather than reflecting true non-steady-state behaviour, these divergences may be due to inaccurate estimates of inputs and/or outputs, particularly dry deposition and hydrological fluxes. Hydrological fluxes, and thus elemental leaching, are difficult to measure in many forest ecosystems where total watershed analyses are not possible (Likens *et al.*, 1977).

Sulphur may be retained by either geochemical or biochemical mechanisms. The geochemical mechanism includes sulphate adsorption (e.g. Walker Branch, Chestnut Oak; Johnson, 1984; Johnson *et al.*, 1982) as well as sulphate mineral formation (e.g. Solling Spruce; Meiwes and Khanna, 1981). The biochemical mechanisms of microbial immobilization and plant uptake have been shown to be important in regulating sulphate leaching for sites with lower rates of atmospheric inputs (David, Mitchell and Schindler, 1984; David, Mitchell and Scott, 1987; Swank, Fitzgerald and Strickland, 1984).

9.4.7 ORGANIC SULPHUR TRANSFORMATIONS IN THE SOIL

Sulphur mineralization (the transformation of organic S to sulphate) and immobilization (the transformation of sulphate to organic S) occur concomitantly in all soil horizons. Many investigators have attempted to measure the transformation rates, pools of sulphur involved, and net alterations in pools with respect to sulphur input/output budgets by both laboratory and field experiments. Various conceptual and mathematical models have also been developed to evaluate how these processes might affect soil sulphur pools and sulphate retention.

Most work on sulphur mineralization–immobilization of forest soils has been performed in temperate regions of North America. These processes have been quantified using laboratory incubations for Spodosols in the northeastern US (David *et al.*, 1983; David, Mitchell and Schindler, 1984; David, Mitchell and Scott, 1987; Fuller *et al.*, 1986a; Schindler *et al.*, 1986). Soils at Coweeta (North Carolina) have also been studied in detail (Fitzgerald and Andrew, 1984, 1985; Fitzgerald and Johnson, 1982; Fitzgerald *et al.*, 1983; Fitzgerald, Strickland and Swank, 1982; Fitzgerald, Watwood and Rose, 1985; Strickland and Fitzgerald, 1984, 1985, 1986; Strickland, Fitzgerald and Swank, 1984, 1986a, 1986b; Strickland *et al.*, 1987; Swank, Fitzgerald and Ash, 1984; Watwood and Fitzgerald, 1988; Watwood *et al.*, 1988). Other investigations have included work in Europe (Christophersen and Wright, 1981; Farrell *et al.*, 1984), Canada (Maynard, Germida and Addison, 1986), Australia (Freney, Melville and Williams, 1971), New Zealand (McLaren, Keer and Swift, 1985) and other regions of the US (Houghton and Rose, 1976; Jarvis, Lang and Wieder, 1987; Wieder, Lang and Granus, 1987).

From these studies some generalizations can be made. Sulphate can be directly incorporated into ester sulphate or C-bonded sulphur in oxic forest soils. Many investigators have shown these processes to be microbial and regulated by various environmental factors (Fitzgerald *et al.*, 1983, 1985; Strickland, Fitzgerald and Swank, 1984, 1986a, 1986b). For most trees, C-bonded sulphur, but not ester sulphate, is synthesized; this C-bonded sulphur is added to forest soils by both root and leaf litter (David, Mitchell and Scott, 1987). Both ester sulphate of microbial origin and C-bonded sulphur can be mineralized, releasing sulphate into the soil solution.

There is presently little information on the changes of organic sulphur constituents under field conditions, especially for periods extending to many years. There is a need to distinguish between net and gross mineralization and immobilization rates. Generally, measurements using radioisotopes over short incubation intervals (i.e. days) approximate gross rates while evaluations using stable isotopes evaluate changes in sulphur pools over extended periods (i.e. years). Quantification of those processes that lead to differential distribution of different isotopes of sulphur may aid in the evaluation of long-term net rates (Rundel, Ehleringer and Nagy, 1988).

9.4.8 ORGANIC SULPHUR IN SOIL SOLUTION

Organic sulphur can also undergo transfer in a soil solution. David, Mitchell and Scott (1987) found that large amounts of C-bonded sulphate and ester sulphate were leached from the O and E horizons of a Spodosol at the Huntington Forest. Because ester sulphate is not input to the forest floor via precipiation, litter or roots, it must be formed in situ. David *et al.* assumed steady-state conditions at the site, where formation of ester sulphate would equal leaching losses (0.22 g (S) m^{-2} a^{-1}; 6.9 mmol m^{-2} a^{-1}). This would imply that a minimum of 25% of ester sulphate in the forest floor turns over annually. Carbon-bonded sulphur in the same forest floor was primarily a result of litter and root inputs (David, Mitchell and Scott, 1987). Although immobilization of sulphate into C-bonded sulphur occurred, there was net mineralization of C-bonded sulphur in the forest floor. The large C-bonded sulphur pool in the mineral soil can probably be attributed to the accumulation of humified organic material derived from direct root inputs and illuviation.

Leaching of organic sulphur from the microbially active forest floor provides an important source of organic sulphur which accumulates in mineral horizons (Schoenau and Bettany, 1987). Further support of this transfer is from laboratory studies by Schindler and Mitchell (1987). $^{35}SO_4{}^{2-}$ was added to soil taken from the O and B horizons from the Huntington Forest, and it was rapidly incorporated into C-bonded sulphur in O horizons. The amount of organic sulphur (as C-bonded S) leached from the Oa horizon soil exceeded the net release of water-soluble sulphate. However, most of the isotope (92–93%) added to the O horizon soil was leached from the soil and thus not incorporated. In another study, Mitchell *et al.* (1989), using a mass balance approach, demonstrated that leaching of organic sulphur in solution from the forest floor is an important mechanism in the formation of organic S in B horizons. They suggested that approximately 2500 years of leaching could account for the accumulation of organic sulphur in the mineral soil.

9.5 CONTROLS ON TRANSFORMATIONS IN SOIL: ENERGETICS AND KINETICS

One of the most useful conceptual models of sulphur transformations in soil was developed by McGill and Cole (1981). This model linked microbial, C, N, and S transformations in organic matter, and treated the organic S constituents, ester sulphate and C-bonded sulphur, separately. These authors hypothesized that mineralization (or hydrolysis) of ester sulphate was regulated by the biological demand for sulphur (biochemical mineralization), whereas C-bonded sulphur mineralization (or oxidation) was more closely associated with mineralization of C and N for energy needs (biological mineralization). Although this model includes sulphate, it does not treat adsorption and desorption reactions. Also, the model of McGill and Cole (1981) predicts that the synthesis of enzymes which cleave sulphate esters (sulphatases) is a function of microbial sulphur demand. Although formation of some such enzymes (arylsulphatases) is repressed by elevated sulphate concentrations, synthesis mechanisms of others (alkylsulphatases) are induced in response to microbial C and energy requirements (Dodgson, White and Fitzgerald, 1982), thus complicating this conceptual approach.

Bettany and Stewart (1983) and Maynard, Stewart and Bettany (1984) expanded the McGill and Cole model by dividing organic sulphur fractions (both C-bonded sulphur and ester sulphate) into labile, clay-protected and stable organic constituents. They also included labile inorganic sulphur, solution sulphate, and stable inorganic sulphur (adsorbed and occluded sulphate and precipitated sulphide). This model suggests that only small pools of organic sulphur are included in rapid transformations, and that there is a large stable organic pool that is not labile.

A few investigators have evaluated the McGill and Cole conceptual model using laboratory experiments. David *et al.* (1983) examined mineralization–immobilization in organic and mineral horizons from the hardwood site at the Huntington Forest. In the organic-rich Oa and Bh horizons, C-bonded sulphur mineralization was associated with high mineralization rates of organic C. However, in the Bs1 and Bs2 horizons, which were low in organic matter, sulphate desorption and ester sulphate mineralization were of greater importance as sources of sulphate, which suppports the McGill and Cole model. Because soluble sulphate was being leached from the soil, ester sulphate mineralization increased in the C horizons, which had insufficient rates of C-bonded sulphur mineralization to provide enough sulphur for microbial growth. In the Oa and Bh horizons mineralization of C-bonded sulphur was sufficient, and thus ester sulphate was not mineralized.

Further work on the Huntington soil with radioisotopes has shown that ester sulphate turnover is more rapid than that of C-bonded sulphur (Schindler *et al.*, 1986). Also, McLaren, Keer and Swift (1985) proposed that only a small fraction of soil sulphur, most likely microbial biomass, is involved in short-term sulphur cycling. Hunt, Stewart and Cole (1986) developed a simulation model based on McGill and Cole concepts. The model was tested using data from ^{35}S laboratory experiments; the results supported the hypothesis that excess sulphate is stored as ester sulphate and that both C—S bonded and ester sulphate sulphur were rapidly mineralized.

9.6 INORGANIC SULPHUR TRANSFORMATIONS

9.6.1 GENERAL DESCRIPTION OF SULPHATE ADSORPTION

The ability of many soils to retain sulphate by inorganic adsorption mechanisms has long been recognized (Ensminger, 1954; Kamprath, Nelson and Fitts, 1956; Tikhova, 1958; Chao, Harward and Fang, 1962), and adsorption of sulphate may regulate sulphur leaching in some forested ecosystems (Cole and Johnson, 1977; Johnson *et al.*, 1983; Fuller, David and Driscoll, 1985; Gaston, Mansell and Rhue, 1986). The exact nature of sulphate adsorption in soils is chemically complex. However, adsorption is generally considered to take place by two mechanisms, one of which results in an equivalent release of another anion, typically OH^-. The other mechanism results in neutral adsorption of sulphate along with accompanying cations and has been designated 'selective' or 'salt' adsorption. The multiple negative charge and the bridging that can occur due to sulphate being divalent are important for its co-adsorption with cations.

Sulphate retention has generally been shown to release OH^- at low soil pH (e.g. Rajan, 1978; Fuller, David and Driscoll, 1985; Hodges and Johnson, 1987) which suggests an adsorption mechanism where the negative charge of sulphate adsorbed on to an Al or Fe hydroxide surface is balanced by displacement of OH^-. Several researchers (as reviewed by Rajan, 1978) have found $SO_4^{2-} : OH^-$ stoichiometry close to 1 : 1 for sulphate adsorption. However, other studies (Hue, Adams and Evans, 1985; Zhang, Zhang and Yu, 1987) found sulphate retention was much greater than OH^- release, indicating at least some of the sulphate adsorption was not driven by reactions that alter soil pH.

Calculations of OH^- release during sulphate desorption are generally based on pH measurements (Hue, Adams and Evans, 1985). However, since the salt added to increase sulphate concentration may depress pH (sometimes termed the 'salt effect'), the OH^- released may be underestimated. Even after adjusting these estimates for ionic strength, it was noted that an Ultisol (Alabama, USA) did not show stoichiometric release of OH^- at pH levels > 5.6. Sulphate adsorption also resulted in increased exchangeable cations and cation exchange capacity, indicating co-adsorption of cations with sulphate. Such results indicate that sulphate adsorption may take place by mechanisms that result in a gain in negative charge of the Al or Fe oxide surface and water displacement.

The neutral adsorption mechanism is supported by observations that soils with low pHs release higher levels of OH^- during sulphate adsorption than soils with higher pHs, and that soils with high pH have no anion adsorption capacity (Hue, Adams and Evans, 1985).

In addition to the mechanism of inorganic surface adsorption, mineral formation can explain some sulphate retention in soils, particularly when pH is low and Al high (Nordstrom, 1982). The reaction of gibbsite with sulphate results in the precipitation of jurbanite (Equation 9.1) (Van Breemen, 1973; Evans and Zelazny, 1987) or basaluminite (Equation 9.2) (Adams and Rawajfih, 1977; Adams and Hajek, 1978; Wolt, 1981) and the displacement of equivalent quantities of OH^-:

$$Al(OH)_3 + SO_4^{2-} = Al(OH)SO_4 + 2OH^- \quad (9.1)$$

$$4Al(OH)_3 + SO_4^{2-} = Al_4(OH)_{10}\,SO_4 + 2OH^- \quad (9.2)$$

It is not possible to distinguish reactions (9.1) and (9.2) from simple surface adsorption based on the generation of OH^-.

9.6.2 FACTORS AFFECTING SULPHATE ADSORPTION

Several factors enhance sulphate adsorption including: (1) decreasing soil pH; (2) increasing clay content or the quantity of A1 and Fe oxides; and (3) decreasing organic matter. The relative importance of each of these factors varies with soil type. Researchers have often concentrated on a single factor in evaluating its effect on sulphate adsorption, but these factors often are correlated with each other. Experimental procedures that alter a particular soil property may also change other properties, and unless all changes are carefully evaluated, their actual effect on sulphate adsorption might not be noted.

9.6.2.1 Effect of pH

Adsorption of sulphate typically increases with decreasing soil pH (Kamprath, Nelson and Fitts, 1956; Elkins and Ensminger, 1971; Gebhardt and Coleman, 1974; Couto, Lathwell and Bouldin, 1979; Singh, 1984b; Bergseth, 1985; Zhang, Zhang and Yu, 1987), and several researchers have found SO_4^{2-} adsorption to be correlated to surface positive charge (Van Raij and Peech, 1972; Marsh, Tillman and Syers, 1987). If soil is acidified by naturally formed organic acids or strong mineral acids of anthropogenic origin (i.e. acidic deposition), sulphate adsorption will tend to increase, as expected when anion adsorption is due to exchange. As H^+ is incorporated on to the surface, the mineral surface becomes positively charged and thus can retain additional anions (Figure 9.9). In the case of multivalent anions such sulphate, a bridge can form between the anion and the surface, and there is a potential for covalent bonds to form by loss of water (Goldberg and Sposito, 1984). Anions that form covalent bonds are tightly held and may change the original surface from an anion to a cation exchanger.

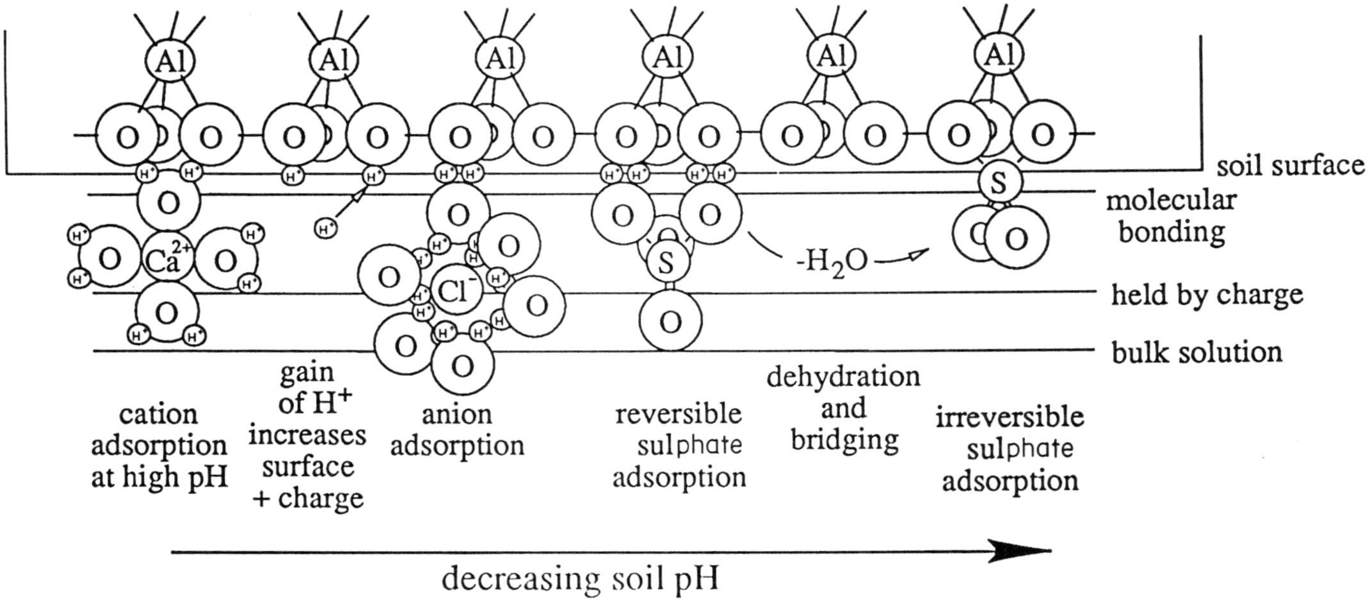

Figure 9.9. Potential model for bridging of sulphate from reversibly adsorbed to irreversibly adsorbed

9.6.2.2 Fe and Al Oxides

Oxides and hydroxide coatings of Fe and Al are primary factors which determine sulphate adsorption of soil (Adams and Rawajfih, 1977; Singh, 1984a; Rajan, 1978; Parfitt and Smart, 1978). Nodvin, Driscoll and Likens (1986a) have shown for several soils that there is pH at which adsorption reaches a maximum, and lower pH levels typically result in solubilization of surface coatings of Fe and Al. Thus, low pH may destroy the sulphate adsorption surfaces and reduce potential sulphate retention sites unless formation of a new surface occurs. In comparing a variety of soils, there is a significant and high correlation between sulphate adsorption and oxalate extractable Al (an index of organically bound and amorphous Al) for Inceptisols and Spodosols, but not for Ultisols (Figure 9.10). Thus, no single factors, such as oxalate extractable Al, has been found which can predict sulphate adsorption for a broad range of soils due to the complex interactions among organic matter, pH, and extractable Al and Fe (Parfitt and Smart, 1978; Johnson and Todd, 1983; Dethier *et al.*, 1988).

9.6.2.3 Organic Matter

Organic matter as a competing ligand may decrease sulphate adsorption. In addition, organic acids such as citrate and oxalate may solubilize of Fe and Al

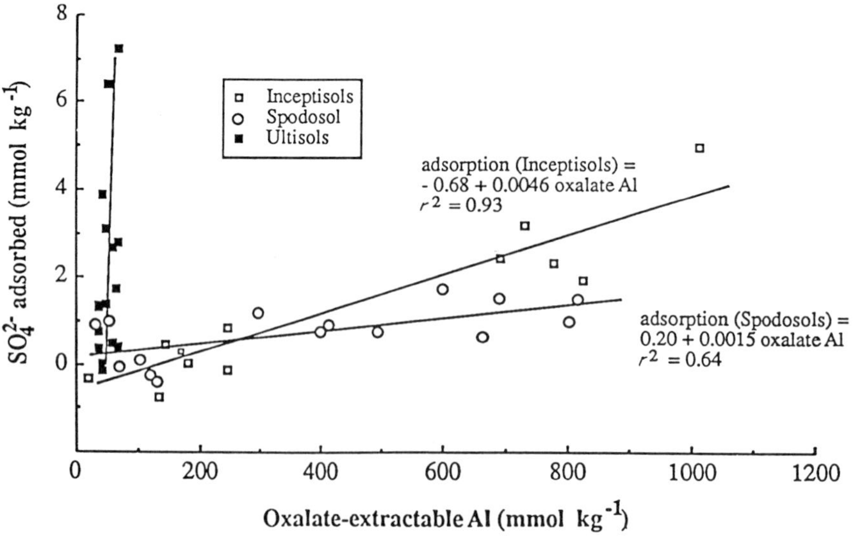

Figure 9.10. Relationship between sulphate adsorption and oxalate-extractable Al for a variety of soils (from Harrison, Johnson and Todd, 1989)

(Schnitzer, 1969; Bartlett and Riego, 1972; Hue, Craddock and Adams, 1988). These two factors may explain the reduction in sulphate adsorption with increasing organic matter concentration in soil. However, the role of soil organic matter in affecting sulphate adsorption may be easily confounded with other factors (e.g. surface area, presence of Fe and Al oxides) which may increase organic matter as well as sulphate retention. For similar soils there is generally a negative relationship between organic matter concentration and sulphate retention. For example, within a single pedon, the surface soil typically has the highest level of organic matter and lowest adsorption.

9.6.3 MODELS FOR DESCRIBING SULPHATE ADSORPTION

The most commonly utilized model for describing adsorption is based on the Langmuir (1918) equation typically written as follows:

$$C / S_S = 1 / M^*b + C / M \tag{9.3}$$

where C is the concentration of sulphate in solution, M is the adsorption capacity, b is a coefficient related to the bonding energy of sulphate adsorption, and S_S is amount of sulphate adsorbed per unit area. When C/S_S is plotted as a function of C, then $1/S_S$ will be the slope and $1/M^*b$ the intercept. The use of this equation has been criticized (Harter and Smith, 1981; Nodvin, Driscoll and Likens, 1986a), because such a plot will almost always result in an apparent significant relationship between C/S_S versus C, because C is a variable located on both axes.

A modification of the Langmuir equation, which uses the initial mass of sulphate added, rather than the final concentration, has resulted in an improved description of sorption in several soils in New Hampshire (Nodvin, Driscoll and Likens, 1986a). Other equations that have been utilized include the Freundlich equation (Kuo and Lotse, 1973; Bolan *et al.*, 1986), and the Temkin equation (Mead, 1981).

9.6.4 REVERSIBILITY OF SULPHATE ADSORPTION

The degree to which sulphate adsorption is reversible is an important consideration in projecting recovery of acidified surface waters and soils following decreases in atmospheric deposition of sulphur (Galloway, Norton and Church, 1983; Reuss and Johnson, 1985, 1986; Harrison, Johnson and Todd, 1989). Reductions of sulphate in forest soil leachates in response to decreases in sulphur inputs will be prolonged if sulphate desorbs (Figure 9.11). Because sorption reactions are kinetic, there is increased adsorption or desorption with increasing time of equilibration between the exchanger and solution. Many researchers consider most adsorption reactions to be reversible given sufficient time for desorption to occur (Barrow and Shaw, 1977), but it is

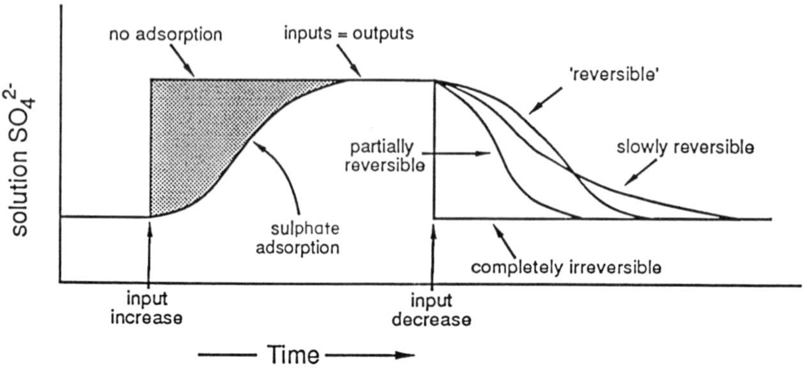

Figure 9.11. Paths of soil solution sulphate concentration with and without adsorption following deposition increases, and showing a variety of sulphate reversibility responses following deposition decreases (from Harrison, Johnson and Todd, 1989)

important to recognize that the residence time of solutes in forest soils may range from minutes to days. Especially under conditions of high flow, solute concentrations may not reflect equilibrium conditions such as those predicted by adsorption isotherms.

The results of sulphate adsorption reversibility experiments are as varied as the observations of sulphate adsorption. Findings have included: (1) release of more sulphate than was absorbed (mobilization of other sulphur pools; (2) complete or near-complete reversibility (all adsorbed sulphate released; (3) partial reversibility of adsorbed sulphate; and (4) little or no reversibility of sulphate adsorption. Most adsorption and desorption studies use batch techniques, but other variables such as time, intensity of shaking and extractant vary considerably. Extractants have included distilled and deionized water or various salts, such as orthophosphate, to increase sulphate displacement. The following discussion is divided into the four possible observations with regard to adsorption reversibility.

9.6.4.1 More Sulphate Desorbed than Adsorbed

Researchers have not often reported observing more sulphate released by desorption than was retained. This may be due to the preponderance of studies using soils with high sulphate adsorption capacities; such soils generally show more retention than release. However, other soils which have been exposed to high sulphur loading and have a relatively small sulphate retention capacity may show more release than adsorption. This net release could be due to the extractant removing previously insoluble sulphate or to the mineralization of organic sulphur. For instance, Cuoto, Lathwell and Bouldin (1979) quantified sulphate adsorption in the laboratory and found a high percentage of desorption for several soils (over 100% in the case of an Acrohumox B2 horizon). They speculated that these high values may have been due to desorption of sulphate which was originally adsorbed at high solution concentrations, but these results could also have been due to the mineralization of organic sulphur which was not determined in their study.

9.6.4.2 Complete or Nearly-complete Reversibility

Chao, Harward and Fang (1962) found that sulphate adsorbed by 11 of 15 soil horizons (Oregon, USA) was readily desorbed by water. The previously adsorbed sulphate was much more readily extracted with KH_2PO_4 solution since phosphate would effectively displace inner and outer sphere sulphate by its higher negative valence and greater charge density (Figure 9.9).

9.6.4.3 Partial Reversibility of Adsorbed Sulphate

Four of the soil horizons that Chao, Harward and Fang (1962) studied exhibited 'appreciable' adsorption. For these soils, added $^{35}SO_4^{2-}$ was leached below a 4 cm depth in the soil column with an application of 20 cm of water, but there was considerable degree of sulphate adsorption. However, the 'adsorption' may also have been due to isotopic dilution caused by non-radioactive sulphate being desorbed or mineralized which would greatly exceed the concentration of $^{35}SO_4^{2-}$ added. Further studies of two of these sulphate-adsorbing soils showed that sulphate adsorption was reversible (> 70%) only after an extended period of extraction (four successive extractions for 2 h with deionized water).

Hodges and Johnson (1987) studied adsorption reversibility by miscible displacement and estimated that complete desorption of previously adsorbed sulphate would require a period of 10 to 20 times greater than the initial adsorption time. In a study of sulphate adsorption on kaolinite (Aylmore, Karim and Quirk, 1967), sulphate adsorption was mostly reversible (> 50%), but sulphate adsorbed on to Fe and Al oxides was essentially irreversible (< 50%). It would be expected that sulphate adsorbed on to amorphous Al and Fe oxides might show greater irreversibility compared to that adsorbed on kaolinite; there would be a higher potential for the sulphate ion to enter the inner sphere beneath the amorphous surface of Al and Fe (Figure 9.9) than the crystalline structure of kaolinite.

Weaver, Khanna and Beese (1985) found that one of four soil horizons (10–20 cm; sampling down to 30 cm) showed irreversible sulphate adsorption. The particular horizon showing irreversible adsorption had the lowest pH and a lower original sulphate concentration than soil horizons above it. This would be expected since soil properties that lead to higher initial sulphate adsorption appear to decrease adsorption reversibility. In another study of sulphate adsorption reversibility, 10–30% of the total of previously adsorbed sulphate was extracted with dilute HCl (pH 3.5), 20 to 40% with water (pH 5.6), 50 to 60% with 0.01 mol $Ca(NO_3)_2$ and nearly 100% of adsorbed sulphate with 0.01 mol $Ca(H_2PO_4)_2$ by repeated extractions of several soils (Singh, 1984b). Soil samples included iron-podzols (highest sulphate adsorption) and brown earths (lowest sulphate adsorption). In yet another study, 33 to 76% of adsorbed sulphate was desorbed with four KH_2PO_4 (16 mmol P) extractions from clay and sandy loam soils from the West Indies (Haque and Walmsley, 1973).

Nodvin, Driscoll and Likens (1986a, 1986b) and Fuller *et al.* (1987) noted that the subsurface horizons of some northeastern US spodosols adsorbed up to 0.5 μmol sulphate per gram of soil (16 μg (S) g^{-1}). Little of the sulphate retained in these horizons was desorbed by water, but desorption was higher with phosphate extraction, ranging from 0.3 to 2.1 μmol g^{-1} (10 to 67 μg (S) g^{-1}); the higher values indicate there was net 'desorption'.

Bornhemisza and Llanos (1967) observed leaching of $^{35}SO_4{}^{2-}$ by water in columns of three Costa Rican soils, including the A and B horizons of a Regosol, a pumice-originated alluvial soil, and a Latosol. They found that ^{35}S did not move to any large degree in the Latosol and the alluvial soil, but that the Regosol showed little retention. When orthophosphate was intermixed with sulphate applications, the added sulphate was more mobile due to displacement by phosphate. However, in the subsoil of the Latosol, sulphate was retained despite the additions of phosphate.

9.6.4.4 Little or No Reversibility of Sulphate Adsorption

As mentioned above, Aylmore, Karim and Quirk (1967) found sulphate adsorption on kaolinite to be mostly reversible, but sulphate adsorbed on Fe and Al oxides was essentially irreversible. Khanna and Beese (1978) found that leaching of chloride and nitrate through a podzolic acid brown soil released little previously adsorbed sulphate. These monovalent anions would not be expected to compete to a high degree with sulphate for retention sites due to their lower charge densities; however, the high concentrations of these monovalent ions used by Khanna and Beese should have displaced any sulphate retained in the outer sphere of the retention surface (Figure 9.9) by mass action. This would be analogous to displacing exchangeable Ca^{2+} by adding KCl to a soil. These observations tend to support inner sphere retention of sulphate as the primary mechanism for irreversible adsorption.

9.7 MODELS OF SULPHUR IN FOREST SOILS

Models of sulphur include conceptual formulations, mass balance calculations, chemical equilibrium and surface complexation models, and process level representation. Such models tend to fall into two classes: those which assume chemical equilibrium, and those which assume kinetic control.

9.7.1 CHEMICAL EQUILIBRIUM MODELS

It is difficult to quantify changes of sulphate minerals using conventional techniques such as X-ray diffraction or infrared methods due to the complex and heterogeneous structure of soil. Typically, therefore, the potential for mineral formation and control of solution chemistry is evaluated by estimating the chemical activity of ions in solution, and comparing ionic activity products of known sulphate minerals with accepted solubility products of reference minerals. This technique offers distinct advantages in evaluating the chemistry of solutions passing through forest soils and can be used to predict which minerals control solution composition.

The estimation of ionic activities and their formation or dissolution potential is based on relatively simple chemical principles (Adams, 1971). However, since solutions present in terrestrial systems typically contain a variety of chemical species, the calculation of ionic activities may be quite complex. Numerous computer programs have been developed that facilitate the calculation of ionic activities; the more commonly utilized programs include MINEQL (Westall, Zachary and Morel, 1985), WATEQ (Truesdell and Jones, 1974), and GEOCHEM (Sposito and Mattigod, 1979). Most computer programs utilize an analysis of the chemical composition of a solution to estimate ionic activities with either the Debye–Huckel, or Davie's equation (Adams, 1971; Truesdell and Jones, 1974). Ionic activity products are calculated and compared to solubility products of reference minerals. For instance, some studies of soils subject to high levels of acid sulphate additions have shown that the ionic-activity product of the mineral jurbanite was constant over a range of soil conditions, and thus could be controlling sulphate solubility in these systems (Van Breemen, 1973; Evans and Zelazny, 1987; Khanna *et al.*, 1987; Von Freiesleben, 1988).

9.7.2 KINETIC AND PROCESS MODELS

Reuss, Christophersen and Seip (1986) reviewed four freshwater and soil acidification models. These models, which describe sulphur retention as a function of sulphate adsorption and desorption, in turn are dependent on sulphate concentration for a given soil type. Mitchell and Fuller (1988) examined a broader range of models with specific attention to the role of sulphur dynamics within the soil. Many of these models focus on sulphate adsorption, generally depicting sulphate adsorption as a partition coefficient or as a simple isotherm in which adsorption is dependent solely on concentration; complete reversibility is assumed. Such models have been used to predict the acidification of surface waters with varying degrees of success (Chen *et al.*, 1983; Christophersen and Wright, 1981; Christophersen, Seip and Wright, 1982; Christophersen *et al.*, 1984; Cosby, Hornberger and Galloway, 1985; Cosby *et al.*, 1985a, 1985b, 1986; Cosby, Whitehead and Neale, 1986; Gherini *et al.*, 1985; Seip *et al.*, 1985; Wright, 1984).

Although process level models have been formulated which include organic sulphur transformations (Fuller *et al.*, 1986a; Hunt, Stewart and Cole, 1986), these have generally not been used to model sulphur dynamics of forest ecosystems because the information required for simulating organic sulphur dynamics is unavailable or it has been assumed that organic sulphur dynamics plays a minor role in regulating sulphur flux through the ecosystem.

9.8 CONCLUSIONS AND MAJOR UNANSWERED QUESTIONS

Our understanding of the biogeochemistry of sulphur in forested ecosystems has improved markedly during the past decade. Many recent studies have focused on ascertaining the impacts of acidic deposition on forest vegetation, soils and surface waters. Sulphur dynamics affect the flux of H^+ and sulphate, both of which interact with other elemental cycles (especially N and basic cations), and affect the acidification of soils and downstream aquatic ecosystems.

There is extensive information on the sulphur constituents and fluxes for temperate forest ecosystems, but little for tropical areas (see appendix to this chapter). These latter areas may be particularly important contributors to trace sulphur gases in the troposphere (see Andreae and Jaeschke, this volume).

On regional as well as global scales, forest ecosystems are being subjected to changes in atmospheric loadings of sulphur as well as other anthropogenic effects, including deforestation and climatic alterations. Predicting how these changes will affect both organic and inorganic sulphur pools of forest ecosystems, and how these alterations will influence gaseous and leaching losses, should be the general focus of future research on sulphur biogeochemistry.

We suggest that there are four major areas which must be emphasized in future research on sulphur dynamics of forest ecosystems:

(1) Gaseous emission from forest vegetation, particularly in tropicals forests.
(2) Measurements of dry and cloud water deposition.
(3) Determination of the reversibility of sulphur retention–adsorption in soil.
(4) Evaluation of changes in organic sulphur constituents over extended periods.

REFERENCES

Adams, F. (1971). Ionic concentrations and activities in soil solutions. *Soil Sci. Soc. Am. Proc.*, **35**, 420–6.

Adams, F. and Hajek, B. F. (1978). Effects of solution sulfate, hydroxide, and potassium concentrations on the crystallization of alunite, basaluminite, and gibbsite from dilute aluminium solutions. *Soil Sci.*, **126**, 169–73.

Adams, F. and Rawajfih, Z. (1977). Basaluminite and alunite: A possible cause of sulfate retention by acid soils. *Soil Sci. Soc. Am. J.*, **41**, 686–92.

Andreae, M. O. and Jaeschke, W. A. This volume.

Aneja, V. P. and Cooper, W. J. (1989). Biogenic sulfur emissions: a review. In: Saltzman, E. S. and Cooper, W. J. (Eds). *Biogenic Sulfur in the Environment*. ACS Symposium Series (New Orleans, Louisiana, 1987), American Chemical Society, Washington DC, pp. 2–13.

Aspiras, R. B., Kenny, D. R. and Chesters, G. (1972). Determination of reduced and inorganic sulfur forms as sulfide by zinc-hydrochloric acid distillation. *Anal. Lett.*, **5**, 425–32.

Aylmore, L. A. G., Karim, M. and Quirk, J. P. (1967). Adsorption and desorption of sulfate ions by soil constituents. *Soil Sci.*, **103**, 10–15.

Barrow, N. J. and Shaw, T. C. (1977). The slow reactions between soil and anions: 7: effect of time and temperature on contact between an adsorbing soil and sulphate. *Soil Sci.*, **124**, 347–54.

Bartel-Ortiz, L. and David, M. B. (1988). Sulfur constituents and transformations in upland and floodplain forest soils. *Can. J. For. Res.*, **18**, 1106–12.

Bartlett, R. J. and Riego, D. C. (1972). Effect of chelation on the toxicity of aluminum. *Plant Soil*, **37**, 419–23.

Bergseth, H. (1985). Selektierungsvermogen eines eisenoxidehydroxids und einiger tonminerale gegenuber phohphat- und sulfationen. *Acta Agric. Scand.*, **35**, 375–88.

Bettany, J. R. and Stewart, J. W. B. (1983). Dynamics of organic sulphur in semi-arid agroecosystems. In: More, A. I. (Ed.). *Sulphur- 82*: *Proceedings of the International Conference*, Vol. II. British Sulphur Corporation, London, pp. 767–85.

Bolan, N. S., Scotter, D. R., Syers, J. K. and Tillman, R. W. (1986). The effect of adsorption on sulfate leaching. *Soil Sci. Soc. Am. J.*, **50**, 1419–24.

Bornhemisza, E. and Llanos, R. (1967). Sulfate movement, adsorption, and desorption in three Costa Rican soils. *Soil Sci. Soc. Am. Proc.*, **31**, 356–60.

Chao, T. T., Harward, M. E. and Fang, S. C. (1962). Movement of ^{53}S tagged sulfate through soil columns. *Soil Sci. Soc. Am. Proc.*, **26**, 27–32.

Chen, C. W., Gherini, S. A., Hudson, R. J. M. and Dean, J. D. (1983). *The Integrated Lake-Watershed Acidification Study*. Vol. 1: *Model Principles and Application Procedures*. EPRI Report EA-3221, Vol. 1, Research Project 1109-5. 214pp.

Christophersen, N., Seip, H. M. and Wright, R. F. (1982). A model for steamwater chemistry at Birkenes, Norway. *Water Resources Res.*, **18**, 977–97.

Christophersen, N. and Wright, R. F. (1981). Sulfate budget and a model for sulfate concentration at Birkenes, a small forested catchment in southernmost Norway. *Water Resources Res.*, **17**, 377–89.

Christophersen, N., Dymbe, L. H., Johannssen, M. and Seip, H. M. (1984). A model for sulphate in streamwater at Storgama, Southern Norway. *Ecol. Model.*, **21**, 35–61.

Cogbill, C. V. and Likens, G. E. (1974). Acid precipitation in the northeastern United States. *Water Resources Res.*, **10**, 1133–7.

Cole, D. W. and Johnson, D. W. (1977). Atmospheric sulfate additions and cation leaching in a Douglas-fir ecosystem. *Water Resources Res.*, **13**, 313–17.

Cosby, B. J., Hornberger, G. M. and Galloway, J. N. (1985). Modelling the effects of acid deposition: assessment of a lumped parameter model of soil water and streamwater chemistry. *Water Resources Res.*, **21**, 51–63.

Cosby, B. J., Whitehead, P. G. and Neale, R. (1986). A preliminary model of long-term changes in stream acidity in southwestern Scotland. *J. Hydrol.*, **84**, 381–401.

Cosby, B. J., Hornberger, G. M., Galloway, J. N. and Wright, R. F. (1985). Time scales of catchment acidification. *Environ. Sci. Technol.*, **19**, 1144–9.

Cosby, B. J., Wright, R. F., Hornberger, G. M. and Galloway, J. N. (1985b). Modelling the effects of acid deposition: estimation of long-term water quality responses in a small forested catchment. *Water Resources Res.*, **21**, 1591–1601.

Cosby, B. J., Hornberger, G. M., Rastetter, E. B., Galloway, J. N. and Wright, R. F. (1986). Estimating catchment water quality response to acid deposition using mathematical models of soil ion exchange processes. *Geoderma*, **38**, 77–95.

Couto, W., Lathwell, D. J. and Bouldin, D. R. (1979). Sulfate adsorption by two oxisols and an alfisol of the tropics. *Soil Sci.*, **127**, 108–16.

Cronan, C. S. and Schofield, C. L. (1979). Aluminum leaching response to acid precipitation: effects on high-elevation watersheds in the Northeast. *Science*, **204**, 304–6.

David, M. B., Mitchell, M. J. and Nakas, J. P. (1982). Organic and inorganic sulfur constituents of a forest soil and their relationship to microbial activity. *Soil Sci. Soc. Am. J.*, **46**, 847–52.

David, M. B., Mitchell, M. J. and Schindler, S. C. (1984). Dynamics of organic and inorganic sulfur constituents in hardwood forest soils. In: Stone, E. L. (1984). *Forest Soils and Treatment Impacts. Proceedings of the Sixth North American Soils Conference*. The University of Tennessee, Knoxville, Tenn., pp. 221–45.

David, M. B., Mitchell, M. J. and Scott, T. J. (1987). Importance of biological processes in the sulfur budget of a northern hardwood ecosystem. *Biol. Fertil. Soils*, **5**, 258–64.

David, M. B., Schindler, S. C., Mitchell, M. J. and Strick, J. E. (1983). Importance of organic and inorganic sulfur to mineralization processes in a forest soil. *Soil Biol. Biochem.*, **15**, 671–7.

Dethier, D. P., Jones, S. B., Feist, T. P. and Ricker, J. E. (1988). Relations among sulfate, aluminum, iron, dissolved organic carbon, and pH in upland forest soils of northwestern Massachusetts. *Soil Sci. Soc. Am. J.*, **52**, 506–12.

Dewalle, D. R., Sharpe, W. R. and Edwards, P. J. (1988). Biogeochemistry of two Appalachian deciduous forest sites in relation to episodic stream acidification. *Water Air Soil Pollut.*, **40**, 143–56.

Dodgson, K. S., White, G. F. and Fitzgerald, J. W. (1982). *Sulfatases of Microbial Origin*, Vol. 2. CRC Press, Boca Raton, Florida, 199pp.

Driscoll, C. T. and Schecher, W. D. (1988). Aluminum in the environment. In: Sigel, H. and Sigel, A. (Eds). *Metal Ions in Biological Systems*. Vol. 24, *Aluminum and its Role in Biology*. Marcel Dekker, New York, pp. 59–122.

Elkins, D. M. and Ensminger, L. E. (1971). Effect of soil pH on the availability of adsorbed sulfate. *Soil Sci. Soc. Am. Proc.*, **35**, 931–4.

Ensminger, L. E. (1954). Some factors affecting the adsorption of sulfate by Alabama soils. *Soil Sci. Soc. Am. Proc.*, **18**, 259–64.

Evans, A. and Zelazny, L. W. (1987). Effects of sulfate additions on the status of exchangeable Al in a Cecil soil. *Soil Sci.*, **143**, 410–17.

Everett, A. G., Retzsch, W. C., Kramer, J. P., Montunez, I. P. and Duhaime, P. F.

(1983). Hydrochemical characteristics of Adirondack waters influenced by terrestrial environments. In: Kotik, M. D. (Ed.). *The Second National Symposium on Acid Rain*, 18pp.

Farrell, E. P., Nilsson, S. I., Wiklander, G. and Tamm, C. O. (1984). Distribution of sulphur fractions in lysimeters previously treated with sulphuric acid, NPK fertilizer and a combination of acid and fertilizer. *For. Ecol. Manage.*, **8**, 41–57.

Farwell, S. O., Sherrard, A. E., Pack, M. R. and Adams, D. F. (1979). Sulfur compounds volatilized from soils at different moisture contents. *Soil Biol. Biochem.*, **11**, 4111–15.

Fitzgerald, J. W. (1976). Sulfate ester formation and hydrolysis: a potentially important yet often ignored aspect of sulfur cycle of aerobic soils. *Bacteriol. Rev.*, **40**, 698–721.

Fitzgerald, J. W. (1978). Naturally occurring organosulfur compounds in soil. In: Nriagu, J. O. (Ed.). *Sulfur in the Environment*. Part II. *Ecological Impacts*. Wiley, New York, pp. 391–443.

Fitzgerald, J. W. and Andrew, T. L. (1984). Mineralization of methionine sulphur in soils and forest floor layers. *Soil Biol. Biochem.*, **16**, 565–70.

Fitzgerald, J. W. and Andrew, T. L. (1985). Metabolism of methionine in forest floor layers and soil: influence of sterilization and antibiotics. *Soil Biol. Biochem.*, **17**, 881–3.

Fitzgerald, J. W. and Johnson, D. W. (1982). Transformations of sulphate in forested and agricultural lands. In: More, A. I. (Ed.). *Sulphur-82*: *Proceedings of the International Conference*, Vol. II. British Sulphur Corporation, London, pp. 411–26.

Fitzgerald, J. W., Strickland, T. C. and Swank, W. T. (1982). Metabolic fate of inorganic sulphate in soil samples from undisturbed and managed forest ecosystems. *Soil Biol. Biochem.*, **14**, 529–36.

Fitzgerald, J. W., Watwood, M. E. and Rose, F. A. (1985). Forest floor and soil arylsulphatase: hydrolysis of tyrosine sulphate, an environmentally relevant substrate for the enzyme. *Soil Biol. Biochem.*, **17**, 885–7.

Fitzgerald, J. W., Ash, J. T., Strickland, T. C. and Swank, W. T. (1983). Formation of organic sulfur in forest soils: a biologically mediated process. *Can. J. For. Res.*, **13**, 1077–82.

Foster, N. W. (1985). Acid precipitation and soil solution chemistry within a maple-birch forest in Canada. *For. Ecol. Manage.*, **12**, 215–31.

Foster, N. W., Morrison, I. K. and Nicolson, J. A. (1986). Acid deposition and ion leaching from a podzolic soil under hardwood forest. *Water Air Soil Pollut.*, **31**, 879–89.

Foster, N. W., Nicolson, J. A. and Hazlett, P. W. (1989). Temporal variation in nitrate and nutrient cations in drainage waters from a deciduous forest. *J. Environ. Qual.*, **18**, 238–44.

Freney, J. R. (1961). Some observations on the nature of organic sulphur compounds in soil. *Aust. J. Agric. Res.*, **12**, 424–32.

Freney, J. R. (1967). Sulfur-containing organics. In: McLaren, A. D. and Peterson, G. H. (Eds). *Soil Biochemistry*. Marcel Dekker, New York, pp. 229–59.

Freney, J. R., Melville, G. E. and Williams, C. H. (1971). Organic sulphur fractions labelled by addition of ^{35}S-sulphate to soil. *Soil Biol. Biochem.*, **3**, 133–41.

Freney, J. R., Melville, G. E. and Williams, C. H. (1975). Soil organic matter fractions as sources of plant-available sulphur. *Soil Biol. Biochem.*, **7**, 217–21.

Freney, J. R., Stevenson, F. J. and Beavers, A. H. (1972). Sulfur-containing amino acids in soil hydrolysates. *Soil Sci.*, **114**, 468–76.

Freney, J. R. and Williams, C. H. (1983). The sulfur cycle in soil. In: Ivanov, M. V. and Freney, J. R. (Eds). *The Global Biogeochemical Sulphur Cycle*. Wiley, New York.

Fuller, R. D., David, M. B. and Driscoll, C. T. (1985). Sulfate adsorption relationships in forested spodosols of the northeastern US. *Soil Sci. Soc. Am. J.*, **49**, 1034–40.

Fuller, R. D., Driscoll, C. T., Schindler, S. C. and Mitchell, M. J. (1986a). A simulation model of sulfur transformations in forested Spodosols. *Biogeochemistry*, **2**, 313–28.

Fuller, R. D., Mitchell, M. J., Krouse, H. R., Wyskowski, B. J. and Driscoll, C. T. (1986b). Stable sulfur isotope ratios as a tool for interpreting ecosystem sulfur dynamics. *Water Air Soil Pollut.*, **28**, 163–71.

Fuller, R. D., Driscoll, C. T., Lawrence, G. B. and Nodvin, S. C. (1987). Processes regulating sulfate flux after whole-tree harvesting. *Nature*, **325**, 707–10.

Galloway, J. N., Norton, S. A. and Church, M. R. (1983). Freshwater acidification from atmospheric deposition of sulfuric acid: A conceptual model. *Environ. Sci. Technol.*, **17**, 541a–5a.

Garten, C. T. (1988). Fate and distribution of sulfur-35 in yellow poplar and red maple trees. *Oecologia*, **76**, 43–50.

Garten, C. T., Bondietti, E. A. and Lomax, R. D. (1988). Contribution of foliar leaching and dry deposition to sulfate in net throughfall below deciduous trees. *Atmos. Environ.*, **22**, 1425–32.

Gaston, L. A., Mansell, R. S. and Rhue, R. D. (1986). Sulfate mobility in acid soils and implications with respect to cation leaching: A Review. *Soil Crop Sci. Soc. Fla. Proc.*, **45**, 67–72.

Gebhardt, H. and Coleman, N. T. (1974). Anion adsorption by allophanic tropical soils: II. sulfate adsorption. *Soil Sci. Soc. Am. Proc.*, **38**, 259–62.

Gherini, S. A., Mok, L., Hudson, R. J. M., Davis, G. F., Chen, C. W. and Goldstein, R. A. (1985). The ILWAS model: Formulation and application. *Water Air Soil Pollut.*, **26**, 425–59.

Gibson, J. H., Galloway, J. N., Schofield, C., McFee, W., Johnson, R., McCarley, S., Dise, N. and Herzog, D. (1983). *Rocky Mountain Acidification Study*. US Fish and Wildlife Service, Division of Biological Services, Eastern Energy and Land Use Team, FWS/OBS-80/40.17.

Glass, N. R., Arnold, D. E., Galloway, J. N., Hendry, G. R., Lee, J. J., McFee, W. W., Norton, S. A., Powers, C. F., Rambo, D. L. and Schofield, C. L. (1982). Effects of acid precipitation. *Environ. Sci. Technol.*, **16**, 162–9.

Goldberg, S. and Sposito, G. (1984). A chemical model of phosphate adsorption by soils: II. Noncalcareous soils. *Soil Sci. Soc. Am. J.*, **48**, 779–83.

Haines, B. (1983). Forest ecosystem SO_4-S input–output discrepancies and acid rain: are they related? *Oikos*, **41**, 139–43.

Haines, B. This volume.

Haque, I. and Walmsley, D. (1973). Adsorption and desorption of sulfate in some soils of the West Indies. *Geoderma*, **9**, 269–78.

Harrison, R. B., Johnson, D. W. and Todd, D. E. (1989). Sulfate adsorption and desorption reversibility in a variety of forest soils. *J. Environ. Qual.*, **18**, 419–26.

Harter, R. D. and Smith, G. (1981). Langmuir equation and alternate methods of studying 'adsorption' reactions in soils. In: Dowdy, R. H., Baker, D. E., Ryan, J. A. and Volk, V. V. (Eds). *Chemistry in the Soil Environment*. American Society of Agronomy, Madison WI.

Harwood, J. L. (1980). Sulfolipids. In: Stumpf, P. R. and Conn, E. E. (Eds). *The Biochemistry of Plants*, Vol. 4. Academic Press, New York, pp. 301–20.

Harwood, J. L. and Nicholls, R. G. (1979). The plant sulfolipid—a major component of the sulphur cycle. *Biochem. Soc. Trans.*, **7**, 440–7.

Hicks, B. B. (1986). Measuring dry deposition: a re-assessment of the state of the art. *Water Air Soil Pollut.*, **30**, 75–90.

Hodges, S. C. and Johnson, G. C. (1987). Kinetics of sulfate adsorption and desorption by Cecil soil using miscible displacement. *Soil Sci. Soc. Am. J.*, **51**, 323–31.

Houghton, C. and Rose, F. A. (1976). Liberation of sulfate from sulfate esters by soils. *Appl. Environ. Microbiol.*, **31**, 969–76.

Hue, N. V., Adams, F. and Evans, C. E. (1985). Sulfate retention by an acid BE horizon of an Ultisol. *Soil Sci. Soc. Am. J.*, **49**, 1196–200.

Hue, N. V., Craddock, G. R. and Adams, F. (1988). Effect of organic acids on aluminum toxicity in subsoils. *Soil Sci. Soc. Am. J.*, **50**, 28–34.

Hunt, W. B., Stewart, J. W. B. and Cole, C. V. (1986). Concepts of sulfur, carbon, and nitrogen transformations in soil: evaluation by simulation modeling. *Biogeochemistry*, **2**, 163–77.

Jarvis, B. W., Lang, G. E. and Wieder, R. K. (1987). Arylsulphatase activity in peat exposed to acid precipitation. *Soil Biol. Biochem.*, **19**, 107–109.

Johnson, D. W. (1984). Sulfur cycling in forests. *Biogeochemistry*, **1**, 29–43.

Johnson, D. W. and Todd, D. E. (1983). Relationships among iron, aluminum, carbon, and sulfate in a variety of forest soils. *Soil Sci. Soc. Am. J.*, **47**, 792–800.

Johnson, D. W., Turner, J. and Kelly, J. M. (1982). The effects of acid rain on forest nutrient status. *Water Resources Res.*, **18**, 449–61.

Johnson, D. W., Hornbeck, J. W., Kelly, J. M., Swank, W. T. and Todd, D. E. (1980). Regional pattern of soil sulfate accumulation: relevance to ecosystem sulfur budgets. In: Shriner, D. S., Richmond, C. R. and Lindberg, S. E. (Eds). *Atmospheric Sulfur Deposition, Environmental Impact and Health Effects*. Ann Arbor Science Publishers, Ann Arbor, Michigan, pp. 507–19.

Johnson, D. W., Henderson, G. S., Huff, D. D., Lindberg, S. E., Richter, D. D., Shriner, D. S., Todd, D. E. and Turner, J. (1982). Cycling of organic and inorganic sulfur in a chestnut oak forest. *Oecologia*, **54**, 141–8.

Johnson, D. W., van Miegroet, H., Cole, D. W. and Richter, D. D. (1983). Contributions of acid deposition and natural processes to cation leaching from forest soils: a review. *J. Air Pollut. Control Assoc.*, **33**, 1036–41.

Johnson, D. W., Richter, D. D., Lovett, G. M. and Lindberg, S. E. (1985). The effects of atmospheric deposition on potassium, calcium, and magnesium cycling in two deciduous forests. *Can. J. For. Res.*, **15**, 773–82.

Johnson, D. W., Richter, D. D., Van Miegroet, H., Cole, D. W. and Kelly, J. M. (1986). Sulfur cycling in five forested ecosystems. *Water Air Soil Pollut.*, **30**, 965–79.

Johnson, N. M., Likens, G. E., Bormann, F. H., Fisher, D. W. and Pierce, R. S. (1968). Rate of chemical weathering of silicate minerals in New Hampshire. *Geochim. Cosmochim. Acta*, **32**, 531–45.

Kamprath, E. J., Nelson, W. L. and Fitts, J. W. (1956). The effect of pH, sulfate and phosphate concentrations on the adsorption of sulfate of soils. *Soil Sci. Soc. Am. Proc.*, **26**, 463–6.

Karkanis, M. (1976). The circulation of sulphur in the forest ecosystem Tilo-Carpinetum in the northern part of Puszcza Niepolomicka near Ispina. *Fragmenta Floristica et Geobotanica*, **22**, 351–63.

Khanna, P. K. and Beese, F. (1978). The behavior of sulfate salt input in podzolic brown earth. *Soil Sci.*, **125**, 16–22.

Khanna, P. K., Prenzel, J., Meiwes, K. J., Ulrich, B. and Matzner, E. (1987). Dynamics of sulfate retention by acid forest soils in an acidic deposition environment. *Soil Sci. Soc. Am. J.*, **51**, 446–52.

Kowalenko, C. G. (1978). Organic nitrogen, phosphorus and sulfur in soils. In: Schnitzer, M. and Khan, S. U. (Eds). *Soil Organic Matter. Developments in Soil Science 8*. Elsevier, Amsterdam, pp. 95–136.
Krouse, H. R. and Tabatabai, M. A. (1986). Stable sulfur isotopes. *Sulfur Agric.*, **27**, 169–205.
Kuo, S. and Lotse, E. G. (1973). Kinetics of phosphate adsorption and desorption by hematite and gibbsite. *Soil Sci.*, **116**, 400–8.
Lambert, M. J. (1986). Sulphur and nitrogen nutrition and their interactive effects on *Dothistroma* infection in *Pinus radiata*. *Can. J. For. Res.*, **16**, 1055–62.
Lambert, M. J. and Turner, J. (1977). Dieback in high site quality *Pinus radiata* stands—the role of sulphur and boron deficiencies. *N.Z. J. For. Sci.*, **7**, 333–48.
Lambert, M. J., Turner, J. and Edwards, D. W. (1976). *Effects of Sulphur Deficiency in Forests. Proceedings of the 16th IUFRO World Congress*, Oslo, Norway, November 1976. IUFRO, Oslo.
Landers, D. H., David, M. B. and Mitchell, M. J. (1983). Analysis of organic and inorganic sulfur constituents in sediments, soils and water. *Intern. J. Anal. Chem.*, **14**, 245–56.
Langmuir, I. (1918). The adsorption of gases on plane surfaces of glass, mica, and platinum. *J. Am. Chem. Soc.*, **40**, 1361–82.
Likens, G. E., Bormann, F. H., Pierce, R. S., Eaton, J. S. and Johnson, N. M. (1977). *Biogeochemistry of a Forested Ecosystem*. Springer-Verlag, New York, 146pp.
Likens, G. E., Bormann, F. H., Pierce, R. S., Eaton, J. S. and Munn, R. E. (1984). Long-term trends in precipitation chemistry at Hubbard Brook, New Hampshire. *Atmos. Environ.*, **18**, 2641–7.
Likens, G. E., Bormann, F. H., Hedin, L. O., Driscoll, C. T. and Eaton, J. S. (1990). Dry deposition of sulfur: a 23 record for the Hubbard Brook Forest Ecosystem. *Tellus*. **42B**, 319–29.
Lindberg, S. E. and Garten, C. T. (1988). Sources of sulphur in forest canopy throughfall. *Nature*, **336**, 148–51.
Lindberg, S. E., Lovett, G. M., Richter, D. D. and Johnson, D. W. (1986). Atmospheric deposition and canopy interactions of major ions in a forest. *Science*, **231**, 41–145.
Lovelock, J. E., Maggs, R. J. and Rasmussen, R. A. (1972). Atmospheric dimethyl sulfur and the natural sulfur cycle. *Nature*, **237**, 452–3.
Lovett, G. M., Reiners, W. A. and Olson, R. K. (1982). Cloud droplet deposition in subalpine balsam fir forests: hydrological and chemical inputs. *Science*, **218**, 1303–4.
Lowe, L. E. and DeLong, W. A. (1963). Carbon bonded sulphur in selected Quebec soils. *Can. J. Soil Sci.*, **43**, 151–5.
McGill, W. B. and Cole, C. V. (1981). Comparative aspects of cycling of organic C, N, S and P through soil organic matter. *Geoderma*, **26**, 267–86.
McLaren, R. G., Keer, J. I. and Swift, R. S. (1985). Sulphur transformation in soils using sulphur-25 labelling. *Soil Biol. Biochem.*, **17**, 73–9.
Marsh, K. B., Tillman, R. W. and Syers, J. K. (1987). Charge relationships of sulfate sorption by soils. *Soil Sci. Soc. Am. J.*, **51**, 318–23.
Maynard, D. G., Germida, J. J. and Addison, P. A. (1986). The effect of elemental sulfur on certain chemical and biological properties of surface organic horizons of a forest soil. *Can. J. For. Res.*, **16**, 1050–4.
Maynard, D. G., Stewart, J. W. B. and Bettany, J. R. (1984). Sulfur cycling in grassland and parkland soils. *Biogeochemistry*, **1**, 97–111.
Mead, J. A. (1981). A comparison of the Langmuir, Freundlich and Temkin Equa-

tions to describe phosphate adsorption properties of soils. *Aust. J. Soil Res.*, **19**, 333–42.

Meiwes, K. J. and Khanna, P. K. (1981). Distribution and cycling of sulphur in the vegetation of two forest ecosystems in an acid rain environment. *Plant Soil*, **60**, 369–75.

Mitchell, C. C. and Blue, W. G. (1981). The sulfur fertility status of Florida soils I. Sulfur distribution in Spodosols, Entisols, and Ultisols. *Soil Crop Sci. Soc. Fla. Proc.*, **40**, 71–6.

Mitchell, M. J. and Fuller, R. D. (1988). Models of sulfur dynamics in forest and grassland ecosystems with emphasis on soil processes. *Biogeochemistry*, **5**, 133–63.

Mitchell, M. J., David, M. B., Maynard, D. G. and Telang, S. A. (1986). Sulfur constituents in soils and streams of a watershed in the Rocky Mountains of Alberta. *Can. J. For. Res.*, **16**, 315–20.

Mitchell, M. J., Driscoll, C. T., Fuller, R. D., David, M. B. and Likens, G. E. (1989). Effect of whole-tree harvesting on the sulfur dynamics of a forest soil. *Soil Sci. Soc. Am. J.*, **53**, 933–40.

Mollitor, A. V. and Raynal, D. J. (1983). Acid precipitation and ionic movements in Adirondack forest soils. *Soil Sci. Soc. Am. J.*, **46**, 137–41.

Murphy, M. D. (1980). Essential micronutrients III: sulphur. In: Davies, B. E. (Ed.). *Applied Soil Trace Elements*. Wiley, New York, pp. 235–8.

Nakas, J. P. (1986). Role of the microflora in terrestrial sulfur cycling. In: Mitchell, M. J. and Nakas, J. P. (Eds). *Microfloral and Faunal Interactions in Natural and Agro-Ecosystems*. Martinus Nijhoff, Dordrecht, pp. 285–316.

Neary, A. J., Mistry, E. and Vanderstar, L. (1987). Sulphate relationships in some central Ontario forest soils. *Can. J. Soil Sci.*, **67**, 341–52.

Nodvin, S. C., Driscoll, C. T. and Likens, G. E. (1986a). The effect of pH on sulfate adsorption by a forest soil. *Soil Sci.*, **142**, 69–75.

Nodvin, S. C., Driscoll, C. T. and Likens, G. E. (1986b). Simple partitioning of anions and dissolved organic carbon in a forest soil. *Soil Sci.*, **142**, 27–35.

Nordstrom, S. A. (1982). The effect of sulfate on aluminum concentrations in natural waters: stability relations in the system Al_2O_3—SO_3—H_2O at 298 K. *Geochim. Cosmochim. Acta*, **46**, 681–92.

Nriagu, J. O. and Krouse, H. R. This volume.

Nys, C. and Ranger, J. (1988). Influence d'une substitution d'espèce sur le fonctionnement biogeochimique de l'ecosysteme foret l'exemple du cycle du souffre. *Ann. Sci. For. (Paris)*, **45**, 169–88.

Pack, D. H. (1980). Precipitation chemistry patterns: a two-network data set. *Science*, **208**, 1143–5.

Parfitt, R. L. and Smart, R. S. C. (1978). The mechanism of sulfate adsorption on iron oxides. *Soil Sci. Soc. Am. J.*, **42**, 48–50.

Parker, G. G. (1983). Throughfall and stemflow in the forest nutrient cycle. *Adv. Ecol. Res.*, **13**, 57–133.

Rajan, S. S. S. (1978). Sulfate adsorbed on hydrous alumina, ligands displaced, and changes in surface charge. *Soil Sci. Soc. Am. J.*, **42**, 39–44.

Raynal, D. J., Raleigh, F. S. and Mollitor, A. V. (1985). Atmospheric deposition and ionic movement in Adirondack forests. In: Adams, D. D. and Page, W. P. (Eds). *Acid Deposition*. Plenum Press, New York.

Rennenberg, H. (1984). The fate of excess sulfur in higher plants. *Annu. Rev. Plant Physiol.*, **35**, 121–53.

Rennenberg, H. (1989). Synthesis and emission of hydrogen sulfide by higher plants. In: Saltzman, E. S. and Cooper, W. J. (Eds). *Biogenic Sulfur in the Environment*.

ACS Symposium (New Orleans, Louisiana, 1987). American Chemical Society, Washington DC, pp. 44–57.

Reuss, J. O., Christophersen, M. and Seip, H. M. (1986). A critique of models for freshwater and soil acidification. *Water Air Soil Pollut.*, **30**, 909–30.

Reuss, J. O. and Johnson, D. W. (1985). Effect of soil processes on the acidification of water by acid deposition. *J. Environ. Qual.*, **12**, 26–31.

Reuss, J. O. and Johnson, D. W. (1986). *Acid Deposition and the Acidification of Soils and Waters*. Springer-Verlag, New York, 119–pp.

Rundel, P. W., Ehleringer, J. R. and Nagy, K. A. (1988). *Stable Isotopes in Ecological Research*. Springer-Verlag, New York, 525pp.

Schindler, S. C. and Mitchell, M. J. (1987). Dynamics of ^{35}S in horizons and leachates from a hardwood forest spodosol. *Soil Biol. Biochem.*, **19**, 531–8.

Schindler, S. C., Mitchell, M. J., Scott, T. J., Fuller, R. D. and Driscoll, C. T. (1986). Incorporation of ^{35}S-sulfate into inorganic and organic constituents of two forest soils. *Soil Sci. Soc. Am. J.*, **50**, 457–62.

Schnitzer, M. (1969). Reactions between fulvic acid, a soil humic compound, and inorganic soil constituents. *Soil Sci. Soc. Am. Proc.*, **33**, 75–81.

Schoenau, J. J. and Bettany, J. R. (1987). Organic matter leaching as a component of carbon, nitrogen, phosphorus, and sulfur cycles in a forest, grassland, and gleyed soil. *Soil Sci. Soc. Am. J.*, **51**, 646–51.

Seip, H. M., Seip, R., Dillon, P. J. and Grosbois, E. D. (1985). Model of sulphate concentration in a small stream in the Harp Lake catchment, Ontario. *Can. J. Fish. Aquat. Sci.*, **42**, 927–37.

Shepard, J., Mitchell, M. J., Scott, T. J., Zhang, Y. M. and Raynal, D. J. (1989). Measurements of wet and dry deposition in a northern hardwood forest. *Water Air Soil Pollut.*, **48**, 225–38.

Singh, B. R. (1984a). Sulfate sorption by acid forest soils: 2. Sulfate adsorption isotherms with and without organic matter and oxides of aluminum and iron. *Soil Sci.*, **138**, 294–7.

Singh, B. R. (1984b). Sulfate sorption by acid forest soils: 3. Desorption of sulfate from adsorbed surfaces as a function of time, desorbing ion, pH, and amount of adsorption. *Soil Sci.*, **138**, 346–53.

Smith, R. A. and Alexander, R. B. (1986). Correlations between stream sulphate and regional SO_2 emissions. *Nature*, **322**, 722–4.

Sposito, G. and Mattigod, S. (1979). GEOCHEM: A computer program for the calculation of chemical equilibria in soil solutions and other natural water systems. Kearney Foundation of Soil Science, Univ. of Calif., Riverside.

Stottlemyer, R. and Hanson, D. G. Jr (1989). Atmospheric deposition and ionic concentrations in forest soils of Isle Royale National Park, Michigan. *Soil Sci. Soc. Am. J.*, **53**, 270–4.

Strick, J. E., Schindler, S. C., David, M. B., Mitchell, M. J. and Nakas, J. P. (1982). Importance of organic sulfur constituents and microbial activity to sulfur transformations in an Adirondack forest soil. *Northeast. Environ. Sci.*, **1**, 161–9.

Strickland, T. C. and Fitzgerald, J. W. (1984). Formation and mineralization of organic sulfur in forest soils. *Biogeochemistry*, **1**, 79–95.

Strickland, T. C. and Fitzgerald, J. W. (1985). Incorporation of sulphate-sulphur into organic matter extracts of litter and soil: involvement of ATP sulphurylase. *Soil Biol. Biochem.*, **17**, 779–84.

Strickland, T. C. and Fitzgerald, J. W. (1986). Organosulphur recalcitrance in soil and litter from a hardwood forest. *Soil Biol. Biochem.*, **18**, 661–2.

Strickland, T. C., Fitzgerald, J. W. and Swank, W. T. (1984). Mobilization of recently formed forest soil organic sulfur. *Can. J. For. Res.*, **14**, 63–7.

Strickland, T. C., Fitzgerald, J. W. and Swank, W. T. (1986a). In situ measurements of sulfate incorporation into forest floor and soil organic matter. *Can. J. For. Res.*, **16**, 549–53.

Strickland, T. C., Fitzgerald, J. W. and Swank, W. T. (1986b). In situ mobilization of ^{35}S-labelled organic sulphur in litter and soil from a hardwood forest. *Soil Biol. Biochem.*, **18**, 463–8.

Strickland, T. C., Fitzgerald, J. W., Ash, J. T. and Swank, W. T. (1987). Organic sulfur transformations and sulfur pool sizes in soil and litter from a southern Appalachian hardwood forest. *Soil Sci.*, **143**, 453–8.

Swank, W. T., Fitzgerald, J. W. and Ash, J. T. (1984). Microbial transformation of sulfate in forest soils. *Science*, **223**, 182–4.

Swank, W. T., Fitzgerald, J. W. and Strickland, T. C. (1984). Transformations of sulfur in forest floor and soil of a forest ecosystem. In: Johansson, I. (Ed.). Hydrological and hydrogeochemical mechanisms and model approaches to the acidification of ecological systems., International Hydrological Programme, Uppsala, pp. 137–45.

Tabatabai, M. A. and Bremner, J. M. (1972). Forms of sulfur, and carbon, nitrogen and sulfur relationships in Iowa soils. *Soil Sci.*, **114**, 380–6.

Tikhova, E. P. (1958). Sulfate ion absorption by soil colloids. *Tr. Voronezh. Gos. Univ.*, **45**, 29–32.

Truesdell, A. H. and Jones, B. F. (1974). WATEQ, a computer program for calculating chemical equilibria of natural waters. *J. Res. US Geol. Surv.*, **2**, 233–48.

Turner, J., Johnson, D. W. and Lambert, M. J. (1980). Sulphur cycling in a Douglas-fir forest and its modification by nitrogen applications. *Acta Oecol. Oecal. Plant.*, **15**, 27–35.

Turner, J. and Lambert, M. J. (1980). Sulfur nutrition of forests. In: Shriner, D. S., Richmond, C. R. and Lindberg, S. E. (Eds). *Atmospheric Sulfur Deposition, Environmental Impact and Health Effects*. Ann Arbor Science Publishers, Ann Arbor, Michigan, pp. 321–34.

Turner, J., Lambert, M. J. and Gessel, S. P. (1977). Use of foliage sulphate concentrations to predict response to urea application by Douglas-Fir. *Can. J. For. Res.*, **7**, 476–80.

Ulrich, B., Mayer, R. and Khanna, R. K. (1980). Chemical changes due to acid precipitation in a loess-derived soil in central Europe. *Soil Sci.*, **130**, 193–9.

Van Breemen, N. (1973). Dissolved aluminum in acid sulfate soils and in acid mine waters. *Soil Sci. Soc. Am. Proc.*, **37**, 694–7.

Van Loon, G. W., Hay, G. W. and Goh, R. (1987). Analysis of sulfur-containing components of a soil treated with simulated acid rain. *Water Air Soil Pollut.*, **34**, 233–40.

Van Raij, B. and Peech, M. (1972). Electro-chemical properties of some oxisols and alfisols of the tropics. *Soil Sci. Soc. Am. Proc.*, **36**, 587–93.

Von Freiesleben, N. E. (1988). Sulfate mobility in an acid Danish forest soil. *J. Environ. Qual.*, **17**, 278–84.

Watwood, M. E. and Fitzgerald, J. W. (1988). Sulfur transformations in forest litter and soil: Results of laboratory and field incubations. *Soil Sci. Soc. Am. J.*, **52**, 1478–83.

Watwood, M. E., Fitzgerald, J. W., Swank, W. T. and Blood, E. R. (1988). Factors involved in potential sulfur accumulation in litter and soil form a coastal pine forest. *Biogeochemistry*, **6**, 3–19.

Weaver, G. T., Khanna, P. K. and Beese, F. (1985). Retention and transport of sulfate in a slightly acid forest soil. *Soil Sci. Soc. Am. J.*, **49**, 746–50.

Westall, J. C., Zachary, J. L. and Morel, F. M. M. (1985). MINEQL. A computer program for the calculation of chemical equilibrium composition of aqueous systems. Water Quality Laboratory, Department of Civil Engineering, Massachusetts Institute of Technology. Technical Note No. 18.

Westberg, H. (1984). Estimation of biogenic sulfur emissions from the continental U.S. In: Aneja, V. P. (Ed.). *Environmental Impact of Natural Emissions*. Air Pollution Control Association, Pittsburgh, PA, pp. 41–53.

Whelpdale, D. M. This volume.

Wieder, R. K., Lang, G. E. and Granus, V. A. (1987). Sulfur transformation in *Sphagnum*-derived peat during incubation. *Soil Biol. Biochem.*, **19**, 101–6.

Wiklander, L. (1973). The acidification of soil by acid precipitation. *Grundforbattring*, **26**, 155–64.

Wolt, J. D. (1981). Sulfate retention by acid sulfate-polluted soils in the Copper Basin area of Tennessee. *Soil Sci. Soc. Am. J.*, **45**, 283–7.

Woodman, J. N. and Cowling, E. B. (1987). Airborne chemicals and forest health. *Environ. Sci. Technol.*, **21**, 120–6.

Wright, R. F. (1984). Norwegian models for surface water chemistry: an overview. In: Schnoor, J. L. (Ed.). *Modeling of Total Acid Precipitation Impacts*. Butterworth Publishers, Boston, pp. 73–87.

Wright, R. F. and Johannessen, M. (1980). Input–output budgets of major ions at gauged catchments in Norway. In: Drablos, D. and Tollan, A. (Eds). *Ecological Impact of Acid Precipitation*. SNSF Project. Sandefjord, Norway, pp. 250–1.

Zhang, G. Y., Zhang, X. N. and Yu, T. R. (1987). Adsorption of sulfate and fluoride by variable charge soils. *J. Soil Sci.*, **38**, 29–38.

Zucker, A. and Zech, W. (1985). Sulfur status of four uncultivated soil profiles in northern Bavaria. *Geoderma*, **36**, 229–40.

APPENDIX TO CHAPTER 9: PROJECTED EFFECTS OF TROPICAL DEFORESTATION ON SULPHUR CYCLING

BRUCE HAINES
University of Georgia, Athens GA, USA

Deforestation, the conversion of forest lands to agricultural use, is progressing in tropical forests in response to the rapidly growing human population (Fearnside, 1986; Detwiler and Hall, 1988). Loss of productive capacity from soil erosion, nutrient leaching, and build-up of pests is a consequence of deforestation (Nye and Greeland, 1965; Watters, 1971; Fearnside, 1986). Diminished rainfall is another possible consequence (Salati, 1986). Felled trees may be exported from the site to saw mills and pulp mills or, more commonly, partially burned and left to decompose on the site. Burning removes mostly leaves and small branches, some of the trunks, and very little of the root material. Sulphur, nitrogen, carbon, and other constituents from the forest can be mobilized by fire into the atmosphere as particulates, aerosols, and gases, or to the soil as ash. Following partial burning, tree trunks decompose, gradually yielding some elements to gas phase and some into both organic and inorganic dissolved phases. The proportions of nutrient stock lost to gas, aerosol, particulate, ash, dissolved organic and dissolved inorganic fractions do not appear to have been quantified for any tropical forest. Removal of the energy-absorbing forest canopy brings the kinetic energy of the rain drops directly to the soil surface, greatly accelerating soil erosion and with it some of the sulphur capital from the system.

Potential effects of deforestation on sulphur cycling are outlined in this appendix. Quantitative predictions are not yet possible because of an almost complete lack of data. Carbon is the most thoroughly quantified element in the context of deforestation, and the quantity of carbon stored in and mobilized out of forests through deforestation serves as a starting point in this speculative paper. Sulphur storage and mobilization are then estimated from the C : S ratio in trees. Although the quantity and composition of sulphur gas species and aerosols mobilized to the atmosphere by deforestation, the residence time of those sulphur species in the atmosphere, and the deposition rates have received some attention, many important questions remain to be resolved (Andreae and Jaeschke, this volume).

The sulphur cycle of an intact tropical forest provides a conceptual base line to analyse the potential effects of deforestation on sulphur cycling. Inputs to

Sulphur Cycling on the Continents
Edited by R.W. Howarth, J.W.B. Stewart and M.V. Ivanov

the forest include inorganic and organic sulphur in wet and dry deposition, as sampled with wet- and dry-fall precipitation gauges as well as aerosols impacting on the forest canopy and sulphur gases absorbed by the plant leaves. Among these possible inputs, only wet plus dry precipitation or wet only precipitation have been routinely measured in any tropical ecosystem. Sulphur cycling in the forest involves wash-off of dry deposits from external leaf surfaces as well as leaching from inside the leaves in forest throughfall solution (Mitchell, David and Harrison, this volume). Sulphur gases are emitted from the leaves of some rainforest trees (Haines *et al.*, 1987; Andreae and Jaeschke, this volume). Falling trunks, leaves, twigs, plant reproductive parts, and animal faeces and remains carry sulphur to the forest floor. In the forest floor, the decomposition of litter from the forest canopy as well as from roots generates organic sulphur and sulphate. Mobilized sulphate may be transformed by microbes into carbon-bonded and ester-bonded forms in the litter and soil (Fitzgerald *et al.*, 1988), incorporated into microbial biomass, adsorbed on to the soil (Reuss and Johnson, 1986), taken up by plant roots, or leached to surface or groundwaters. Most sulphur taken up by plant roots is used in synthesis of new plant tissue, but some is converted by some rainforest trees of the family Mimosoideae to the toxic gas CS_2 (Haines *et al.*, 1987; Haines, Black and Bayer, 1989) and to other volatile sulphur gases (Andreae and Jaeschke, this volume). Attempts to partition sulphur quantitatively into all of these pathways have been discussed for temperate forest systems (Mitchell, David and Harrison, this volume), but not for intact tropical forests or cut-and-burned tropical forests. The magnitude of sulphur loss via soil erosion from intact forests and cut-and-burned forests has not been quantified at any site as yet.

Estimates of carbon released by tropical deforestation for 1980 are summarized by Detwiler and Hall (1988). These are for low and for high estimates of land-use change, evaluated both on the basis of timber volume and on destructive sampling. Estimates are also given for the release of carbon from the soil (Table 9A.1). These data can be used to estimate sulphur release by assuming that carbon comprised 50% of the biomass and that the plant biomass is 0.2% sulphur. Values for sulphur as a percentage of total dry weight of tropical trees range from 0.07 to 0.36% (Rodin and Bazilevich, 1967; Ewel *et al.*, 1981). The Detwiler and Hall (1988) estimates of carbon release based on vegetation volume range from 310 to 740 Tg (C) a^{-1}; while their estimates based on destructive sampling range from 560 to 1300 Tg (C) a^{-1}. Emissions from the soil range from 110 to 250 Tg (C) a^{-1}. From the vegetation volume data, sulphur loss (both to the atmosphere and to downstream ecosystems) is calculated to range between 1.2 and 2.9 Tg (S) a^{-1} (0.038 to 0.09 Tmol a^{-1}), while estimates from destructive sampling range from 2.2 to 5.2 Tg (S) a^{-1} (0.069 to 0.16 Tmol a^{-1}).

Table 9A.1. Estimated tropical deforestation sulphur loss (DSL), *c.* 1980 based on estimated of net carbon release from vegetation (Veg) and from soils (Soil) summarized by Detwiler and Hall, 1988. Estimation was done for both vegetation volume (Volume) and from destructive sampling (Destructive)

Source	Carbon release 10^6 t a^{-1}		Sulphur release (DSL) 10^6 t a^{-1}	
	Veg	Soil	Veg	Soil
Low estimate of land-use change (Seiler and Crutzen)				
Volume	310	110	1.24	?
Destructive	560	110	2.24	?
High estimate of land-use change (Seiler and Crutzen)				
Volume	520	240	2.08	?
Destructive	910	240	3.64	?
Estimated land-use change (FAO)				
Volume	310	160	1.24	?
Estimated land-use change (Lanly)				
Volume	530	230	2.12	?
Destructive	930	230	3.72	?
Estimated land-use change (Myers)				
Volume	740	250	2.96	?
Destructive	1300	250	5.2	?

When a tropical forest is cut and burned most of the leaves, twigs, litter and some of the surface root mat, but only some small fraction of the tree trunks, will be consumed by the fire. In a shifting agricultural system in Costa Rica, Central America, Ewel *et al.* (1981) found that of the total initial element storage in the vegetation, burning volatilized 31% of C, 22% of N, and 49% of S. If we assume that the total sulphur loss from tropical forests due to deforestation is between 1.2 and 5.2 Tg (S) a^{-1} (0.038 to 0.16 Tmol a^{-1}) and that 49% of this is volatized to the atmosphere, then sulphur emission to the atmosphere from tropical forest burning is estimated as 0.58 to 2.6 Tg (S) a^{-1} (0.018 to 0.08 Tmol a^{-1}). This estimate is somewhat lower than the estimate of Andreae and Jaeschke (this volume) that 2.8 Tg (S) a^{-1} (0.088 Tmol a^{-1}) is emitted to the atmosphere from biomass burning world-wide, including burning of tropical grasslands and burning in temperate regions. The general agreement of these estimates seems good, particularly since the estimate of Andreae and Jaeschke (this volume) is derived from actual measurements of sulphur in the atmosphere over Amazonia.

Globally, the emission of sulphur to the atmosphere from biomass burning is comparable to the biogenic flux from terrestrial ecosystems (Andreae and Jaeschke, this volume), and this is probably true for tropical rainforests as well. Both of these fluxes are probably much smaller than either industrial sources or biogenic fluxes from the oceans (Andreae and Jaeschke, this volume). None the less, the emission of sulphur to the atmosphere from burning tropical forests can be locally important: dry season burning around Lake Valencia, Venezuela, and subsequent wash-out delivered sufficient nitrate and sulphate to cause acid rain of $pH < 4.0$ (Lewis, 1981).

The fate of sulphur in particulates and in solution entering the soil following burning and during later decomposition of residues has received almost no study in tropical forests. In their study in Costa Rica, Ewel *et al.* (1981) found that the leaching and wind of the first post-burn rainstorm removed 6% of the total initial S. Erosion of soil, accelerated by tropical deforestation, must result in loss of some sulphur. However, while there have been numerous studies of soil erosion and its negative impact on productivity in tropics (Lal, 1984, 1986, 1988; Pimentel *et al.*, 1987), none has actually quantified sulphur loss. Some of the sulphur removed from previously forested sites by leaching and erosion is expected to appear in rivers, but this potential effect has yet to be detected. In a compilation of water and sulphur runoff data from 54 large river basins, Husar and Husar (1985) concluded that the spatial resolution of the data was insufficient to detect man-induced sulphur flows via leaching and erosion.

Burning of savannas and subsequent transport of sulphur through the atmosphere to rainforests has been proposed as a mechanism of sulphur impoverishment in the savanna and sulphur enrichment in the forest into which sulphur is deposited (Fox and Blair, 1986). Similar mechanisms could by hypothesized to operate between cut-and-burned tracts of rainforest and intact rainforest. Such sulphur losses can lead to sulphur limitation (Ewel *et al.*, 1981), particularly if forests are frequently burned.

The net loss of sulphur from forests can be analysed in quantitative terms by comparing sulphur losses upon burning to sulphur inputs in deposition. This approach relies on the assumption that in intact forests, sulphur losses through biogenic emissions and other processes are small compared to inputs in deposition (see Andreae and Jaeschke, this volume; Mitchell, David and Harrison, this volume). Using the land-use data in Detwiler and Hall (1988) and total sulphur loss values in Table 9A.1, sulphur losses per unit area burned can be calculated. These losses are estimated as 1.1 to 6.3 g (S) m^{-2}

a^{-1} burned (34 to 190 mmol m^{-2} a^{-1}). The sulphur inputs in bulk precipitation (as sulphate) have been reported for several tropical forests; organic sulphur inputs in precipitation have not been quantified. For the Amazon, sulphate inputs are in the range of 0.12 to 0.4 g (S) m^{-2} a^{-1} (4. to 13 mmol m^{-2} a^{-1}; Galloway, 1985; Stallard and Edmond, 1981). Thus, for the Amazon, the ratio of sulphur losses during deforestation to input from precipitation ranges from 3 to 63. In the most simple sense, this ratio is the time in years required for wet atmospheric deposition sulphur inputs to replace sulphur mobilized by deforestation. There is obviously reason to be concerned that frequent burning could lead to long-term decreases in forest productivity where sulphur losses are high and inputs are low.

REFERENCES

Andreae, M. O. and Jaeschke, W. A. Chapter 3 in this volume.

Detwiler, R. P. and Hall, C. A. S. (1988). Tropical forests and the global carbon cycle. *Science*, **239**, 42–7.

Ewel, J., Berish, C., Brown, B., Price, N. and Raich, J. (1981). Slash and burn impacts on a Costa Rican wet forest site. *Ecology*, **62**(3), 816–29.

Fearnside, P. M. (1986). *Human Carrying Capacity of the Brazilian Rain Forest*. Columbia University Press, New York, 293pp.

Fitzgerald, J. W., Swank, W. T., Strickland, T. C., Ash, J. T., Hale, D. D., Andrew, T. L. and Watwood, M. E. (1989). Sulphur pools and transformations in litter and surface soil of a hardwood forest. In: Swank, W. T. and Crossley, D. A. Jr (Eds). *Forest Hydrology and Ecology at Coweeta*. Springer-Verlag, Berlin, pp. 245–53.

Fox, R. L. and Blair, G. J. (1986). Plant responses to sulphur in tropical soils. In: Tabatabai, M. A., Beaton, J. D. and Fox, R. L. (Eds). *Sulphur in Agriculture. Agronomy Monograph No. 27*. American Society of Agronomy, Crop Science Society of America, Soil Science Society of America, Madison, WI, pp. 405–34.

Galloway, J. N. (1985). The deposition of sulphur and nitrogen from the remote atmosphere. In: Galloway, J. N., Charlson, R. J., Andreae, M. O., Rodhe, H. and Marston, M. S. (Eds). *The Biogeochemical Cycling of Sulphur and Nitrogen in the Remote Atmosphere*. D. Reidel Publishing Co., Dordrecht, pp. 143–75.

Haines, B., Black, M. and Bayer, C. (1989). Sulphur emissions from roots of the rain forest tree *Stryphnodendron excelsum*. In: Saltzman, E. S. and Cooper, W. J. (Eds). *Biogenic Sulphur in the Environment*. ACS Symposium Series, American Chemical Society, Washington DC, pp. 58–69.

Haines, B., Black, M., Fail, J. Jr, McHargue, L. and Howell, G. (1987). Potential sulphur gas emissions from a tropical rainforest and a southern Appalachian deciduous forest. In: Hutchinson, T. C. and Meema, K. M. (Eds). *Effects of Atmopheric Pollutants on Forest, Wetlands and Agricultural Ecosystems*. Springer-Verlag, Berlin, pp. 599–610.

Husar, R. B. and Husar, J. D. (1985). Regional river sulphur runoff. *J. Geophys. Res.*, **90**, (C1), 1115–25.

Lal, R. (1984). Soil erosion from tropical arable alands and its control. *Adv. Agron.*, **37**, 183–248.

Lal, R. (1986). Conversion of tropical rainforest: agronomic potential and ecological consequences. *Adv. Agron.*, **39**, 173–264.

Lal, R. (1988). Soil erosion by wind and water: Problems and prospects. In: Lal, R. (Ed.). *Soil Erosion Research Methods*. Soil and Water Conservation Society. Ankeny, Iowa USA, pp. 1–6.

Lewis, W. M. Jr (1981). Precipitation chemistry and nutrient loading by precipitation in a tropical watershed. *Water Resources Res.*, **17**(1), 169–81.

Mitchell, M. J., David, M. B. and Harrison, R. B. This volume.

Nye, P. H. and Greenland, D. J. (1965). The soil under shifting cultivation. *Technical Communication No. 51*, Commonwealth Bureau of Soils, Harpenden, England.

Pimentel, D., Allen, J., Beers, A., Guinand, L., Linder, R., McLaughlin, P., Meer, B., Musonda, D., Perdue, D., Poisson, S., Siebert, S., Stoner, K., Salazar, R. and Hawkins, A. (1987). World agriculture and soil erosion. *Bioscience*, **37**, 277–83.

Reuss, J. O. and Johnson, D. W. (1986). Acid deposition and the acidification of soils and waters. *Ecological Studies 59*. Springer-Verlag, New York.

Rodin, L. E. and Bazilevich, N. I. (1967). *Production and Mineral Cycling in Terrestrial Vegetation*. Transl. by Fogg, G. E. Oliver and Boyd, Edinburgh, 288pp.

Salati, E. (1986). The forest and the hydrological cycle. In: Dickinson, R. E. (Ed.). *The Geophysiology of Amazonia, Vegetation and Climate Interactions*. Wiley, New York, pp. 273–96.

Stallard, R. F. and Edmond, J. M. (1981). Geochemistry of the Amazon 1. Precipitation chemistry and the marine contribution to the dissolved load at the time of peak discharge. *J. Geophys. Res.*, **86**(C10), 9844–58.

Watters, R. F. (1971). *Shifting Cultivation in Latin America*. Food and Agriculture Organization of the United Nations, Rome, 305pp.

10 Sulphur Cycling in Upland Agricultural Systems

J. J. SCHOENAU and J. J. GERMIDA
Department of Soil Science, University of Saskatchewan, Saskatoon, Canada

10.1 INTRODUCTION

In upland agricultural systems (oxic soils), the major transformations of S are mineralization, immobilization, and oxidation. Such transformations often result in losses or gains of S in the soil–plant system through processes such as leaching and S gas evolution and absorption. As a result, the S load in adjacent hydrospheric and atmospheric systems is altered. S transformations can be greatly affected by small changes in the environment which, in turn, can cause large shifts in the size of the S pools. To predict accurately the effect of a perturbation such as global warming on S load, detailed knowledge of S forms and transformations in the system is required. In the past, progress in elucidating the soil–plant S cycle has been limited by the complex nature of soil S pools, in particular soil organic S, and the lack of suitable investigation techniques. Analytical difficulties have always been a major limitation in advancing knowledge of S cycling. For example, our progress in understanding soil organic S turnover has stagnated somewhat in recent years as we continue to rely largely on characterization methods developed 30 years ago which separate organic S into only two broad groups: HI-reducible S and C-bonded S.

A knowledge of process response to environmental factors is also critical to accurate prediction of S flux at all scales of interest. Only when the relationship between an environmental parameter and a process such as S mineralization is clearly defined can fluxes for all subunits in a body of interest be accurately predicted and combined to arrive at a flux estimate for the whole region. The predictive tools used to estimate flux are usually mathematical models of a process and must account for the complex interactive effect of environmental parameters such as temperature and moisture.

This chapter addresses the S cycle in upland agricultural systems by considering the forms and transformations of S and their relationship to other components of the global S cycle, in particular the atmosphere. Stable S

Sulphur Cycling on the Continents
Edited by R.W. Howarth, J.W.B. Stewart and M.V. Ivanov

isotope techniques and process modelling are introduced as current approaches to identifying and quantifying S forms and fluxes.

10.2 THE NATURE OF SULPHUR IN UPLAND AGRICULTURAL SYSTEMS

In well drained, non-saline surface soils, more than 90% of the S is organic, and this sulphur is the main source of plant available sulphur (sulphate) in soil (Bettany and Stewart, 1983; Freney, 1986). Organic S is separated into two major groups based on susceptibility to reduction by reducing agents: (1) HI-reducible S, and (2) C-bonded S. The HI-reducible fraction consists of organic S that is directly reducible to H_2S by hydriodic acid. It includes S atoms which are not directly bonded to C but are linked to C via an oxygen, nitrogen, or sulphur atom (e.g. C—O—SO_3^- linkages) (Freney, 1961, 1967, 1986; Williams, 1975; Biederbeck, 1978). The HI-reducible fraction, commonly referred to as organic sulphate, is thought to consist primarily of sulphate esters as well as thioglucosides and sulphamates. Organic sulphates are easily converted to inorganic sulphate by mild physical and chemical treatments such as wetting and drying and by treatments with acids and bases (Spencer and Freney, 1960; Freney, Melville and Williams, 1969). Organic sulphates are thought to be associated mainly with the low molecular weight fraction of soil organic matter (Bettany, Stewart and Halstead, 1973; Schoenau and Bettany, 1987), and are considered to be the most labile fraction. These conclusions were based on the ease of hydrolysis and predominance in the non-recalcitrant fractions of soil organic matter (Freney, 1967; Freney, Melville and Williams, 1975; Fitzgerald, 1976, 1978; Bettany, Stewart and Saggar, 1979). Organic sulphates comprise 30 to 70% of the total organic S in surface soils (Bettany, Stewart and Halstead, 1973).

Even though the organic sulphate fraction constitutes an important fraction of soil organic sulphur, its origins and mode of formation remain unclear. In plant residues, more than 90% of the S usually exists in the C-bonded form, whereas upwards of 70% of the humus S derived from such residues may be comprised of organic sulphates (Bettany, Stewart and Saggar, 1979; Roberts and Bettany, 1985). It is reasonable to suggest a microbial role in the production of such compounds during degradation. However, whether the organic sulphates are metabolic by-products, components of dead cells, or products formed by reaction of inorganic sulphate with humic constituents, remains a point of debate.

The C-bonded S fraction consists of S which is directly bonded to C, in the forms of S-containing amino acids in free and combined states, and sulphonates (Freney, 1967; Freney, Melville and Williams, 1970; Biederbeck, 1978). C-bonded S can be distinguished analytically from the organic sulphate fraction because it is not directly reducible to H_2S by hydriodic acid. C-bonded S is calculated as the difference between total organic S and organic sulphate. A portion of C-bonded S, believed to be mainly amino acid S, can be directly measured by reduction to H_2S with Raney Ni (Lowe and DeLong, 1963; Freney, Melville and Williams, 1970, 1975). The residual fraction is highly resistant to chemical degradation, and is probably not a significant source of available S (Biederbeck, 1978). The C-bonded S fraction is believed to be more associated with the highly condensed aromatic humus core and a more stable organic S form that organic sulphate (Bettany, Stewart and Halstead, 1973; Bettany Stewart and Saggar, 1979). However, a portion of the C-bonded S fraction, possibly consisting of simple amino acid S, contributed significantly to the soluble sulphate pool in short-term tracer experiments (Freney, Melville and Williams, 1975; Maynard, Stewart and Bettany, 1985) and field studies (McLachlan and DeMarco, 1975).

Soil microbial biomass is defined as the living part of soil organic matter. The microbial biomass sulphur (MB-S) in agricultural soils comprises about 2–3% of the total S. However, detailed information on the role of MB is limited as measurement methods have only recently been developed for this fraction. Saggar, Bettany and Stewart (1982) proposed a method involving fumigation with chloroform and extraction with 0.01 M $CaCl_2$ to measure MB-S in soil. Using a modification of this method, Gupta (1989) determined that the quantities of MB-S in Saskatchewan agricultural soils ranged from 2.5 to 12.5 μg (S) g^{-1} (0.08 to 0.39 $\mu mol\ g^{-1}$), accounting for 1–4% of the total organic S.

Despite its small size, the MB is a highly active fraction that acts as the driving force behind mineralization–immobilization and oxidation–reduction processes. Microorganisms rapidly assimilate simple, free organics such as low molecular weight ester sulphates and amino acids, and also inorganic sulphate when the S in the substrate is insufficient to meet microbial requirements (immobilization). Organic S, assimilated in excess of microbial requirements, is mineralized by intracellular enzymes and released as sulphate (biological mineralization) (Biederbeck, 1978). Microorganisms can also secrete extracellular sulphatase enzymes such as arylsulphatases and other sulphohydrolases which are capable of hydrolysing organic sulphate functional groups in soil organic matter (biochemical mineralization) (Tabatabai and Bremner, 1970; Gupta, 1989). Mineralizable organic S found in soils appears to be closely associated with the microbial biomass present as binding agents in macroaggregates (Gupta and Germida, 1988a). Net mineralization of microbial biomass S is reported to be dependent on both microbial and soil C : S ratios as well as percentage organic sulphate (Gupta, 1989). For example, native pasture soils with a wide C : S ratio exhibited low net release of microbial sulphur due to re-assimilation of released microbial sulphur by the new flush of microbial growth.

A conceptual model for the various microbial components and processes of the sulphur cycle in soil is illustrated in Figure 10.1. The MB plays a central role as a sink and source for sulphate in the soil, with the interactions between various microbial populations being an important factor regulating sulphate levels. Gupta and Germida (1989) have recently observed that amoebal grazing of bacteria enhances the biochemical and biological mineralization of MB-S in soil. As well, addition of elemental S fertilizers has been reported to decrease populations of predatory protozoa, suggesting reduced nutrient turnover via microbial predation as a result of elemental S fertilization (Gupta and Germida, 1988b). Such findings point out the need to consider microbial ecology and population dynamics when attempting to model S turnover in soils.

Evidence for downward leaching of low molecular weight, S-rich organics during soil formation has led to the suggestion that organic S leaching is an important long-term translocation process in the soil–plant S cycle (Schoenau and Bettany, 1987; Roberts, Bettany and Stewart, 1989). The removal of labile, S-rich organics from surface horizons by leaching may influence S turnover rates and act as another S export mechanism if the organics are lost entirely from the profile.

While organic S is the main reservoir of S in the soil, inorganic S in the form of sulphate is the highly dynamic plant available pool. Inorganic sulphur may occur in agricultural soils as sulphate and as compounds of lower oxidation states such as sulphide, polysulphide, sulphite, thiosulphate, and elemental S. In well-drained, well-aerated soils, the most common form is sulphate and the amounts of reduced S are generally less than $1\,\mu g$ (S) g^{-1} ($0.03\,\mu mol\ g^{-1}$) (Freney, 1961; Williams, 1975). Sulphate may occur in soils as water-soluble salts, adsorbed sulphate, or insoluble forms. In surface soils of neutral to alkaline pH, sulphate exists mainly in the soil solution. In acid soils, considerable quantities of sulphate may be adsorbed to the mineral fraction of the soil. All of the water-soluble sulphate and most of the adsorbed sulphate is available for plant uptake (Freney and Spencer, 1960).

In highly leached soils, sulphate adsorption is an important mechanism for retaining sulphate in the profile. Sulphate adsorption is negligible above pH 6.5 and increases with decreasing pH (Kamprath, Nelson and Fitts, 1956). In soils containing large amounts of iron and aluminum oxides, sulphate adsorption is generally greater (Chao, Harward and Fang, 1962; Mitchell, David and Harrison, this volume). Sulphate adsorption occurs at many energetically different reaction sites so that sulphate adsorption may vary both within and among soils depending on the inorganic constituents present (Williams, 1975).

Insoluble sulphate can exist in the soil as barium and strontium sulphates, sulphates associated with calcium carbonate, and iron and aluminum sulphates (Williams, 1975). Sulphate as a co-crystallized impurity in calcium carbonate can account for up to 95% of the total S in calcareous soils (Williams and

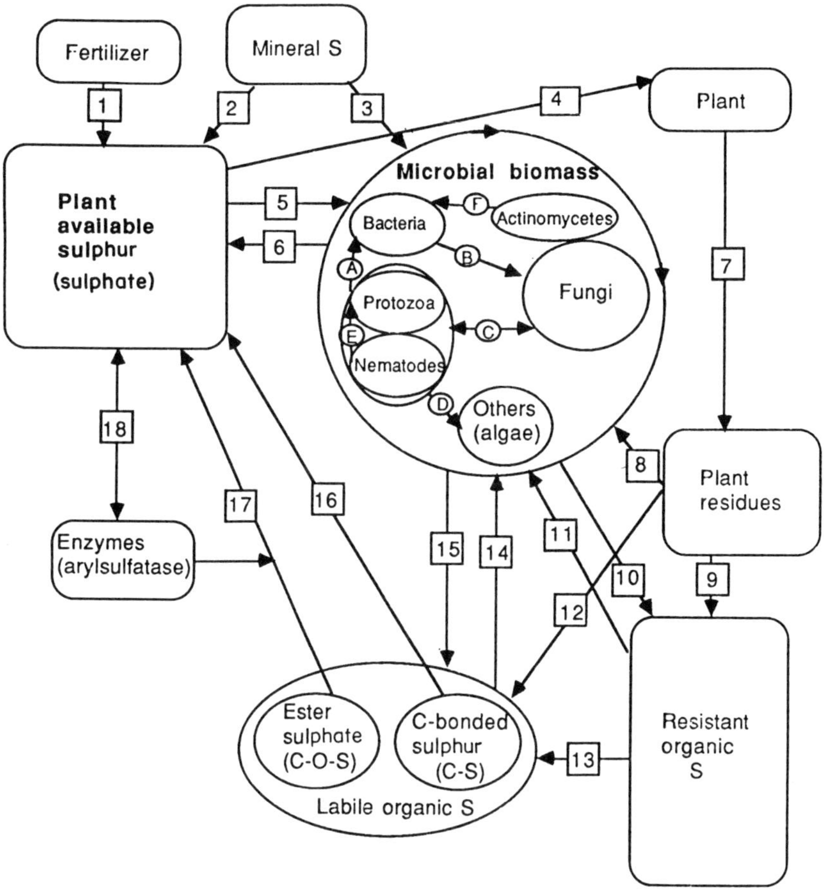

Figure 10.1. A conceptual model for the various microbial components and processes of the sulphur cycle in soil (Gupta, 1989). 1. Fertilizer dissolution and oxidation of elemental S fertilizers. 2. Oxidation of reduced S in minerals and release of sulphate. 3. Microbial assimilation of S from minerals. 4. Plant uptake of sulphate. 5. Microbial assimilation and immobilization of sulphate. 6. Mineralization. 7. Dead and decomposing root material and residual straw. 8. Microbial decomposition of plant material. 9. Humification and/or stabilization of resistant plant residues. 10. Humification and/or stabilization of resistant microbial residues. 11. Microbial utilization of resistant organic S. 12. Release of labile S compounds during plant residue decomposition. 13. Microbial decomposition and organic matter turnover due to other environmental factors such as drying and wetting, freeze and thaw cycles. 14. Microbial assimilation of S from labile organic S compounds. 15. Release of labile microbial S due to grazing by predators. 16. Biological mineralization. 17. Biochemical or enzymatic mineralization. 18. End-product controlled activity of sulphatase enzymes. *Microbial biomass*: (A) Protozoa and nematodes feeding on bacteria; (B) Bacteria feeding on damaged fungal mycelium and on compounds released from dead hyphae; (C) Protozoa and nematodes feeding on fungi; (D) Protozoa grazing on algae; (E) Nematodes grazing on protozoa; and (F) Actinomycetes feeding on bacteria

Steinbergs, 1964). This sulphate is suggested to make little, if any, contribution to the plant available sulphate pool (Williams and Steinbergs, 1964).

In the subsurface horizons of soils with restricted downward leaching, large accumulations of sulphate may occur in the form of gypsum ($CaSO_4 \cdot 2H_2O$) or other water-soluble salts. Gypsum is a slightly soluble salt and produces a high sulphate activity in solution. Consequently, soils developed under arid or semi-arid environments with accumulations of gypsum in the rooting zone are seldom found to be S deficient (Bettany, Janzen and Stewart, 1983). Excessive quantities of sulphate salts within the rooting zone can increase osmotic potential to the point where plant growth is severely restricted. In contrast, in humid climates or under irrigation with good drainage, sulphate may be lost entirely from the soil profile by deep leaching, with reported losses as high as 30 g (S) m^{-2} a^{-1} (1 mol m^{-2} a^{-1}) (Harward and Reisenauer, 1966).

In poorly drained soils, large quantities of inorganic S may be present in reduced forms, mainly sulphides. Sulphides are usually found in subsoils below the water table. Re-oxidation of sulphide to sulphate under aerobic conditions can result in the formation of acid-sulphate soils (Fleming and Alexander, 1961).

Elemental S is commonly used as a fertilizer in S-deficient agricultural systems, but can also be a soil pollutant in the form of wind-blown dust from stockpiles near sour gas processing plants (Maynard, Germida and Addison, 1986). The elemental S is oxidized to sulphate by the soil microbial population. This conversion is necessary to render the S plant available, with the rate of oxidation being a major factor influencing the effectiveness of elemental S fertilizer. Oxidation of S occurs readily in some soils, but chemical, physical, and biological factors limit oxidation rates in other soils (Janzen and Bettany, 1987). Temperature and moisture affect microbial activity and consequently oxidation rate. Very slow rates are observed in cold, dry soils. Decreasing the particle size of the elemental S results in higher oxidation rates due to the increased surface area.

The composition of the microbial community plays an important role in regulating S oxidation. Autotrophic S oxidizers were once believed to entirely dominate S oxidation in soils, and are very active in certain environments (Lee, Watkinson and Lauren, 1988). However, enumeration of autotrophic and heterotrophic elemental S oxidizers in Saskatchewan soils has revealed heterotrophs to be the most abundant sulphur oxidizers (Germida, 1985; Lawrence and Germida, 1988). Between 3 and 37% of the total heterotrophic population of these soils were found to be capable of oxidizing elemental sulphur to thiosulphate. Heterotrophic sulphur oxidizers tended to increase with increasing pH, total S, and clay content, whereas populations of autotrophic thiosulphate oxidizers were negatively correlated with these factors (Lawrence and Germida, 1988). Nevertheless, it remains difficult to partition accurately the rates of sulphur oxidation between heterotrophic and autotrophic oxidation. The plant rhizosphere is a potentially important site for S oxidation and reduction but S fluxes in crop rhizospheres are not well understood. Recent work by Grayston and Germida (1990) indicated that the rhizosphere supports greater sulphur oxidation with larger and more diverse populations of sulphur-oxidizing heterotrophs than non-rhizosphere soil.

10.3 SOIL–ATMOSPHERE INTERACTIONS

In the last 20 years, the exchange of sulphur gases between soils, plants, and the atmosphere has received increased attention. A few studies (Banwart and Bremner, 1976; Minami and Fukushi, 1981) have shown that microbial breakdown of S-containing amino acids added to the soil can give rise to gaseous S compounds such as dimethyl sulphide, methyl mercaptan, dimethyl disulphide, carbonyl sulphide, and carbon disulphide. However, the contribution of indigenous organic S in soil to gas evolution during microbial decomposition is unknown. Recent work by Melillo and Stuedler (1989) has shown that addition of nitrogen to forest soils resulted in increased emission of carbonyl sulphide and carbon disulphide gases. Presumably this was due to increased production of S-containing amino acids, which were then transformed by the soil microbial population into volatile S compounds. It is highly likely that emissions of volatile S compounds are emitted from agricultural soils during residue breakdown in a fashion similar to that observed in forest soils. However, the lack of field measurements in cropping systems makes this difficult to corroborate.

The emission of volatile S compounds from plants has recently been proposed as an important constituent of the biogeochemical S cycle (Rennenberg, 1989). The importance of such emissions to the global atmospheric cycle is still uncertain but is probably not great (Andreae and Jaeschke, this volume). With regard to agroecosystems, Grundon and Asher (1988) have suggested losses of volatile S from fields of alfalfa and wheat to be in the order of tens of milligrams (S) m^{-2} a^{-1} (in the order of 1 mmol m^{-2} a^{-1}), whereas cotton may be losing upwards of hundreds of milligrams (S) m^{-2} a^{-1} (in the order of 10 mmol m^{-2} a^{-1}). These are typical of values reported for other types of vegetation, such as trees in tropical forests (Andreae and Jaeschke, this volume). The production of H_2S and other volatile S compounds by plants is a light-dependent process which increases in response to root injury and S stress (Winner *et al.*, 1981). Stress may be induced in the plants by high sulphate concentrations in the solution bathing the roots, or by exposure of the plants to high concentrations of SO_2. This phenomenon may be important in soils containing high concentrations of sulphate, such as many saline soils, and where repeated injury to the roots occurs as a result of cultivation.

Atmospheric inputs of S to agroecosystems are difficult to quantify due to large variations from location to location (Tabatabai, 1984). Tabatabai (1984) has concluded that the amounts of S added in precipitation in rural areas of North America (50 to 1000 mg (S) m^{-2} a^{-1}; 1.6 to 31 mmol m^{-2} a^{-1}) are important in crop production to maintain the soil S balance, particularly in those areas where use of S fertilizers is low.

Figure 10.2 shows rough estimates of some of the S fluxes for western Canadian agroecosystems. Considerable uncertainty surrounds many of the

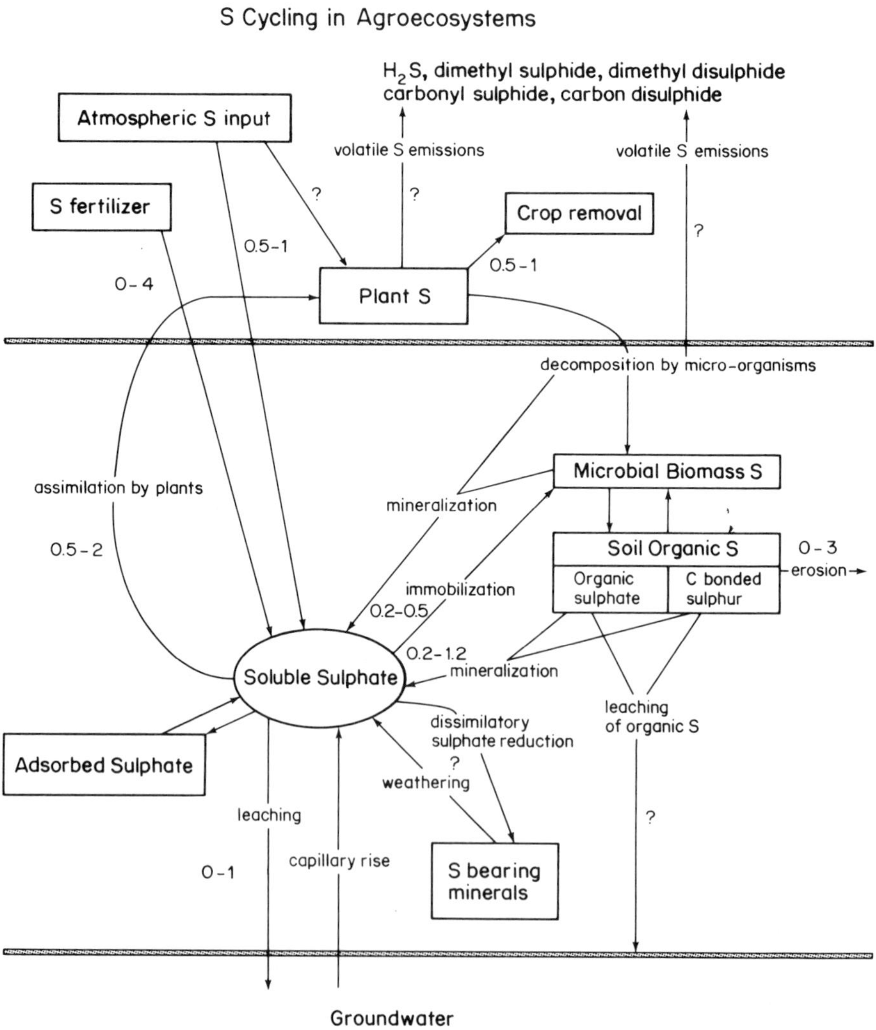

Figure 10.2. Conceptual S cycle in agroecosystems. Numbers represent flux estimates (g (S) m^{-2} a^{-1}) for S transformations in western Canadian soils

estimates. For example, we were unable to assign even approximate values to gaseous S emissions. This points out the difficulty in constructing S budgets for agroecosystems with present information and thus the direction that future work must take.

10.4 STABLE ISOTOPE INVESTIGATIONS

Naturally occurring variations in stable S isotope ratios (^{34}S / ^{32}S) are measured and expressed relative to a standard using the 'delta' ($\delta^{34}S$) notation (Nriagu and Krouse, this volume). Samples which are enriched in ^{34}S relative to the meteoritic standard bear a positive $\delta^{34}S$ while samples depleted in ^{34}S have negative $\delta^{34}S$ values.

Variations in ^{34}S / ^{32}S ratios are created as a result of isotopic fractionation in chemical and biological transformations of S. In many reactions, the lighter isotope (^{32}S) reacts preferentially, forming a product depleted in the heavy isotope (^{34}S) while the remaining reactant is enriched in ^{34}S (Krouse and Tabatabai, 1986; Nriagu and Krouse, this volume). The degree to which isotope abundances are altered is often specific to a reaction and can be used as an indirect indicator of a process occurring in a system. For example, sulphate reduction is associated with large depletions of ^{34}S in the product sulphide. The $\delta^{34}S$ technique has frequently been used in geological studies to assist in tracing the origin of petroleum and ore deposits and in environmental studies which have attempted to trace the fate of S emissions from industrial sources (Krouse and Case, 1981, 1983; Krouse, Legge and Brown, 1984; Nriagu and Krouse, this volume).

In one of the earliest $\delta^{34}S$ studies involving soils, differences in S isotope composition were noted between different soil organic S fractions (Lowe, Sasaki and Krouse, 1971). Fuller *et al.* (1986) observed that the B horizon organic S in a podzolic soil profile had a $\delta^{34}S$ at least 5 ‰ less than the horizon above or below it, which they attributed to isotope fractionation in biotic S transformations. Enrichments in ^{34}S of up to 3 ‰ have been observed in the organic sulphate fraction relative to other S fractions in western Canadian soils (Table 10.1) (Schoenau and Bettany, 1989). This probably reflects the high lability and turnover of organic sulphate (Figure 10.3). ^{34}S enrichment may thus offer a new means of identifying and tracing labile organic S compounds in soil.

Variations in ^{34}S abundance may reveal soil–plant–atmosphere interactions. The total S in wheat straw grown on saline, sulphate-saturated cultivated grassland soils was found to be enriched in ^{34}S by about 3 ‰ relative to its sulphate source (Schoenau and Bettany, 1989) (Table 10.1). Winner *et al.* (1981) and Krouse, Legge and Brown (1984) have noted that plants which are exposed to high levels of S, either in the atmosphere or the solution bathing the roots, become enriched in ^{34}S compared to surrounding S sources due to

Table 10.1. $\delta^{34}S$ values of S fractions in a western Canadian grassland soil (Schoenau and Bettany, 1989)

Soluble sulphate	Organic sulphate	C-bonded S	Plant S
		Cultivated	
−3.8	−3.1	−5.5	−1.7 (wheat straw)
		Native Prairie	
−3.0	−0.1	−3.5	−2.7 (blue grama grass)

the emission of ^{34}S-depleted H_2S from the plant as a stress relief mechanism. The enrichment of ^{34}S in the wheat plants thus provides indirect field evidence for the emission of ^{34}S-depleted gaseous S compounds in response to S stress invoked by the high concentration of sulphate.

The $\delta^{34}S$ approach can provide clues about the origin and fate of sulphur in a soil. In sulphate co-precipitated with soil carbonates, Schoenau and

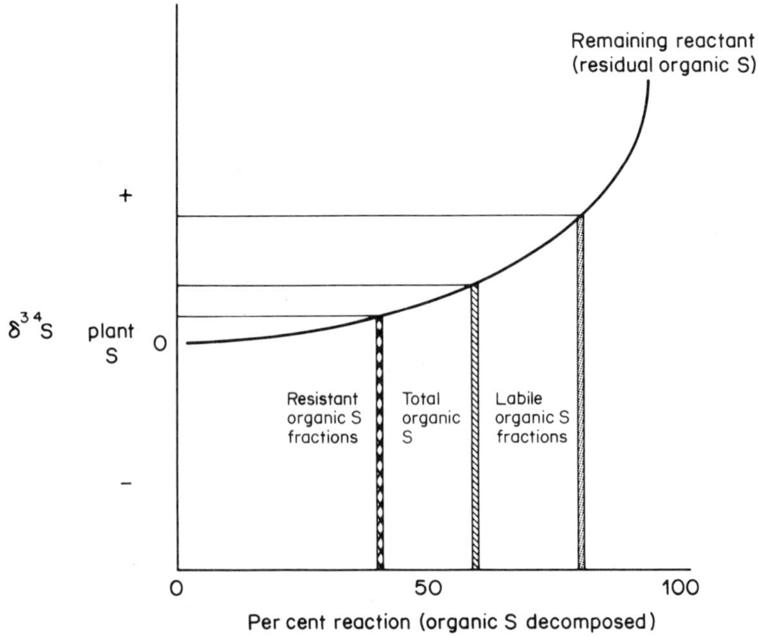

Figure 10.3 Predicted isotope composition of S fractions during mineralization of a plant residue with an initial $\delta^{34}S$ of 0‰ (Schoenau and Bettany, 1989). Rapidly decomposed (labile) S fractions are characterized by ^{34}S enrichment (more positive $\delta^{34}S$) relative to recalcitrant fractions

Bettany (1989) observed $\delta^{34}S$ values that were much more positive than any of the possible S inputs during soil formation. This led to the suggestion that much of the sulphate in the carbonate minerals was of oceanic origin (highly positive $\delta^{34}S$), being precipitated with the carbonates during the evaporation of ancient inland seas. $\delta^{34}S$ values have been used to evaluate the downward movement of manure S applied to a grassland soil (Chae and Krouse, 1986). The $\delta^{34}S$ values of the C-bonded organic S fraction indicated penetration of manure S in this form down to the water table.

As demonstrated above, ^{34}S natural abundance variations are a useful tool in elucidating the soil–plant S cycle. As more information is obtained on the nature of stable S isotope fractionation, the utility of the $\delta^{34}S$ technique will undoubtedly increase. The recent availability of compounds artificially enriched or depleted in ^{34}S offers a host of new research opportunities. For example, using ^{34}S-labelled fertilizer materials and organic residues, it is now possible to trace the long-term fate of S additions in the field.

10.5 MODELLING OF MINERALIZATION–IMMOBILIZATION PROCESSES

Mineralization–immobilization reactions regulate the supply of available S (solution sulphate) for biomass production. Sulphur mineralization–immobilization processes are strongly linked to the cycling of C, N, and P and thus are often investigated simultaneously (Stewart, 1984). In order to construct mathematical models which can be used to predict the net mineralization of S in different soils over a growing season, detailed information is required on how S mineralization rates change with time and environment. A first-order kinetic model derived from curve fitting to laboratory or field mineralization data has been used to predict N and S mineralization as a function of time (Stanford and Smith, 1972; Ellert and Bettany, 1988). This model is of the following form:

$$I_{ct} = M_o\,(1 - e^{-kt})$$

where I_{ct} is the cumulative inorganic S or N mineralized after time t (mg kg^{-1}), t is the time from start of incubation (weeks), M_o is the potentially mineralizable organic S or N (may be determined by chemical extraction or extrapolation of mineralization curve), and k is the first-order rate constant (week^{-1}).

Mathematical models of S cycling require temperature and moisture response functions to adjust mineralization rates for field conditions (Coughenour *et al.*, 1980). If the rate constant as a function of temperature, moisture, and other environmental parameters is known, then the above expression may be used to predict S mineralization under a wide variety of conditions.

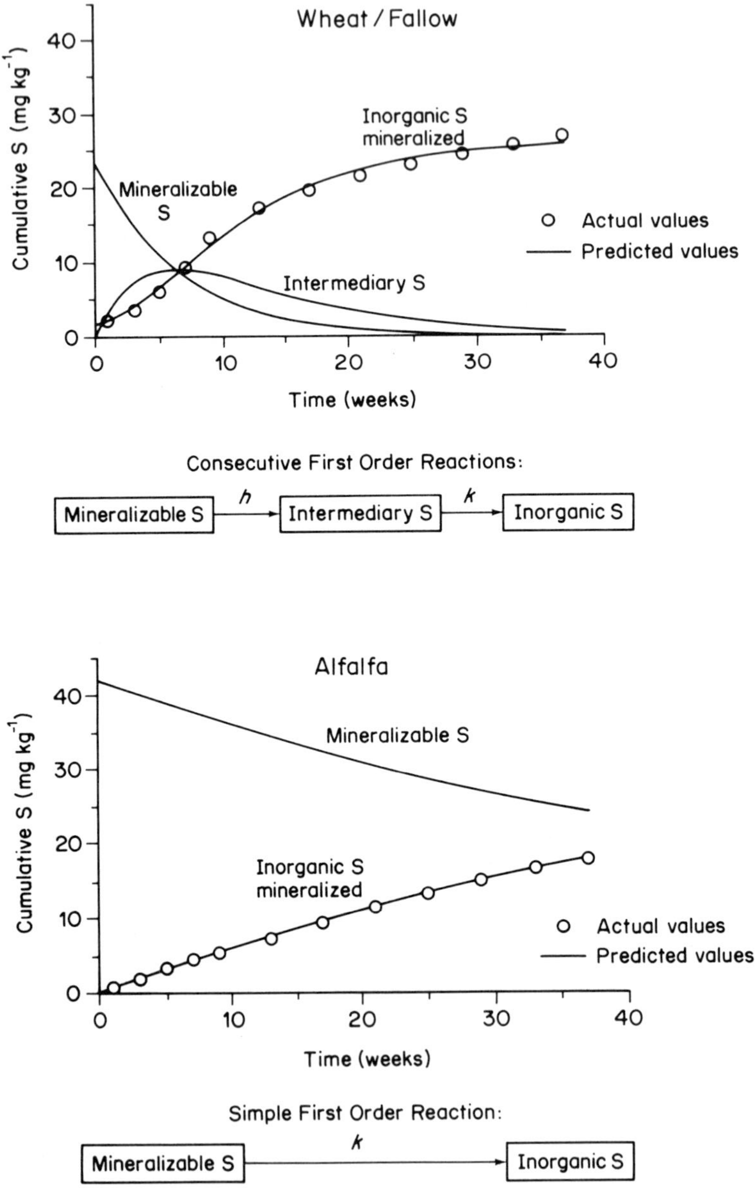

Figure 10.4. Cumulative sulphate-S mineralized from a wheat–fallow and perennial forage (alfalfa) field indicating S release and changes in mineralizable fractions predicted by best-fitting kinetic models. Included are schematic representations of the corresponding kinetic models

However, simple first-order expressions do not always adequately describe mineralization patterns observed in the field. For example, S mineralization in Saskatchewan soils cropped to perennial forage (alfalfa) has been observed to follow first-order kinetics, while soils cropped under a wheat/fallow rotation showed an initial lag during the first two weeks of a 37-week incubation (Ellert and Bettany, 1988) (Figure 10.4). Similarly, S mineralization rates in native forest soils in Saskatchewan have shown much smaller responses to further temperature increases when temperature was warm (> 20° C) than cultivated soils, reflecting possible differences in substrate composition and/or the microbial community (B. Ellert, unpub. data).

Our understanding of mineralization–immobilization processes is far from complete. Temperature and moisture may influence the size and composition of the mineralizable fraction, which in itself is not yet well defined. If advances are to be made in understanding organic S turnover, improvements in characterization of organic S are vital. The separation of organic S into the two large, chemically 'fuzzy' pools of organic sulphate and carbon-bonded S is no longer sufficient for interpretation of the sophisticated incubation and tracer experiments now utilized. Attempts must be made to distinguish analytically the organic S pools of definite chemical and biological function. The recent development of the 'microbial biomass S pool' (Saggar, Bettany and Stewart, 1982; Gupta, 1989) is a good step in this direction.

10.6 CONCLUSIONS

At present, we have a basic understanding of the forms and amounts of S in agricultural soils and the processes controlling the supply of S to plants. The exchange of sulphur gases between the soil–plant system and the atmosphere is less well documented and understood. The current challenge is to predict accurately the impact of man-induced or natural changes on S loading in all components of the biosphere. Prerequisites to this are as follows:

(1) To obtain a better definition of S forms and transformations in the soil–plant S cycle accomplished through use of new approaches such as microbial biomass and stable isotope investigations.
(2) To make more and better field measurements of soil–atmosphere S exchange from agroecosystems.
(3) To understand how key processes in the S cycle respond to environmental factors, e.g. construction of models of mineralization and volatilization processes that include temperature, moisture, substrate, and microbial composition response functions.

(4) To develop better approaches for extrapolating data from a few field sites or the laboratory to entire ecosystems, e.g. studies which emphasize how smaller components of a system combine to form the whole.

REFERENCES

Andreae, M. O. and Jaeschke, W. A. This volume.

Banwart, W. L. and Bremner, J. M. (1976). Evolution of volatile sulfur compounds from soils treated with sulfur-containing organic materials. *Soil Biol. Biochem.*, **8**, 439–43.

Bettany, J. R., Janzen, H. H. and Stewart, J. W. B. (1983). Sulfur deficiency in the prairie provinces of Canada. In: More, A. I. (Ed.). *Proceedings of the International Sulphur '82 Conference*, Nov. 1982, London, England, Vol. II. The British Sulphur Corporation, pp. 787–800.

Bettany, J. R. and Stewart, J. W. B. (1983). Sulphur cycling in soils. In: More, A. I. (Ed.). *Proceedings of the International Sulphur '82 Conference*, Nov. 1982, London, England, Vol. II. The British Sulphur Corporation, pp. 767–86.

Bettany, J. R., Stewart, J. W. B. and Halstead, E. H. (1973). Sulfur fractions and carbon, nitrogen, and sulfur relationships in grassland, forest, and associated transitional soils. *Soil Sci. Soc. Am. Proc.*, **37**, 915–18.

Bettany, J. R., Stewart, J. W. B. and Saggar, S. (1979). The nature and forms of sulfur in organic matter fractions of soils selected along an environmental gradient. *Soil Sci. Soc. Am. J.*, **43**, 481–5.

Biederbeck, B. P. (1978). Soil organic sulfur and fertility. In: Schnitzer, M. and Khan, S. U. (Eds). *Soil Organic Matter*. Elsevier, New York, pp. 273–310.

Chae, Y. M. and Krouse, H. R. (1986). Alteration of sulfur-34 natural abundance in soil by application of feedlot manure. *Soil Sci. Soc. Am. J.*, **50**, 1425–30.

Chao, T. T., Harward, M. E. and Fang, S. C. (1962). Adsorption and desorption phenomena of sulfate ions in soils. *Soil Sci. Soc. Am. Proc.*, **26**, 234–7.

Coughenour, M. B., Parton, W. J., Lauenroth, W. K., Dodd, J. L. and Woodmansee, R. G. (1980). Simulation of a grassland sulfur cycle. *Ecol. Model.*, **9**, 179–213.

Ellert, B. H. and Bettany, J. R. (1988). Comparison of kinetic models for describing net sulfur and nitrogen mineralization. *Soil Sci. Soc. Am. J.*, **52**, 1692–702.

Fitzgerald, J. W. (1976). Sulfate ester formation and hydrolysis: a potentially important yet often ignored aspect of the sulfur cycle of aerobic soils. *Bacteriol. Rev.*, **40**, 698–721.

Fitzgerald, J. W. (1978). Naturally occurring organosulfur compounds in soils. In: Nriagu, J. O. (Ed.). *Sulfur in the Environment*. Part II, *Ecological Impacts*. Wiley, New York, pp. 391–443.

Fleming, J. F. and Alexander, L. T. (1961). Sulfur acidity in South Carolina tidal marsh soils. *Soil Sci. Soc. Am. Proc.*, **25**, 94–5.

Freney, J. R. (1961). Some observations on the nature of organic sulphur compounds in soil. *Aust. J. Agric. Res.*, **12**, 424–32.

Freney, J. R. (1967). Sulfur-containing organics. In: McLaren, A. D. and Peterson, G. H. (Eds). *Soil Biochemistry*. Marcel Dekker, New York, pp. 220–59.

Freney, J. R. (1986). Forms and reactions of organic sulfur compounds in soils. In: Tabatabai, M. A. (Ed.). *Sulfur in Agriculture*. Agronomy Monograph No. 27. ASA-CSSA-SSSA, Madison, pp. 207–31.

Freney, J. R., Melville, G. E. and Williams, C. H. (1969). Extraction, chemical nature and properties of soil organic sulphur. *J. Sci. Food. Agric.*, **20**, 440–5.

Freney, J. R., Melville, G. E. and Williams, C. H. (1970). The determination of carbon bonded sulfur in soil. *Soil Sci.*, **109**, 310–18.

Freney, J. R., Melville, G. E. and Williams, C. H. (1975). Soil organic matter fractions as sources of plant-available sulphur. *Soil Biol. Biochem.*, **7**, 217–21.

Freney, J. R. and Spencer, K. (1960). Soil sulphate changes in the presence and absence of growing plants. *Aust. J. Agric. Res.*, **11**, 339–45.

Fuller, R. D., Mitchell, M. J., Krouse, H. R., Wyskowski, B. J. and Driscoll, C. T. (1986). Stable sulfur isotope ratios as a tool for interpreting ecosystem sulfur dynamics. *Water Air Soil Pollut.*, **28**, 163–71.

Germida, J. J. (1985). Modified sulphur-containing media for studying sulphur-oxidizing microorganisms. In: Caldwell, D. E., Brierley, J. A. and Brierley, C. L. (Eds). *Planetary Ecology*. Van Norstrand Reinhold, New York, pp. 333–44.

Grayston, S. J. and Germida, J. J. (1990). Influence of crop rhizosphere on populations and activity of heterotrophic sulfur-oxidizing microorganisms. *Soil Biol. Biochem.*, **22**, 457–63.

Grundon, N. J. and Asher, C. J. (1988). Volatile losses of sulfur from intact plants. *J. Plant Nutr.*, **11**, 563–76.

Gupta, V. V. S. R. (1989). Microbial biomass sulfur and biochemical mineralization of sulfur in soil. *Ph.D. Thesis*, University of Saskatchewan, Saskatoon, Saskatchewan, Canada.

Gupta, V. V. S. R. and Germida, J. J. (1988a). Distribution of microbial biomass and its activity in different soil aggregate size classes as affected by cultivation. *Soil Biol. Biochem.*, **20**, 777–86.

Gupta, V. V. S. R. and Germida, J. J. (1988b). Populations of predatory protozoa in field soils after 5 years of elemental S fertilizer application. *Soil Biol. Biochem.*, **20**, 787–91.

Gupta, V. V. S. R. and Germida, J. J. (1989). Influence of bacterial–amoebal interactions on sulfur transformations in soil. *Soil Biol. Biochem.*, **21**, 921–30.

Harward, M. E. and Reisenauer, H. M. (1966). Movement and reactions of inorganic soil sulfur. *Soil Sci.*, **101**, 326–35.

Janzen, H. H. and Bettany, J. R. (1987). The effect of temperature and water potential on sulfur oxidation in soils. *Soil Sci.*, **144**, 81–9.

Kamprath, E. J., Nelson, W. L. and Fitts, J. W. (1956). The effect of pH, sulfate and phosphate concentrations on the adsorption of sulfate by soils. *Soil Sci. Soc. Am. Proc.*, **20**, 463–6.

Krouse, H. R. and Case, J. W. (1981). Sulfur isotope ratios in water, air, soil and vegetation near Teepee Creek gas plant, Alberta. *Water Air Soil Pollut.*, **15**, 11–28.

Krouse, H. R. and Case, J. W. (1983). Sulphur isotope abundances in the environment and their relation to long term sour gas flaring near Valleyview, Alberta. Prep. for Alberta Environment, Research Management Division and Pollution Control Division, Air Quality Control Branch by Department of Physics, University of Calgary, RMD Report. 110pp.

Krouse, H. R., Legge, A. and Brown, H. M. (1984). Sulphur gas emissions in the boreal forest: The West Whitecourt Case Study V: Stable Sulphur Isotopes. *Water Air Soil Pollut.*, **22**, 321–47.

Krouse, H. R. and Tabatabai, M. A. (1986). Stable S isotopes. In: Tabatabai, M. A. (Ed.). *Sulfur in Agriculture Agronomy Mongraph No. 27*. ASA-CSSA-SSSA, Madison, pp. 165–205.

Lawrence, J. R. and Germida, J. J. (1988). Most-probable-number procedure to enumerate S°-oxidizing, thiosulfate producing heterotrophs in soil. *Soil Biol. Biochem.*, **20**, 577–8.

Lee, A., Watkinson, J. H. and Lauren, D. R. (1988). Factors affecting oxidation rates of elemental sulphur in a soil under a ryegrass dominant sward. *Soil Biol. Biochem.*, **20**, 809–16.

Lowe, L. E. and DeLong, W. A. (1963). Carbon bonded S in selected Quebec soils. *Can. J. Soil Sci.*, **43**, 151–5.

Lowe, L. E., Sasaki, A. and Krouse, H. R. (1971). Variations of sulphur-34 : sulphur-32 ratios in soil fractions in western Canada. *Can. J. Soil Sci.*, **51**, 129–31.

McLachlan, K. D. and DeMarco, D. G. (1975). Changes in soil sulphur fractions with fertilizer additions and cropping treatments. *Aust. J. Soil Res.*, **13**, 169–76.

Maynard, D. G., Germida, J. J. and Addison, P. A. (1986). The effect of elemental sulfur on certain chemical and biological properties of surface organic horizons of a forest soil. *Can. J. For. Res.*, **16**, 1050–4.

Maynard, D. G., Stewart, J. W. B. and Bettany, J. R. (1985). The effects of plants on soil sulfur transformations. *Soil Biol. Biochem.*, **17**, 127–34.

Melillo, J. M. and Stuedler, P. A. (1989). The effect of nitrogen fertilization on the COS and CS_2 emission from temperate forest soils. *J. Atmos. Chem.*, **9**(4), 411–18.

Minami, K. and Fukushi, K. S. (1981). Volatilization of carbonyl sulfide from paddy soils treated with sulfur-containing substances. *Soil Sci. Plant Nutr.*, **27**, 339–45.

Mitchell, M. J., David, M. B. and Harrison, R. B. This volume.

Nriagu, J. O. and Krouse, H. R. This volume.

Rennenberg, H. (1989). Synthesis and emission of hydrogen sulfide by higher plants. In: Saltzman, E. S. and Cooper, W. J. (Eds). *Biogenic Sulfur in the Environment.* American Chemical Society, Washington, pp. 45–57.

Roberts, T. L. and Bettany, J. R. (1985). The influence of topography on the nature and distribution of soil sulfur across a narrow environmental gradient. *Can. J. Soil Sci.*, **65**, 419–34.

Roberts, T. L., Bettany, J. R. and Stewart, J. W. B. (1989). A hierarchical approach to the study of organic C, N, P, and S in Western Canadian soils. *Can. J. Soil Sci.*, **69**, 739–49.

Saggar, S., Bettany, J. R. and Stewart, J. W. B. (1982). Measurement of microbial biomass sulfur in soil. *Soil Biol. Biochem.*, **13**, 493–8.

Schoenau, J. J. and Bettany, J. R. (1987). Organic matter leaching as a component of carbon, nitrogen, phosphorus, and sulfur cycles in a forest, grassland, and gleyed soil. *Soil Sci. Soc. Am. J.*, **51**, 646–51.

Schoenau, J. J. and Bettany, J. R. (1989). ^{34}S natural abundance variations in prairie and boreal forest soils. *J. Soil Sci.*, **40**, 397–413.

Spencer, K. and Freney, J. R. (1960). A comparison of several procedures for estimating the sulphur status of soils. *Aust. J. Agric. Res.*, **11**, 948–59.

Stanford, G. and Smith, S. J. (1972). Nitrogen mineralization potentials of soils. *Soil Sci. Soc. Am. Proc.*, **36**, 465–72.

Stewart, J. W. B. (1984). Interrelation of carbon, nitrogen, sulfur, and phosphorus cycles during decomposition processes in soil. In: Reddy, C. A. and Klug, M. J. (Eds). *Current Perspectives in Microbial Ecology. Proceedings of the 3rd International Symposium*, American Society of Microbiology, Washington DC, pp. 442–6.

Tabatabai, M. A. (1984). Importance of sulphur in crop production. *Biogeochemistry*, **1**, 45–62.

Tabatabai, M. A. and Bremner, J. M. (1970). Arylsulfatase activity of soils. *Soil Sci. Soc. Am. Proc.*, **34**, 225–9.

Williams, C. H. (1975). The chemical nature of sulphur compounds in soils. In: McLachlan, K. D. (Ed.). *Sulphur in Australasian Agriculture*. Sydney Univ. Press, Sydney, pp. 21–30.

Williams, C. H. and Steinbergs, A. (1964). The evaluation of plant-available sulphur in soils. II. The availability of adsorbed and insoluble sulphates. *Plant Soil*, **21**, 50–62.

Winner, W. E., Smith, C. L., Kach, G. W., Mooney, H. A., Bewley, J. E. and Krouse, H. R. (1981). Rates of emission of H_2S from plants and patterns of stable sulfur isotope fractionation. *Nature*, **289**, 672–3.

11 Sulphur Cycling in Rice Wetlands

ROD D. B. LEFROY
University of New England, Armidale NSW, Australia

CEZAR P. MAMARIL
International Rice Research Institute, Los Baños, Laguna, Philippines

GRAEME J. BLAIR
University of New England, Armidale NSW, Australia

and

P. J. GONZALES
International Rice Research Institute, Los Baños, Laguna, Philippines

11.1 INTRODUCTION

Most Asian countries have adopted intensified rice-based cropping systems to increase food production for a rapidly growing population. Crop intensification and increased yield have significantly increased the output of nutrients and, where there has been an imbalance between outputs and inputs, has resulted in declining soil fertility and an increase in the incidence of deficiencies of certain plant nutrients, including sulphur.

To understand the cycling of sulphur in rice wetlands, it is essential that the geographical distribution and characteristics of these wetlands be appreciated, along with the important biological, physical, and chemical changes which occur as a result of flooding and which influence the transformation and thus the cycling of sulphur. Many of the characteristics of rice wetlands and the biological, physical, and chemical processes which occur in them are similar to those of natural wetlands (Giblin and Wieder, this volume). However, there are some significant differences which largely result from their distribution and the management practices employed. For instance, land preparation for rice involves 'puddling', a management practice aimed at establishing an impermeable layer at lower soil depths while thoroughly mixing the upper soil layers to provide a uniform seed bed and sufficient rooting depth. In many areas, the heavy texture of the soil makes cultivation difficult during the dry season, and puddling starts after the onset of flooding.

Sulphur Cycling on the Continents
Edited by R.W. Howarth, J.W.B. Stewart and M.V. Ivanov

11.2 RICE WETLAND DISTRIBUTION

Wetlands have been defined as areas where saturation with water is the dominant factor determining the nature of soil development and the types of plant and animal communities (Guthrie, 1985). On this basis the US Department of Agriculture has classified about 10% of the world's total land area as wetlands. These range from frozen wetlands in Alaska, Canada, and the USSR, to the coastal and alluvial plains of the temperate and tropical zones.

In their undrained state, these wetlands support only a limited number of agriculturally important species. Sago (*Metroxylon* spp.) and taro (*Colocasia* spp.) are two species which are of local importance in limited areas, but rice (*Oryza sativa*) is the only major food crop of wetland soils. It has been estimated that rice constitutes nearly 20% of the world's food grain production (Stangel, 1979) and that 90% of this production takes place in monsoonal Asia. Kyuma (1988) suggested that the combination of physiography and monsoonal climate of the region has resulted in the large areas which are suitable for rice cultivation. These areas either receive sufficient rainfall during the monsoonal periods or, in the case of the alluvial plains and deltas of the major continental rivers and the broad coastal lowlands of the islands, receive enough water in addition to rainfall as a result of their extended catchment areas.

Rice is grown under a range of hydrological regimes from the unflooded rainfed, through the rainfed and irrigated flooded, to the deepwater and floating cultivation systems. Neue and Bloom (1989) outlined a comprehensive classification of rice environments by applying water source and floodwater depth as diagnostic criteria. Further differentiation is made by modifiers related to climate, landforms, fieldforms, floodwater, and soil and cropping systems. Greenland and De Datta (1985) estimated that of the 136×10^6 ha planted with rice in 1980, just over 50% was irrigated (Table 11.1). A further 11% was rainfed and considered hydrologically favourable for rice production, while the remaining 38% consisted of rainfed submerged, dryland, deepwater, or floating rice cultivation, which were susceptible to hydrological problems of drought or excessive submergence. Man has significantly altered the hydrology of many of these rice wetlands by draining, terracing, embanking, and irrigating.

It is clear that the wetlands used for rice production differ from a large number of the natural wetlands, not only in their geographical distribution with respect to latitude and altitude, but also in their hydrology. Most soils used for rice cultivation go through at least one natural or management imposed annual wetting and drying cycle, with the dry period ranging from several weeks to eight months. Even those rice soils which are only drained for a short period, and thus may remain saturated, normally receive regular inundation with oxygenated surface water. In contrast, many natural wet-

Table 11.1. Areas of world rice production on the basis of hydrological regime, climatic zone, and cropping system (from Greenland and De Datta, 1985)

	Arca (10^6 ha)	Max water depth
Subtropical		
Irrigated; temperate and terraced highlands	40	5 cm
Tropical		
Irrigated; multiple cropping	9	5 cm
Irrigated; single crop	20	5 cm
Non-irrigated; dryland	19	30 cm
Non-irrigated; deep water and floating	17	0 cm
Non-irrigated; shallow rainfed	31	1–6 m

lands are continuously flooded, or have relatively short dry periods (Giblin and Wieder, this volume). Those natural wetlands which are not continuously flooded, such as inter-tidal salt marshes, differ from most rice wetlands in characteristics other than their hydrology, for instance in their high organic matter content.

11.3 SOIL CHARACTERISTICS

11.3.1 SOIL TYPES

Within the various geographical and hydrological constraints discussed above, rice is grown on a wide range of soils, from all the 10 Soil Taxonomy Orders (USDA, 1975). Those of major importance are the Alfisols, Entisols, Inceptisols, and Ultisols (Moormann, 1978). Soils which are highly weathered and high sorbing, very light textured or organic/peaty soils are seldom of major importance. In contrast many natural wetland soils have organic-C contents of from 5 to 80%.

11.3.2 CHANGES WITH FLOODING

The flooding of a soil sets in motion a series of physical, chemical, and biological processes that have a very marked effect on the properties of soils, and therefore on the ability to support plant growth. Many of the changes that take place during flooding have already been dealt with in a previous chapter (Giblin and Wieder, this volume) and need not be dealt with in great detail here. The limited oxygen diffusion into the soil, the supply of biodegradable

C, the level of other oxidants, and the biological activity in the soil, particularly in the root rhizosphere, are the key to most of the biochemical and chemical processes in flooded soils.

Early laboratory work suggested that the reduction of substances in flooded soils should proceed stepwise in a sequence based on thermodynamic considerations (Ponnamperuma, 1972; Patrick and Reddy, 1978), but such thermodynamic considerations are only partly applicable under field conditions (Neue, 1988). Since most oxidation–reduction reactions in soils are mediated by microorganisms, their presence, population dynamics, and physiological activity are further controllers of the rate and order of the reactions. A thermodynamically possible reaction may not occur due to substrate or pH effects on the mediating microorganism. In addition, an immediate effect of flooding and puddling a soil is to disrupt soil aggregates, disperse soil colloids, and, depending on the floodwater being used, dilute the soil solution. These changes increase the area of reactive surfaces, and thus change the ion exchange and pH buffering capacities (Samosir, Blair and Lefroy, 1988).

The combined effect of pH, oxidation–reduction potential, microbial activity, and reactant and product concentrations is that the sequence of oxidation–reduction reactions does not entirely follow thermodynamic consideration but is set by the kinetics of the reactions. Consequently, within a soil a range of reactions can be occurring at the same bulk soil oxidation–reduction potential. Further, measured E_h potentials correspond only loosely to actual in situ oxidation–reduction potentials (Neue, 1988).

11.3.3 HETEROGENEITY WITHIN RICE WETLANDS

The heterogeneity of a flooded soil, in terms of its redox potential and the concomitant chemical characteristics, is a result of initial heterogeneity in its mineral phase, differential diffusion and mixing of the various redox reaction substrates and products and differences in the distribution of soil organisms, which will affect the chemistry and physical mixing of the soil. One of the most obvious and important aspects of this heterogeneity is the distribution of oxygen. While diffusion of oxygen in water is much lower than in air, significant diffusion into the surface soil does occur; this is further enhanced by the activity of soil microfauna and macrofauna (Roger and Kurihara, 1988). The result is that in a flooded rice field, there is a large range in the oxygen tension of different parts of the system, from an often highly oxidized floodwater and soil surface, to highly reduced subsoil layers. These properties are not fixed but have large diurnal and seasonal fluctuations.

11.3.4 THE EFFECT OF THE RICE PLANT

A further level of heterogeneity is introduced by the plant. Rice roots have well developed aerenchyma which have been shown to enlarge in response to reduced aeration (Katayama, 1961; Armstrong, 1971). The aerenchymatous nature of rice roots enables them to transfer oxygen to the roots from the shoots (van Raalte, 1940; Barber, Ebert and Evans, 1962; John, Limpinuntana and Greenway, 1974). This capacity for rapid diffusion of oxygen through the extracellular spaces of shoots and roots is shared by many plants which are adapted to natural wetlands (Armstrong, 1964). Transfer of oxygen to the roots allows the more efficient oxidative phosphorylation to continue in root cells and also results in the oxidation of the root rhizosphere (Armstrong, 1971). Yu (1985) has reported higher oxygen contents and higher E_h at, or near, the root surface than in soil away from the roots. The spatial difference in E_h as a result of rice root distribution is also reflected in lower levels of reducing substances in the rhizosphere. Yu (1985) cites work by Lui and Yu (1962) which showed an inverse correlation between planting density and concentrations of ferrous iron, manganous manganese, and reduced organic substances at a range of growth stages.

The root also provides substrates for microorganisms in the form of root exudates, changes rhizosphere pH as a result of proton pump activity, and alters soil solution concentrations as a result of nutrient uptake. The result is that within the vertical stratification of oxygenation in a flooded soil, from the oxidized floodwater and soil surface to the reduced subsoil, roots provide a horizontal and vertical discontinuity in oxygen tension and other factors. The nature of these rhizosphere conditions will not only stimulate microbial activity, but also will change the population dynamics of soil microorganisms by providing a niche for aerobes and a change in the physiology of facultative anaerobes.

11.4 SULPHUR CYCLING IN RICE WETLAND SOILS

For many nutrients, the balance between ionic forms, as affected by redox potential and pH, changes the solubility and strength of adsorption/exchange and therefore the amount of that nutrient available for plant uptake. In the case of nitrogen, the ionic form also changes the uptake mechanism. Sulphur, however, is taken up by the roots of most plants in the oxidized sulphate form. There is evidence that some wetland plant species, such as the salt-marsh grass *Spartina alterniflora* can also assimilate sulphide (Carlson and Forrest, 1982; Peterson, Howarth and Garritt, 1986). It is unknown whether or not rice also can assimilate sulphide, but we suspect it is unlikely to be a major source of sulphur to the plant since rice is sensitive to low levels of sulphide in its tissues. The hydrology, chemistry, biology, and physical characteristics of rice wetlands, as discussed above, are all important in sulphur cycling in flooded rice systems as they affect the form of and availability of S to the rice plant.

The level of total S and sulphate-S in wetland rice soils varies widely. The mean concentrations reported for total S in rice wetland soils of a number of different countries range from 83 to 1176 μg (S) g^{-1} dry weight (2.6 to 37 μmol g^{-1}), while the mean sulphate-S levels range from 2 to 210 μg (S) g^{-1} dry weight (0.06 to 6.6 μmol g^{-1}) (Table 11.2). Generally, the S content of rice wetlands is related to soil texture, with light textured soils containing less S than clay soils (Parkpian *et al.*, 1986). The total S content is also correlated to total organic matter content, as has been reported for a large number of Philippine soils (Reyes *et al.*, 1985), with a significant fraction of the total S in the soil being present as organic sulphur (Blair and Lefroy, 1987). The same is generally true of natural freshwater wetlands; their normally higher organic matter content being correlated with higher total sulphur content (Giblin and Wieder, this volume).

Organic sulphur compounds in soil have been grouped on the basis of their chemical bonding or reactivity with certain reducing agents (Tabatabai, 1982). Very few analyses of the groups of organic S in rice wetland soils have been reported, although some measurements of the amount of organic S reduced to H_2S by hydriodic acid have been made; this form of organic S is not bonded directly to C and is believed to be largely in the form of ester sulphates. Furusaka and Freney (cited by Freney, Jacq and Baldensperger, 1982) reported that the mean HI-reducible S in some lowland soils in Japan was 50% of total organic S, while the mean of 12 wetland rice soils in the Philippines was 36% of total organic S (Gaddi, 1988; Cacnio and Mamaril unpub. data). These values are within the range reported for HI-reducible organic S in upland soils of between 30 and 70% of organic S (Freney, Jacq and Baldensperger, 1982). The form of S in flooded soils is dependent on the balance between the various S transformations that can occur in flooded soils. Freney *et al.* have divided these transformations into five categories, namely:

Table 11.2. Total sulphur and sulphate sulphur contents of wetland rice soils from different countries

Country	Number of sites	Total S (μg/g)	SO_4-S (μg/g)	Source
Brazil	3	83	2	Wang, Liem and Mikkelsen (1976)
China	152	260	22	Liu, Chen and Cao (1980)
India	30	1176	210	Hegde, Rao and Biddappa (1980)
Indonesia	9	1006	48	Ismunadji and Zulkarnaini (1978)
Japan	5	1122	45	Furusaka and Freney (cited in Freney, Jacq and Baldensperger (1982)
Philippines	2	305	3	Mamaril and Cacnio unpub. data
	30	446	23	Islam and Ponnamperuma (1982)
Thailand	3	174	20	Parkpian *et al.* (1986)

(1) Immobilization or assimilation of S into organic compounds
(2) Mineralization or decomposition of organic S compounds
(3) Sulphide production
(4) Production of volatile S compounds
(5) Oxidation of S and inorganic S compounds.

Reactions (1), (2), and (4) are prevalent in both oxic and anoxic zones of flooded soils, reaction (3) occurs principally in anoxic zones and reaction (5) occurs principally in oxic zones (but see Luther and Church, this volume, for a discussion of exceptions). Measurements of sulphate-S in the soil solution under field conditions would therefore vary with time, distance from the root, or root density, and the depth of sampling.

The significant fraction of soil S present as organic sulphur has to be mineralized to sulphate (or perhaps sulphide, if rice behaves like saltmarsh grasses) before it is available to plants. This mineralization depends not only on microbial activity but also on the nature of the organic matter. The mineralization of organic matter in wetland rice soils is enhanced by the annual wetting and drying cycles, generally high soil temperatures, and generally high and balanced nutrient supply. In contrast to many natural wetlands of the temperate zone, there is likely to be less build-up of organic matter that is resistant to decomposition. In addition, after each rice crop there is a substantial input of predominantly readily mineralizable organic matter from floodwater flora and fauna, rice roots and, depending on residue management practices, rice shoots. As production increases, as a result of management practices, so the mineralizable organic matter and therefore sulphur returned to the system is likely to increase.

The decomposition of organic matter that occurs after flooding and puddling is accompanied by a rapid fall in redox potential and changes in pH. Once oxygen and other more energetically favourable electron acceptors have been consumed, at least in a restricted zone, sulphate will be reduced to sulphide. The delay in sulphide formation and its subsequent rate of formation is controlled not only by the amount of oxygen in the system but also by other oxidants such as nitrate and manganese and iron oxides, which are generally thought to be reduced earlier in the soil reduction sequence. Productive rice wetland soils are usually relatively rich in iron oxides and in some cases MnO_2. The redox potential in the bulk soil is consequently poised by the Fe(III)–Fe(II) and MnO_2–Mn(II) couples, well above the potential where rice scientists have generally thought sulphate reduction can occur (Ponnamperuma, 1972; Patrick and Reddy, 1978). Because of this assumption, few workers have actually studied sulphate reduction in flooded rice soils. However, E_h measurements can be a poor indication of whether or not particular microbially mediated reactions occur (Neue, 1988). In salt-marsh soils, high rates of sulphate reduction are found in soils with measured redox potentials as high as +350 mV (Howarth and Teal, 1979; Howes *et al.*, 1981). The measured redox potential remains high because sulphides are rapidly precipitated as pyrite (FeS_2), and the low abundance of more energetically favourable electron acceptors, rather than the measured redox potential, regulates sulphate reduction in these natural wetlands (Howarth and Teal, 1979; Howarth, 1984). High rates of sulphate reduction have also been measured in the surface layers of subtidal marine sediments where measured redox potentials of +400 mV occur (Howarth and Jørgensen, 1984).

When sulphate is reduced, hydrogen sulphide, which is toxic to plants and microorganisms, is formed. If not precipitated as metal sulphides or oxidized in the surface soil and water layers, S can be lost from the system as H_2S gas. It is thought that the presence of significant amounts of H_2S is only likely to occur in organic or very sandy soils which are very low in reducible iron or when very large quantities of organic matter have been incubated in flooded soils; these are conditions rare in productive rice soils. There is very little evidence for significant losses of sulphur as H_2S from rice wetlands (Freney, Jacq and Baldensperger, 1982). Fluxes of H_2S are generally not significant losses of sulphur from natural wetlands as well (Andreae and Jaeschke, this volume; Giblin and Wieder, this volume). However, very few measurements of the flux or controls of H_2S production have been made in rice soils. More effort should be put into actual measurements, especially as there are several reasons to believe that sulphate reduction rates may be greater than once thought.

It can be speculated that losses of H_2S from productive rice soils may be greater than currently thought due to increased losses at night as a result of the nocturnal decline in the degree of oxidation of the rhizosphere, surface soil, and floodwater. In addition, as rice cropping is intensified and larger areas are brought under controlled irrigation, many rice wetland soils are being maintained in anoxic states for longer periods. This potentially increases the degree of soil reduction and thus gaseous S emissions. The loss of S may be further increased by the use of waste water which is high in carbon to irrigate rice fields, as is occurring near some large cities, particularly in China. The planned or incidental application of sulphate may also increase gaseous S losses, especially where more reducing management practices are used.

11.5 THE FATE OF SULPHATE

Sulphate in the floodwater and soil solution of the oxidized layers may be taken up by plants, immobilized into organic matter, sorbed on to anion exchange sites, removed by leaching and runoff or reduced to sulphide. With an active floodwater and surface soil biota, some sulphate will be rapidly incorporated into organic matter, but with the relatively short generation time of many of these organisms the turnover should be rapid. Anything that affects the activity of this wetland biota, such as temperature or floodwater/soil fertility, will clearly affect the rate of sulphate immobilization.

Sorption of sulphate from the floodwater and oxidized surface soil layer will be affected by changes in the anion exchange capacity, the sulphate concentration, and changes in the concentration of competing anions (Mitchell, David and Harrison, this volume). Sulphate sorption capacity declines as pH increases in upland (Kamprath, Nelson and Fitts, 1956; Barrow, 1970; Gebhart and Coleman, 1974; Couto, Lathwell and Bouldin, 1979) and lowland soils (Samosir, Blair and Lefroy, 1988). Thus on acid soils, the pH increase that occurs with flooding would decrease the sorption capacity, while flooding of alkaline soils would increase sorption capacity. Changes in sorption capacity with changes in redox will depend on the nature of the major sorption sites. The reduction of ferric iron compounds to the soluble ferrous forms, along with the effects of puddling can weaken soil structure, thus increasing the anion adsorption capacity by exposing more exchange sites. Similarly, the formation of colloidal or amorphous iron compounds from the solubilized ferrous iron compounds would increase sorption capacity. Increases in colloidal iron, as measured in an ammonium acetate extract, have been found to occur across a range of pH conditions, as redox potential declined (Samosir, Blair and Lefroy, 1988). Exchange capacity can also be increased by the interaction of organic matter breakdown products, such as humic and fulvic acid, with iron oxides (Parfitt, Fraser and Farmer, 1977; Tipping, 1981). Alternatively, the solubilizing of ferric oxide coatings may remove all the associated positive charges and reveal negatively charged clay surfaces, thus reducing anion adsorption. The net effect on exchange capacity will be a balance between these processes.

Whatever the change in anion exchange capacity, the net effect on sulphate sorption will also depend on changes in the concentration of potentially competing anions. As the redox potential falls and ferric ions are reduced, the levels of iron and phosphate in solution increase dramatically (Patrick, Gotoh and Williams, 1973; Chang, 1976; Samosir, Blair and Lefroy, 1988). The increased phosphate in solution is likely to increase the sulphate in solution by competition with sulphate for adsorption sites (Parfitt, 1980; Nagarajah, Posner and Quirk, 1970).

At the onset of flooding, CO_2 and bicarbonate accumulate from organic matter decomposition. The bicarbonate ion will also compete with sulphate for sorption sites and so is likely to increase the sulphate in solution. The organic acids that result from organic matter decomposition will also increase competition for adsorption sites (Parfitt, 1980; Nagarajah, Posner and Quirk, 1970). The effect of flooding on solution sulphate concentrations will be a balance between these changes in anion exchange capacity and the concentration of competing ions (Samosir, Blair and Lefroy, 1988). The competitive effects of the increased concentrations of phosphate, organic acids, and bicarbonate are most likely to result in an increase in sulphate in solution in the oxidized layers unless the anion exchange capacity is increased enormously. An increase in sulphate in solution after flooding has been observed by Ponnamperuma, Attanandana and Beye (1973) and Patnaik (1978).

The loss of sulphate in floodwater will depend on the water management system. If significant amounts of floodwater are flowing through the rice fields, significant losses of sulphate and other nutrients may occur. However, if water is added when required to maintain floodwater depth, then losses would be reduced.

Movement of sulphate in the soil solution will largely be controlled by soil structure, although influenced by the degree of puddling and soil macrofauna activity. Movement of sulphate down the profile will generally result in its reduction and subsequent precipitation as sulphide. In some very light textured soils, losses of sulphur may occur as a result of lateral and vertical soil solution movement.

The precipitation of relatively immobile sulphide in the reduced soil layers renders the sulphur temporarily unavailable for plants. As a root grows into a zone where metal sulphides have been precipitated, the oxygenated rhizosphere allows bacteria to oxidize sulphide into the plant available sulphate form. If the sulphide is present deep in the profile it is unlikely to be available during the growing season as rice does not have a deep root system; in fact, after maximum tillering a large portion of the rice roots form a mat on the soil surface (Alberda, 1954). Sulphides deposited deep in the soil profile can be oxidized to sulphate under the unsaturated conditions which occur during fallow periods. This sulphate may then be available for the rice plants at the beginning of the next cropping season, however it may also be leached from the soil as the profile becomes saturated.

11.6 SULPHUR CYCLING IN RICE CROPPING SYSTEMS

11.6.1 SULPHUR UPTAKE AND USE BY RICE

Sulphur is important in rice nutrition for the synthesis of amino acids and proteins, which account for approximately 90% of organic S in the plant. The sulphur requirement of rice varies according to the nitrogen supply. When S becomes limiting, addition of N does not change the yield or protein level of plants (Dijkshoorn and van Wijk, 1967). Sulphur is required early in the growth of rice plants. If it is limiting during early growth, then tiller number and therefore final yield will be reduced (Blair, Momuat and Mamaril, 1979). Although the concentration of sulphur in rice grain is higher than in rice

Table 11.3. Sulphur balance sheet for rice grown on an Aeric Paleaquult at Ubon Ratchatani, Thailand (from Lefroy, Blair and Blair, 1988)

	No fertilizer	Fertilizer
Yield ($t\ ha^{-1}\ a^{-1}$)		
Grain	1.23	1.17
Residue	3.60	4.61
S input ($g\ m^{-1}\ a^{-1}$)		
Fertilizer	0	0.8
Rain	0.36	0.36
S uptake ($g\ m^{-1}\ a^{-1}$)		
Grain	0.15	0.12
Residue	0.27	0.44
S balance ($g\ m^{-1}\ a^{-1}$)		
Nothing returned	−0.06	0.6
Stubble returned	0.21	1.04

straw, there is less re-mobilization of S to the grain than for phosphorus. The uptake of S and its distribution within the plant for rice grown in northeast Thailand can be seen in the sulphur balance sheet in Table 11.3.

If it is assumed that all of the sulphur from the atmosphere is available to the rice crop and that no leaching losses occur then, with no added sulphur fertilizer, the system shows a net loss of 60 mg (S) m^{-2} (0.6 kg ha^{-1}, or 2 mmol m^{-2}) if no residues are returned. Less than half of the sulphur in the above ground part of the plant was in the grain; consequently, the return of all residue to the cropping area reduces the sulphur loss to the point where there was a net gain of 210 mg (S) m^{-2} (6.6 mmol m^{-2}). The addition of 800 mg (S) m^{-2} (8 kg ha^{-1}, or 25 mmol m^{-2}) as gypsum resulted in a net gain of S irrespective of residue management. In the following season, when no further S fertilizer was applied and the previous season's residue was incorporated, a response in grain and straw yield was seen. This suggests the S balances in Table 11.3 are an overestimate, presumably because not all the S from the atmosphere can be used by the rice crop and leaching losses were not taken into account. On a slightly more fertile site in Thailand, yielding about 2.5 t ha^{-1} (250 g m^{-2}) of grain, the total S uptake when no sulphur fertilizer was applied was 770 mg (S) m^{-2} (24 mmol m^{-2}). This site received 340 mg (S) m^{-2} (11 mmol m^{-2}) during the year from the atmosphere. In this case, even when the stubble was returned, there was a net loss of 200 mg (S) m^{-2} (6 mmol m^{-2}). This site showed no response to sulphur and had been cropped for many years; it would therefore appear that either there was a very large soil sulphur pool which was being depleted, or there was an additional input which kept the system in a positive or near-neutral sulphur balance. The most likely additional input is in irrigation water.

11.6.2 INPUTS OF SULPHUR IN IRRIGATION WATER AND RAINFALL

There has been relatively little work that has accurately assessed the role of irrigation water in rice nutrition. Amarasiri and Latheef (1982) estimated that the sulphur input in irrigation water across four sites in Sri Lanka ranged from 1.4 to 4.3 g (S) m^{-2} (44 to 130 mmol m^{-2}) per season and that this was from 2.6 to 17 times as much as received in rainfall. Freney, Jacq and Baldensperger (1982) report from various sources a range of 0.2 to 20.2 mg (S) l^{-1} (6 to 630 μM) in irrigation and river water.

The question of what concentration of sulphur is required in irrigation water is more complex, as it is necessary to consider the plant's uptake system and various site-specific factors, such as the adsorption capacity of the soil and the residence time of the floodwater in the rice field. Thus, there is no consensus among research reports that try to relate sulphur concentration in irrigation waters to crop requirements and production (Yoshida and Chaudhry, 1979; Wang, 1979; Ishizuka and Tanaka, 1959; Blair, Momuat and Mamaril, 1979).

There is also little work on the variation in amounts of sulphur in rainfall or its effectiveness to rice plants. The S in rainfall is often assumed to be largely in the sulphate form and is generally higher in industrialized countries and near industrial centres, volcanoes, swamps, and seas (Blair, Mamaril and Momuat, 1978). Measurements in Korea have shown that the concentration of sulphate in rainfall is inversely related to the amount of precipitation, with very high concentrations in the first 5 mm and lower concentrations with increasing amounts of rainfall (Shin, 1987).

A wide range of sulphur deposition in rainfall has been measured in rice growing countries of Asia (Table 11.4). Total sulphur inputs to these rice systems from the atmosphere may be larger than these values indicate depending on the amount of dry deposition from the atmosphere and how much dry deposition has been included in these measurements. There are virtually no separate measurements of wet and dry deposition of sulphur in the major rice growing areas of the world.

Recent measurements of S in rainfall at approximately 60 sites across Malaysia and Thailand show large differences in the amount of S received and its temporal distribution (Lefroy, unpub. data). As expected, the S inputs were found to vary with proximity to industrial and population centres, the amount of rainfall, proximity to the coast, and the season. While there were

Table 11.4. Rates of sulphur deposition in rainfall in rice growing countries

Country	Sulphur deposition ($g\ (S)\ m^{-2}\ a^{-1}$)	Reference
Philippines	1.4 to 2.5	Mamaril and Gonzales (1987)
Sri Lanka	0.5 to 2.4	Amarasiri and Latheef (1982)
China	1.3 to 2.9	Liu (1987)
Malaysia	0.4 to 2.8	Zahari *et al.* (1983)
Taiwan		
150 km from industrial area	0.6	Shin (1987)
adjacent to industrial area	2.9	Shin (1987)
Korea		
mountains	1.2	Shin (1987)
coastal area	2.8	Shin (1987)

large local inputs from industrial and population centres, the major influence was the marine input during monsoonal periods, particularly near the east coast of peninsular Malaysia during the northeast monsoons. On light textured soils in this region, large responses of rice to sulphur applications were seen at the beginning of the northeast monsoon, a month after transplanting. By maturity these differences had disappeared with the extra sulphur input coming in monsoonal rain estimated to be between 500 mg and 1 g (S) m^{-2} (15 to 31 mmol m^{-2}), an amount sufficient to explain the disappearance of S deficiency.

Sulphur in rainfall declines along a transect from the coast as shown in northern Queensland (Probert, 1976) and in northern New South Wales (Lefroy, unpub. data). Monthly measurements of rainfall and sulphur accession across a 700 km transect from the coast have shown, after 12 consecutive collections, a decline from a total of more than 2 g (S) m^{-2} (63 mmol m^{-2}) at the coast to less than 200 mg (S) m^{-2} (6.3 mmol m^{-2}) 150 km from the coast. At 13 km from the coast, despite similar rainfall to the coastal site, the sulphur accession was down to less than 700 mg (S) m^{-2} (22 mmol m^{-2}).

11.6.3 RESIDUE MANAGEMENT

The balance sheet approach used above indicates the relatively large quantity of sulphur that remains in rice stubble and panicle straw and thus emphasizes the importance of residue management. The management of crop residues is influenced by agronomy, socio-economics, and tradition. In many rice-growing regions, rice residues are an important source of animal fodder. In Indonesia it is estimated that 50% of rice crop residue produced during the dry season is used for animal fodder, but during the wet season only 2% is used, since more forage is available (Ranjhan, 1986). The use of rice residue as animal feed may not represent a total loss of sulphur from the rice wetlands if some animal manure is returned. Other uses, such as for paper production, for mushroom cultivation or for fuel, generally represent more complete losses from the system.

In some areas all the rice residues are incorporated (i.e. added back to the soil). However, there are significant problems associated with incorporating large quantities of residue, especially when cropping intensities are high and there is only a short fallow period between crops. In these and other situations the residues are often burnt. Attempts to simulate the slow burn that typifies the normal practice have shown that between 40 and 60% of the sulphur is lost (Chaitep, 1990). By using the ^{35}S-labelled residues, it was possible to show that similar proportions of the sulphur applied in burnt or unburnt residue were recovered by the plant. However, when the amount of sulphur lost during burning was taken into account, the proportion of sulphur in the original unburnt residue that was recovered by the plant was much lower when the residue was burnt; a larger amount of sulphur remained in the soil after the crop when the unburnt residue was applied (Table 11.5). It appears that any small gains in efficiency of use of S from burnt residue are offset by the substantial losses of S when straw is burnt, assuming that sulphur lost during burning is lost from the system and not returned as ash particles or in rainfall.

Many traditional cropping systems include an upland crop in rotation with rice, especially when there is insufficient water for an additional rice crop. These upland crops are grown for many and often multiple uses, including human and animal feed, fuel, fibre, and green manure for improved physical and chemical soil fertility. With intensification of rice cultivation, often associated with improved irrigation, many of these traditional multiple cropping systems have been replaced by multiple rice crop systems. Recently

Table 11.5. The proportion of S in straw recovered, either in the rice plant or soil, or lost in burning

	Straw	Ash
Per cent S in straw		
Recovered in plant	5.2	2.3
Recovered in soil	94.8	39.4
Lost in burning	—	58.3

there has been more interest in using green manure crops in rice culture (IRRI, 1988), especially as legume germplasm shows greater prospects for multiple use in more sustainable farming systems. The case for green manuring must take into account the effect on rice production, in terms of a possible reduction in cropping intensity and the benefits of improved soil fertility, as well as the changes in nutrient demands of these multiple cropping systems. The sulphur requirements of many of these upland crops, especially the legumes, are higher than for rice, but the sulphur dynamics of these systems with regard to their residue and soil, particularly hydrological, management is not well known and needs further clarification.

11.6.4 SULPHUR FERTILIZERS

Responses of flooded rice to S application in Bangladesh, Burma, Brazil, India, Indonesia, and Taiwan are cited by Freney, Jacq and Baldensperger (1982), with yield responses as high as 278% in Indonesia and 330% in Brazil. Blair, Momuat and Mamaril (1979) reported an average grain yield response of 18.6% at 28 sites in South Sulawesi, Indonesia.

Sulphur is applied in fertilizers as either sulphate or elemental sulphur. Rice plants take up S from solution as sulphate (as mentioned above, there is some evidence that some marsh plants can take up sulphide, but if this is true of rice it is most likely a minor pathway). Consequently, elemental sulphur must be converted to sulphate before it is available; the inhibition of oxidation of elemental sulphur to sulphate will decrease the availability of fertilizer sulphur to the plant. Also, the reduction of sulphate to sulphide will reduce the availability of the sulphur, particularly if the sulphides are precipitated. To be effective, S-containing fertilizers must be placed in a zone of high oxidation–reduction potential such as the oxidized soil layer at the soil–water interface or within the root rhizosphere.

Samosir (1989) found in a pot experiment using ^{35}S-labelled fertilizers, that over 80% of the rice plant's S was derived from a surface application of 4 g (S) m^{-2} (13 mmol m^{-2}), applied as potassium sulphate, yet when the same application was incorporated 7 to 21 cm below the surface, only 34% of the plant's S came from the fertilizer. The lower recovery from deep placement was most likely due to reduction to sulphide and precipitation of the sulphide. However, a lower biomass of roots at depth may also have contributed to this finding. The surface application of 4 g (S) m^{-2} as elemental sulphur was effective as a fertilizer, but somewhat less so than the surface application of sulphate, with 50% of the plant's S being derived from this treatment. The inhibition of oxidation of elemental sulphur when it was placed deep within the soil greatly reduced its effectiveness as a fertilizer, providing only 8% of plant sulphur. Similar reductions in the effectiveness of sulphate and elemental fertilizers when deep placed in pots have been found by Friesen and Chien (1986) and Chien, Hellums and Henao (1987).

In field experiments in Sulawesi, a comparison of sulphur fertilizers showed that sulphate and elemental sulphur were equally effective when surface applied (Blair, Momuat and Mamaril, 1979). Glasshouse trials comparing urea-S, elemental S, ammonium sulphate, gypsum, and S bentonite on 11 wetland soils from the Philippines showed that urea-S produced significantly higher yields than elemental sulphur, ammonium sulphate, or gypsum and that S bentonite was not immediately effective, although it showed a residual effect on succeeding crops (Mamaril and Gonzales, 1987).

The effectiveness of applied S on rice grain yield depends on timing application to coincide with plant requirement as well as placement of fertilizer sulphur in the vicinity of roots. Sulphur is best applied between transplanting or seeding and maximum plant tillering. Application after maximum tillering does not have an appreciable effect on yield since S has a large affect on tiller formation and therefore grain yield via panicle number. In a field experiment in the Philippines using two varieties of different growth duration (110–111 and 111–120 days), S application at 15 days after planting or later significantly reduced the yield of the short duration variety, when compared to an application at transplanting, but did not affect the longer duration variety (Mamaril and Gonzales unpub. data). Application at 30 days after transplanting significantly affected the yield of both varieties.

The work of Blair, Momuat and Mamaril (1979) also showed the importance of timing of sulphur applications. The application of elemental sulphur 20 days before transplanting produced significantly lower yields than the same application at transplanting. This was presumably because of the oxidation of elemental sulphur and its subsequent loss by leaching as sulphate from the paddy or immobilization as sulphides in the soil. Clearly the management of sulphur fertilizer, in terms of source, placement and timing, as well as rate, is of significance to the rice crop and to the environment at large.

11.7 CONCLUSIONS: SULPHUR MANAGEMENT OF RICE

There are many aspects of the sulphur cycle in wetland rice systems, as outlined above, which are influenced by management practices. Such practices not only influence the yield of rice, but also the efficiency of use of sulphur which ultimately affects the economics of production and the environmental consequences of the wetland rice cropping system.

The flooding and puddling of a soil can generally be said to produce an improved medium for the cultivation of rice in terms of its hydrology, nutrition, and weed control. This is particularly true if, as is most often the case with rice, the soil undergoes wetting and drying cycles between crops and substantial amounts of readily decomposable organic matter are incorporated. This general case is true for sulphur as long as highly reduced conditions do not dominate, particularly in soils low in reducible iron. In these cases not only will sulphur availability be reduced by sulphide formation but also damage can be done to the crop as a result of sulphide toxicity, and in extreme cases losses of sulphur to the atmosphere may occur. Other losses of sulphur from rice wetland systems arise from outflow in floodwater and leaching losses in light textured soils. These leaching losses may occur in the saturated state or when sulphides deposited deep in the profile are oxidized and leached during unsaturated fallow periods.

The management of sulphur relies on balancing input with output in crop harvests. In recent years, rice production has been increased in many areas by greater use of fertilizers, the use of high yielding rice varieties, the improvement of management practices, such as irrigation, and the intensification of cropping (Blair and Lefroy, 1987). This increased production has resulted in greater sulphur output. The use of high yielding varieties, with their improved harvest index, as well as changes in residue management, with more use of residues as forage and more burning of residues to facilitate intensive cropping, has resulted in a lower proportion of plant sulphur being returned to the soil. Similarly, the increased use of low S or S-free fertilizers, such as urea and triple superphosphate, has resulted in less sulphur being applied in fertilizers.

In some areas the reduction of industrial sulphur emissions may have reduced sulphur input in rainfall, and further reductions are likely in some areas. However, on the industrial side of agriculture, the increasing demand for fertilizers, both sulphur fertilizers and phosphate fertilizers produced by sulphuric acid processes, may lead to greater emission of SO_2 during elemental sulphur production from sour gas sources. Sales of elemental sulphur are increasing yearly, and the manufacture of fertilizer is the major use.

There is increasing evidence that the result of these changes in the balance between sulphur input and sulphur output is an increased deficiency in sulphur, reducing the yields of rice, and other crops, in certain areas. In terms of the whole area of rice production, the areas currently showing or having the potential to develop sulphur deficiency are probably not large, although on the local scale they may be very important. They are likely to be on lighter textured soils, which are intensively cropped, with little residue return and with low inputs of sulphur in rainfall or irrigation—in other words, not too near the coast where brackish irrigation water or high sulphur in rainfall increases the sulphur inputs. It is important that appropriate management practices be developed for these soils. First, the return of as much plant residue as possible should be encouraged. This may involve the application of ash or animal manure, particularly where the demand for rice residue for fuel, in industry, or for animal forage is high, or the quantity and timing of application of organic matter may result in a very rapid decline in the redox potential of the soil.

Second, appropriate fertilizers and fertilizer application methods should be developed. The use of elemental sulphur in a fertilizer management scheme has the attraction of being a high analysis form of sulphur, and the oxidation rate is dependent on temperature and moisture, as is plant demand. Elemental sulphur can be applied alone, as a powder or in a pelleted form, or with other fertilizers as a coating or a mixture. The application of elemental sulphur in a S–N fertilizer is probably not appropriate since for maximum efficiency nitrogen fertilizers should be deep placed, while deep placed elemental sulphur will be oxidized only slowly. The application of elemental sulphur combined with a phosphate fertilizer, such as triple superphosphate, seems most appropriate, as both are most effective in the more oxidized surface layers and both need to be applied early in crop growth.

Balancing the demands of the wetland rice crop, both at particular stages during the crop and for the crop as a whole, with the inputs of sulphur is important for maximizing yield and the economics of production as well as for achieving a sustainable agriculture system with minimal adverse environmental consequences. Adequate understanding of the sulphur cycle is essential for development of management practices which will achieve these aims. At present there are many unknowns, not the least of these being an adequate knowledge of the soil sulphur transformations and fluxes on a daily and seasonal basis under various cropping systems and management conditions. It appears that agriculturalists now need to adopt some of the methods used by geochemists and ecologists in natural wetlands to understand more fully the dynamics of the sulphur cycle.

REFERENCES

Alberda, T. (1954). Growth and root development of lowland rice and its relation to oxygen supply. *Plant Soil*, **5**, 1–28.

Amarasiri, S. L. and Latheef, M. A. (1982). The sulphur deposition by rainfall in some agriculturally important areas of Sri Lanka. *Trop. Agric.*, **138**, 33–43.

Andreae, M. O. and Jaeschke, W. A. This volume.

Armstrong, W. (1964). Oxygen diffusion from the roots of some British bog species. *Nature*, **204**, 801–2.

Armstrong, W. (1971). Radial oxygen losses from intact rice plants as affected by distance from the apex, respiration and waterlogging. *Physiol. Plant.*, **25**, 192–7.

Barber, D. A., Ebert, M. and Evans, N. T. S. (1962). The movement of ^{15}O through barley and rice plants. *J. Exp. Bot.*, **13**, 397–403.

Barrow, N. J. (1970). Comparison of the adsorption of molybdate, sulfate and phosphate by soils. *Soil Sci.*, **109**, 282–8.

Blair, G. J. and Lefroy, R. D. B. (1987). Sulphur cycling in tropical soils and the agronomic impact of increasing use of S free fertilizers, increased crop production

and burning of crop residue. In: *Proceedings of the Symposium on Fertilizer Sulphur Requirements and Sources in Developing Countries of Asia and the Pacific*. FADINAP, FAO, TSI and ACIAR Bangkok, Thailand, pp. 12–17.
Blair, G. J., Mamaril, C. P. and Momuat, E. O. (1978). Sulfur nutrition of wetland rice. *IRRI Res. Pap. Ser.*, **21**, 29.
Blair, G. J., Momuat, E. O. and Mamaril, C. P. (1979). Sulfur nutrition of rice. II. Effect of source and rate of S on growth and yield under flooded conditions. *Agron. J.*, **71**, 477–80.
Carlson, P. R. and Forrest, J. (1982). Uptake of dissolved sulfide by *Spartina alterniflora*: evidence from natural sulfur isotope abundance ratios. *Science*, **216**, 633–5.
Chang, S. C. (1976). Phosphorus in submerged soils and phosphorus nutrition and fertilization of rice. In: *The Fertility of Paddy Soils and Fertilizer Applications for Rice*. Food and Fertilizer Technology Center, ASPAC, Taipei, Taiwan, pp. 93–116.
Chaitep, W. (1990). The use of fertilisers and crop residues as sources of sulfur in flooded and unflooded rice cropping systems. *Ph.D. Thesis*, University of New England.
Chien, S. H., Hellums, D. T. and Henao, J. (1987). Greenhouse evaluation of elemental sulfur and gypsum for flooded rice. *Soil Sci. Soc. Am. J.*, **51**, 120–3.
Couto, W., Lathwell, D. J. and Bouldin, D. R. (1979). Sulfate sorption by two oxisols and an alfisol of the tropics. *Soil Sci.*, **127**, 108–16.
Dijkshoorn, W. and van Wijk, A. L. (1967). The sulphur requirements of plants as evidenced by the sulphur–nitrogen ratio in the organic matter. A review of published data. *Plant Soil*, **26**, 129–57.
Freney, J. R., Jacq, V. A. and Baldensperger, J. F. (1982). The significance of the biological sulfur cycle in rice production. In: Dommergues, Y. R. and Diem, H. G. (Eds). *Microbiology of Tropical Soils and Plant Productivity. Developments in Plant and Soil Science 5*. Martinus Nijhoff, The Hague, pp. 271–317.
Friesen, D. K. and Chien, S. H. (1986). Sulphur fertilizers and their application: General review and S source evaluation at IFDC. In: *Sulphur in Agricultural Soils*. BARC and TSI, Dhaka, Bangladesh, 1986, pp. 407–31.
Gaddi, V. Q. (1988). Soil sulphur fractions as indices of plant-available sulphur and the influence of presubmergence on sulphur availability in selected rice soils in the Philippines. *MS Thesis*, University of the Philippines, Los Baños.
Gebhart, H. and Coleman, N. T. (1974). Anion adsorption by allophanic tropical soils. II. Sulfate adsorption. *Soil Sci. Soc. Am. Proc.*, **38**, 259–62.
Giblin, A. E. and Wieder, R. K. Chapter 5 in this volume.
Greenland, D. J. and De Datta, S. K. (1985). Constraints to rice production and wetland soil characteristics. In: *Wetland Soils: Characterization, Classification and Utilization*. International Rice Research Institute, Los Baños, Philippines, pp. 23–36.
Guthrie, R. L. (1985). Characterizing and classifying wetland soils in relation to food production. In: *Wetland Soils: Characterization, Classification and Utilization*. International Rice Research Institute, Los Baños, Philippines, pp. 11–20.
Hegde, T. M., Rao, N. N. and Biddappa, C. C. (1980). Evaluation of sulphur status of different rice soils—I. *Mysore J. Agric. Sci.*, **14**, 171–6.
Howarth, R. W. (1984). The ecological significance of sulfur in the energy dynamics of salt marsh and coastal marine sediments. *Biogeochemistry*, **1**, 5–27.
Howarth, R. W. and Jørgensen, B. B. (1984). Formation of ^{35}S-labelled elemental sulfur and pyrite in coastal marine sediments (Limfjorden and Kysing Fjord, Denmark) during short-term $^{35}SO_4^{2-}$ reduction measurements. *Geochim. Cosmochim. Acta*, **48**, 1807–18.

Howarth, R. W. and Teal, J. M. (1979). Sulfate reduction in a New England salt marsh. *Limnol. Oceanogr.*, **24**, 999–1013.

Howes, B. J., Howarth, R. W., Teal, J. M. and Valiela, I. (1981). Oxidation–reduction potentials in a salt marsh: Spatial patterns and interactions with primary production. *Limnol. Oceanogr.*, **26**, 350–60.

IRRI (1988). Green manure in rice farming. *Proceedings of a Symposium on Sustainable Agriculture*. International Rice Research Institute, Los Baños, Philippines, 379pp.

Ishizuka, Y. and Tanaka, A. (1959). Inorganic nutrition of rice plants. IV. Effect of calcium, manganese and sulfur levels in culture solution on yields and chemical composition of the plant. *J. Sci. Soil Man., Jpn*, **30**, 414–16 (in Japanese). Cited in Freney, Jacq and Baldensperger (1982).

Islam, M. M. and Ponnamperuma, F. N. (1982). Soil and plant tests for available sulphur in wetland soils. *Plant Soil*, **68**, 97–113.

Ismunadji, M. and Zulkarnaini, I. (1978). Sulphur deficiency of lowland rice in Indonesia. *Sulphur Agric.*, **2**, 17–19, 22.

John, C. D., Limpinuntana, V. and Greenway, H. (1974). Adaptation of rice to anaerobiosis. *Aust. J. Plant Physiol.*, **1**, 513–20.

Kamprath, E. J., Nelson, W. L. and Fitts, J. W. (1956). The effect of pH, sulfate and phosphate concentrations on the adsorption of sulfate by soils. *Soil Sci. Soc. Am. Proc.*, **20**, 463–6.

Katayama, T. (1961). Studies on the intercellular spaces in rice I. *Proc. Crop Sci. Soc. Jpn.*, **29**, 229–33.

Kyuma, K. (1988). Paddy soils in Japan in comparison with those in tropical Asia. In: *First International Symposium on Paddy Soil Fertility*. Department of Land Development, Bankghen, Thailand, pp. 5–19.

Lefroy, R. D. B., Blair, S. and Blair, G. J. (1988). Phosphorus and sulfur efficiency in tropical cropping systems. ACIAR Project 8328 Final Report. University of New England, Armidale, NSW 2351, Australia.

Liu, C. Q. (1987). Sulphur in the agriculture of China. In: *Proceedings of the Symposium on Fertilizer Sulphur Requirements and Sources in Developing Countries of Asia and the Pacific*. FADINAP, FAO TSI and ACIAR, Bangkok, Thailand, pp. 26–30.

Liu, C. Q., Chen, G. A. and Cao, S. Q. (1980). Sulphur content and distribution on paddy soils of South China. In: *Proceedings of Symposium on Paddy Soils*, pp. 628–34.

Luther, G. W. and Church, T. M. Chapter 6 in this volume.

Mamaril, C. P. and Gonzales, P. B. (1987). Sulphur status and requirements of the Philippines. In: *Proceedings of the Symposium on Fertilizer Sulphur Requirements and Sources in Developing Countries of Asia and the Pacific*. FADINAP, FAO, TSI and ACIAR, Bangkok, Thailand, pp. 67–75.

Mitchell, M. J., David, M. B. and Harrison, R. B. This volume.

Moorman, F. R. (1978). Morphology and classification of soils on which rice is grown. In: *Soils and Rice*. International Rice Research Institute, Los Baños, Philippines, pp. 255–72.

Nagarajah, S., Posner, A. M. and Quirk, J. P. (1970). Competitive adsorption of phosphate with polygalacturonate and other organic anions on kaolinite and oxide surfaces. *Nature (Lond).*, **228**, 83–5.

Neue, H. U. (1988). Wholistic view of chemistry of flooded soil. In: *First International Symposium on Paddy Soil Fertility*. Department of Land Development, Bangkhen, Thailand, pp. 21–53.

Neue, H. U. and Bloom, P. R. (1989). Nutrient kinetics and availability in flooded soils. In: *Proceedings of the International Rice Research Conference 1987*. International Rice Research Institute, Los Baños, Philippines.

Parfitt, R. L. (1980). Chemical properties of variable charge soils. In: Theng, B. K. G. (Ed.). *Soils with Variable Charge*. New Zealand Society of Soil Science, pp. 167–94.

Parfitt, R. L., Fraser, A. R. and Farmer, V. C. (1977). Adsorption on hydrous oxides. III. Fulvic acid and humic acid on goethite, gibbsite and imogolite. *J. Soil Sci.*, **28**, 289–96.

Parkpian, P., Pongsakul, P., Sangtong, P., Cholitkul, W. and Ratanarat, S. (1986). Indigenous soil properties influencing the availability of sulphur in selected Thai soils. In: *Sulphur in Agricultural Soils*. BARC and TSI, Dhaka, Bangladesh, pp. 469–87.

Patnaik, S. (1978). Natural sources of nutrients in rice soils. In: *Soils and Rice*. The International Rice Research Institute, Los Baños, Philippines, pp. 501–19.

Patrick, W. H. Jr, Gotoh, S. and Williams, B. G. (1973). Strengite dissolution in flooded soils and sediments. *Science*, **179**, 564–5.

Patrick, W. H. Jr and Reddy, C. N. (1978). Chemical changes in soils. In: *Soils and Rice*. The International Rice Research Institute, Los Baños, Philippines, pp. 361–37.

Peterson, B. J., Howarth, R. W. and Garritt, R. H. (1986). Sulfur and carbon isotopes as tracers of salt-marsh organic matter flow. *Ecology*, **67**, 865–74.

Ponnamperuma, F. N. (1972). The chemistry of submerged soils. *Adv. Agron.*, **24**, 29–96.

Ponnamperuma, F. N., Attanandana, T. and Beye, G. (1973). Amelioration of three sulphate soils for lowland rice. In: Dost, H. (Ed.). *Acid Sulphate Soils*. International Institute for Land Reclamation and Improvement, Wageningen, The Netherlands, pp. 391–405.

Probert, M. E. (1976). The composition of rainwater at two sites near Townsville, Qld., Australia. *J. Soil Res.*, **14**, 397–402.

Ranjhan, S. K. (1986). Sources of feed for ruminant production in Southeast Asia. In: Blair, G. J., Ivory, D. A. and Evans, T. R. (Eds). *Forages in Southeast Asian and South Pacific Agriculture. ACIAR Proceedings Series No. 12*, Canberra, Australia, pp. 24–8.

Reyes, I. C., Villapando, R. R., Rosales, R. T., Gongales, P. B. and Mamaril, C. P. (1985). Sulphur studies in some Philippine wetland rice soils. Paper presented at Saturday Seminar, International Rice Research Institute, Los Baños, Philippines.

Roger, P. A. and Kurihara, Y. (1988). Floodwater biology of tropical wetlands. In: *First International Symposium on Paddy Soil Fertility*. Department of Land Development, Bangkhen, Thailand, pp. 275–300.

Samosir, S. S. R. (1989). Dynamics of sulfur in flooded soils and the sulfur nutrition of rice plants. *Ph.D. Thesis*, University of New England.

Samosir, S. S. R., Blair, G. J. and Lefroy, R. D. B . (1988). Sulfur transformations in soils undergoing flooding. In: *First International Symposium on Paddy Soil Fertility*. Department of Land Development, Bangkhen, Thailand, pp. 587–600.

Shin, J. S. (1987). Sulphur in Korean agriculture. In: *Proceedings of the Symposium on Fertilizer Sulphur Requirements and Sources in Developing Countries of Asia and the Pacific*. FADINAP, FAO, TSI and ACIAR, Bangkok, Thailand, pp. 76–82.

Stangel, P. J. (1979). Nitrogen requirement and adequacy of supply for rice production. In: *Nitrogen and Rice*. International Rice Research Institute, Los Baños, Philippines, pp. 45–69.

Tabatabai, M. A. (1982). Sulfur. In: *Method of Soil Analysis*, Part 2. *Chemical and*

Microbiological Properties. American Society of Agronomy and Soil Science, Madison, Wisconsin, pp. 501–38.

Tipping, E. (1981). The adsorption of aquatic humic substances by iron oxides. *Geochim. Cosmochim. Acta*, **45**, 191–9.

USDA Soil Conservation Service, Soil Survey Staff (1975). *Soil Taxonomy. A Basic System for Soil Classification for Making and Interpreting Soil Surveys. USDA Agricultural Handbook 436*. US Department of Agriculture, US Government Printing Office, Washington DC.

van Raalte, M. H. (1940). On the oxygen supply of rice roots. *Ann. Jardin Bot. Buitenzorg*, **50**, 99–114.

Wang, C. H. (1979). Sulphur fertilization of rice—diagnostic techniques. *Sulphur Agric.*, **3**, 12–15.

Wang, C. H., Liem, T. H. and Mikkelsen, D. S. (1976). Sulphur deficiency—a limiting factor in rice production in the lower Amazon Basin. I. Development of sulphur deficiency as limiting factor for rice production. *IRI Res. Inst. Bull. No. 47*.

Yoshida, S. and Chaudhry, M. R. (1979). Sulfur nutrition of rice. *Soil Sci. Plant Nutr.*, **25**, 121–34.

Yu, T. R. (1985). Soil and plants. In: Yu, T. R. (Ed.). *Physical Chemistry of Paddy Soils*. Science Press, Beijing, pp. 195–217.

Zahari, A. B., Tham, K. C., Wong, N. C. and Kerridge, P. C. (1983). Sulfur deficiencies in the agriculture of Malaysia. In: Blair, G. J. and Till, A. R. (Eds). *Sulfur in South East Asian and South Pacific Agriculture*. UNE, Indonesia, pp. 155–64.

12 Summary and Conclusions

J. W. B. STEWART
University of Saskatchewan, Saskatoon, Canada

and

R. W. HOWARTH
Cornell University, Ithaca NY, USA

This book reviews and synthesizes information on the processes that govern the cycling of sulphur in the various ecosystems of the continents. Whenever possible, chapters have synthesized information both on rates of sulphur-cycle processes and on the controls of these processes. An understanding of these processes and their controls is essential to determine accurately the effect of anthropogenic sulphur deposition on ecosystem structure and function and to predict the role of altered sulphur cycling in global change. Current evidence indicates that biogenic sources of sulphur to the atmosphere, with the exception of marine sources of DMS, are much smaller than once thought, and so anthropogenic sources of atmospheric sulphur dominate (Chapter 2 by Whelpdale and Chapter 3 by Andreae and Jaeschke). These anthropogenic sulphur emissions have started to decrease in North America and Europe and are predicted to continue to decrease, but large increases in Asia, Africa, and South America seem likely. A knowledge of sulphur cycling at the scale of the ecosystem is required to evaluate the recovery of areas already polluted by acid deposition (SCOPE 19) and the potential impact of further sulphur deposition and acidification in tropical countries (SCOPE 36).

Detailed and relatively accurate information on sulphur processes is generally now available for a variety of specific sites in a variety of types of ecosystems. However, measurements are usually made at a fine areal scale (a few square metres), and a challenge exists in developing the ability to integrate these to regions, to continents, and to global basis (SCOPE 35). One approach may be to synthesize process information into credible mathematical simulation models that accurately mimic the behaviour of sulphur and natural ecosystems. Obviously we have much to do before we can achieve these aims, despite the large amount of study over the past two decades of sulphur and nutrient cycling, element interactions, acid deposition, and

Sulphur Cycling on the Continents
Edited by R.W. Howarth, J.W.B. Stewart and M.V. Ivanov

climate change. None the less, remarkable progress in understanding has been made, and we believe the synthesis in this book of the existing material from workers using different techniques and approaches will be extremely useful. We found that bringing this material together exposed many gaps in information, uneveness in progress, and sometimes a lack of communication between groups looking at the same problem from different angles or in different ecosystems. Thus much was gained from bringing groups together and discussing common problems at the Trent workshop. We hope that the readers of this book will profit from this synthesis.

Sulphur and sulphur compounds are not easy to determine analytically. Thus, from the onset of work on sulphur in soils and sediments, there has been much discussion and debate on analytical methodology. Despite progress in this area (Appendix by Tabatabai), further method development is needed for many questions, such as measuring the diurnal fluctuation of sulphur in the vicinity of a plant root under aerobic or anaerobic conditions, or accurately estimating fluxes of sulphur gases from point sources to the atmosphere. Similarly, the complexity of the interactions that occur in the decomposition of plant material in soils and sediments has not allowed us to determine the chemical nature of much of the sulphur that is held in organic material. Indeed, most methods of studying sulphur transformations rely heavily on isotopic methods to gain estimates of the fluxes of sulphur in such systems and do not attempt to measure the exact quantities of chemical compounds. These methods themselves are complicated, especially with regard to isotopic exchange among pools and the discrimination that occurs in reduction and oxidation processes between the heavier and lighter isotopes of sulphur. However, an understanding of the discrimination processes has allowed this particular technique to advance our knowledge of anthropogenic fluxes of sulphur in the environment (SCOPE 43; Appendix to Chapter 3 by Nriagu and Krause, this volume).

The analytical challenge of sulphur has been further complicated by the number of additional sulphides that have been discovered in the atmosphere during the last 15 years, such as dimethylsulphide, carbonyl sulphide and carbon disulphide. Thus, as our knowledge expands on the existence of natural sources of atmospheric sulphur compounds to the atmosphere, we require much more information on the processes of formation and decomposition that occur, both in the atmosphere and in terrestrial systems (Chapter 3). Andreae and Jaeschke (Chapter 3) summarize the most important naturally emitted sulphur compounds known at the time of writing and present best estimates of these fluxes. They emphasize three uncertainties regarding continental emissions of biogenic sulphur compounds. First, there is the difficulty of being able to determine these at low levels. Second, there are technical problems of either measuring point sources or estimating these across regions. Third, knowledge of the exact processes involved is incomplete, as are all the controls on these processes. Understanding these controls is complicated by the great diversity of interactions of the sulphur cycle with other element cycles (Chapter 4).

Despite problems and complications, substantial progress has been made during the past two decades, especially through intensive studies to understand the effects of anthropogenic deposition on lakes, wetland ecosystems, and upland forests. These studies have emphasized the critical role that sulphur plays in the biogeochemistry of these ecosystems.

Much progress has been made in our knowledge of sulphur cycling in marine and freshwater wetlands (Chapter 5 by Giblin and Wieder). In marine wetlands, the tidal fluxes of sulphur are very large in comparison to the rates of sulphur processing within the sediment, and rapid cycling occurs primarily through inorganic reduced forms of sulphur. Sulphate reduction is the dominant form of decomposition in these systems. Gaseous fluxes are high in comparison to any other ecosystems, but represent only a tiny fraction of a cycle within the wetlands. Also, because of the relatively limited area of these marine wetlands, their influence on the global atmospheric sulphur cycle is limited. The freshwater wetlands have a much lower sulphur input and the cycling of organic sulphur is much more important. One study, which examined the effect of increased sulphur deposition on a freshwater peatland, suggested that it may increase rates of anoxic decomposition and thereby decrease storage of organic carbon in the peat (See Appendix to Chapter 5 by Wieder *et al.*). Increased inputs of sulphur to a wetland also clearly reduce the fluxes of methane from the wetland to the atmosphere. The chemistry of sulphur in wetland systems is complicated due to the wide range of oxidation states of sulphur, -2 to $+6$, the wide variety of both organic and inorganic sulphur compounds found, and the methodological problems already mentioned. Chapter 6 by Luther and Church integrates information on the chemistry of sulphur (emphasizing marine wetlands) and suggests major transfers among inorganic and organic species in natural ecosystems.

The research of the past two decades on sulphur cycling in lakes and other inland water bodies has been similarly encouraging. Freshwater lakes are among the ecosystems most obviously damaged by acid deposition. Research has clearly noted the resulting changes that occur in the sulphur cycle, the attendant changes in acid–base chemistry of lakes, the alteration of interactions with carbon, nitrogen, phosphorus, and other major elements, and the ability of the lakes to partially counteract acidification. Cook and Kelly (Chapter 7) point to the key features in freshwater lakes as being assimilatory reduction and uptake of sulphate in the water column by algae and bacteria, dissimilatory sulphate reduction by bacteria in anoxic zones, reaction of reduced S species with organic matter and reduced metals, and the production of sulphur gases likely to accompany decomposition of sulphur-bearing organic matter.

In Chapter 8, Lein and Ivanov explore another form of alteration of the sulphur cycle in aquatic ecosystems: the intensification of sulphate reduction which accompanies eutrophication in continental and coastal seas. The processes which occur are the same as in the freshwater lakes subject to acidification, but the magnitude and consequences of the processes differ. As with marine wetlands, sulphate reduction dominates the decomposition which occurs in the sediments of seas. Increased inputs of nutrients and organic carbon to these seas increase the rates of sulphate reduction, with a concomitant increase of hydrogen sulphide in the waters and rise in the depth of the anoxic–oxic boundary.

Chapter 9 by Mitchell *et al.* deals with sulphur dynamics in forest ecosystems and summarizes much of the information gained from studying the impacts of acid-sulphate deposition on forest vegetation and soils. The role that sulphur retention or release plays in affecting the flux of acidity to downstream ecosystems is now clear, as is the interaction of sulphur with other elemental cycles, especially nitrogen and basic cations. However, Chapter 9 notes that there are still inconsistencies in the measurement of dry and cloud-water deposition, and that there are still major unanswered questions with regard to sulphur retention and adsorption in soil. For example, the effect of sulphur storage in organic compounds has to be understood in terms of release rates over a longer period: is this organic sulphur held in a recalcitrant organic matter pool or can it be released quickly? The fluxes of sulphur gases from forests also remain poorly studied, particularly for tropical forests and for fluxes directly from forest vegetation.

Significant progress also has been made in understanding sulphur cycling in upland agricultural systems (Chapter 10 by Schoenau and Germida). We now have a basic understanding of the forms and amounts of sulphur in agricultural soils and of the processes controlling the supply of sulphur to plants. Again, the exchange of sulphur gases between the soil and plant system and the atmosphere is less well understood, as is the immediate and long-term effect of sulphur loading from the atmosphere. The authors of Chapter 10 suggest we still need a better understanding of the basic forms of sulphur in the soil, and of transformations through chemical and biological reactions. An emphasis is also needed in extrapolation from square metre to field and to region, and a requirement for more emphasis on the construction of models of mineralization and volatilization processes that include information on controls such as temperature, moisture, substrate and microbial composition response functions.

Managed lowland systems, specifically wetlands used in rice culture, offer many opportunities for expansion of knowledge. In many rice-paddy soils, high yield agriculture has stressed natural reserves of soil sulphur to an extent where sulphur deficiency is becoming a common occurrence (Chapter 11 by Lefroy *et al.*). Partially for this reason, research has concentrated on the supply of sulphur to plants and on the efficiency of fertilizer use, and much of the work reviewed in Chapter 11 concentrates on the efficiency of recovery by plants of applied fertilizer sulphur. In comparison with the other ecosystems discussed in this book, much less information is available on the processes of sulphur cycling within rice paddies. The focus of research on use of fertilizer sulphur provides practical short-term solutions to sulphur limitations, but it may not deal adequately with the long-term supply of this essential element. Also left relatively under-studied are the potential toxic effects of sulphur on rice and the effect of sulphur cycling on methane fluxes from rice paddies. We suggest that there should be a major effort towards research on sulphur processes and their controls in rice-paddy ecosystems, as there has been in other types of ecosystems, including managed upland agricultural systems. Such an emphasis on understanding sulphur cycling is especially needed as there has been a recent trend to re-examine organic amendments or incorporation of legumes in sustainable rice cultivation, a type of green manuring. This will require a better understanding of the dynamics of sulphur, nitrogen, and carbon in these systems. In our discussions at Trent, the synergism between those working on natural wetlands and cultivated wetlands gave rise to hopes that some of the methods recently developed in natural wetland studies could have application in cultivated wetland studies. Similarly, the managed nature of the rice wetlands provided the opportunity for understanding processes under more controlled conditions that could clarify the same processes occurring under natural conditions. This was a most encouraging aspect of the Trent workshop.

Finally, we conclude by noting that society has greatly altered sulphur cycles in a variety of ways, but the full magnitude of these is still unknown. Acid sulphur deposition has damaged forests and lakes, and the challenge has been accepted in industrialized temperate regions to begin to ameliorate the problem and restore damaged areas. Tropical areas are also known to be vulnerable to a greater or lesser extent, but questions arise as to our ability to control the magnitude of anthropogenic sulphur inputs to these systems and to limit their deleterious effects. Eutrophication of continental and coastal seas has increased the abundance of sulphides in the water column, to the detriment of life in these water bodies. And by developing high yield agriculture, we have stressed natural systems of sulphur supply and developed a need to add fertilizer sulphur in a manner that supplies plant needs without altering important soil quality characteristics and properties. Atmospheric sulphur comes principally from anthropogenic sources, and biogenic fluxes of sulphur from terrestrial ecosystems are thought to be relatively small. However, we do not have the capability to predict how they will change in response to added inputs of sulphur or other human influences. The major biogenic source of atmospheric sulphur appears to be oceanic DMS, but the exact magnitude of this flux and its control remain poorly known. We are unable at present to predict how this flux may change in response to climatic change or changing nutrient inputs to the oceans.

Appendix: Methods of Measurement of Sulphur in Soils, Plant Materials and Waters

M. A. TABATABAI
Agronomy Department, Iowa State University, Ames IA, USA

A.1 INTRODUCTION

A variety of organic and inorganic S compounds occur in soil–plant–water systems. In soils of the most humid and semihumid regions, the organic S compounds accounts for > 95% of the total S. The proportion of organic and inorganic S in a soil sample, however, varies widely according to soil type and depth of sampling. Similarly, S occurs in plants in such compounds as sulphate, coenzyme A, methionine, cysteine, thiamine, biotin, thiotic acid, thiocyanates, glutathiones, alkaloid, and others (Beaton, Burns and Platou, 1968) as well as in the intermediates associated with the synthesis of those S compounds. In most plant materials methionine and cysteine account for about 90% of the total S (Allaway and Thompson, 1966).

Although the sulphate ion is predominant in most natural waters, other inorganic S species and organic compounds could become predominant under certain conditions. These range from accumulation of thiosulphate in mine tailing effluents (Wolkoff and Larose, 1975) to accumulation of sulphides and organic S compounds in surface waters originating from peat and other organic deposits (Howarth *et al.*, 1983; Howarth and Merkel, 1984; Lowe and Bustin, 1985, 1989; Wieder, Lang and Granus, 1985; Wieder and Lang, 1988). Most of the methods used for determination of total S and the various S species in soil samples are applicable to analysis of plant materials and water samples. For convenience, the methods used will be discussed according to the type of samples analysed.

Sulphur Cycling on the Continents
Edited by R.W. Howarth, J.W.B. Stewart and M.V. Ivanov

A.2 SULPHUR IN SOILS

A variety of organic and inorganic S compounds occur in soils. Inorganic sulphate may occur in soils as water-soluble salts, adsorbed on to soil colloids, or may occur as insoluble forms. Organic S is the dominant form of S in most soils. Although the chemical nature of organic S in soils has been under investigation for the past four decades, little is known about the specific organic S compounds present in soils. The methods used for determination of the various inorganic forms of S and the different fractions of organic S in soils are described in Sections A.2.2 and A.2.3, respectively.

A.2.1 TOTAL SULPHUR

A.2.1.1 Wet Chemical Methods

Total S in mineral soils may range from < 20 mg kg^{-1} in sandy soils to $>$ 600 mg k^{-1} in heavy texture soils. Organic soils may contain as much as 0.5% S. Most soils, however, contain between 100 and 500 mg (S) kg^{-1}.

The methods available for accurate determination of total S in soils involve two steps: (i) conversion of the various S compounds in soils to one form, either by oxidation to sulphate (dry or wet procedure) or by reduction to sulphide (conversion to sulphate is more common than conversion to sulphide); and (ii) determination of the sulphate or sulphide produced.

Various techniques have been employed to oxidize soil S to sulphate. One of the most widely accepted procedures is fusion with Na_2CO_3 and an oxidizing agent. The procedure recommended by the Association of Official Agricultural Chemists (1955) requires fusion of sample with Na_2CO_3 and Na_2O_2. This method is probably the best for all types of soils, but tedious and operator sensitive. A mixture of Na_2CO_3 and $NaNO_3$ has been used in other procedures (Bertrand and Silberstein, 1927; Robinson, 1930). Sodium carbonate alone may be adequate to oxidize and retain S during fusion (Evans and Rost, 1945). Sodium peroxide has also been used alone (Hart and Peterson, 1911; Steinbergs, 1955). Any procedure used to accomplish fusion before S analysis should be adequate, but inclusion of an oxidizing agent would add assurance of complete oxidation during fusion.

Of the various acid treatments proposed, digestion with $HClO_4$ or a mixture of $HClO_4$ and HNO_3 has been most popular. In all $HClO_4$ digestion procedures, predigestion of soils high in organic matter with HNO_3 before addition of $HClO_4$ is necessary to avoid the danger of explosion and fire. To overcome the unsafe features of boiling $HClO_4$, Tabatabai and Bremner (1970a) developed a method that involves oxidation of soil S with NaOBr. This method is based on wet oxidation of the soil S under alkaline conditions, because wet oxidation procedures are more convenient than ignition techniques for conversion of soil S to sulphate (particularly for soil extracts), and use of alkaline oxidation should not involve the risk of gaseous loss of S (as H_2S, SO_2, or SO_3) associated with use of acid oxidants under certain conditions. Methods based on reduction of total S before its estimation are also available (Little, 1957).

Several methods are available for determination of the oxidized S. In most of the oxidation procedures described, the product is sulphate, which can be determined gravimetrically as $BaSO_4$ (AOAC, 1955), turbidimetrically (Steinbergs, 1955; Bentley, Hoff and Scott, 1955), or colorimetrically after its reduction to H_2S (Johnson and Nishita, 1952). However, methods involving conversion of soil S to SO_2 and determination of the SO_2 evolved are also available (Bloomfield, 1962; Jenkinson, 1968).

Of the various colorimetric methods recommended for determination of sulphate, the procedure developed by Johnson and Nishita (1952) is the most sensitive and accurate method available. A modification of the apparatus developed by Johnson and Nishita is shown in Figure A.1. This method involves reduction of sulphate to H_2S by a reducing mixture containing HI, HCOOH, and H_3PO_2. The H_2S thus liberated is absorbed in a buffer containing $Zn(OAc)_2$ and NaOAc and subsequently treated with *p*-aminodimethylaniline sulphate and ferric ammonium sulphate reagents for methylene blue colour development (Gustafsson, 1962). The intensity of the methylene blue colour is determined colorimetrically at a wavelength of

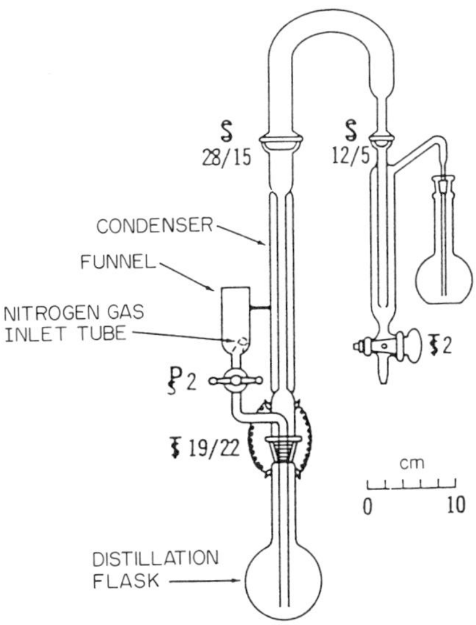

Figure A.1. Modified Johnson–Nishita apparatus

670 nm. The colour developed is stable for 24 hours when the solution is stored in a stoppered flask in subdued light. The conversion of sulphate to H_2S by the method of Johnson and Nishita (1952) after fusion, ignition, digestion, or oxidation of soil sample obviates many problems. This method is capable of completely converting all sulphates (soluble or insoluble) to sulphide (Johnson and Nishita, 1952; Freney, 1958). Methylene blue is formed by reacting a H_2S solution with an acid solution of *p*-aminodimethylaniline in the presence of Fe^{3+}. At first the solution is red, but it changes to blue as the dye is formed. The reaction may be represented as follows:

$$SO_4^{2-} + \text{Reducing mixture} \xrightarrow{\Delta} H_2S \qquad \text{(A.1)}$$

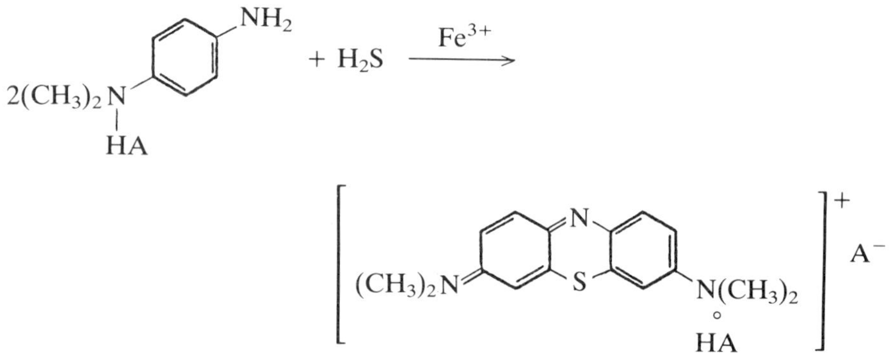

(A.2)

where A^- represents an anion. Although the reaction is not quantitative, the amount of methylene blue produced is quite reproducible if the conditions of colour development are adhered to. St Lorant (1929) obtained 68.7%, Skerrett and Dickes (1961) obtained 64.7% and Gustafsson (1960) obtained 66.7% H_2S conversion to methylene blue.

Of the procedures described, oxidation of soil S to sulphate by NaOBr and analysis of the entire residue for sulphate by the Johnson and Nishita (1952) reduction–distillation procedure are preferred. This is mainly because the methylene blue procedure has the sensitivity, accuracy, and precision needed for total S analysis of soils and soil extracts (Tabatabai and Bremner, 1970a). Other methods (e.g. turbidimetrically) in which extracts of oxidation residues are analysed for sulphate can be subject to serious errors if the soil under analysis contains significant amounts of Ba or other metallic cations that form sparingly soluble salts with sulphate (the Ba contents of most soils are between 100 and 3000 mg kg^{-1}). When the methylene blue method is used for determination of the sulphate formed, the results obtained by NaOBr oxidation procedure were similar to those obtained by using $HClO_4$ digestion and $NaHCO_3$–Ag_2O ignition procedures (Tabatabai and Bremner, 1970b).

Recently, an ion chromatographic method was developed for determination of total S in soils (Tabatabai, Basta and Pirela, 1988). It involves ignition of a mixture of soil and $NaHCO_3$ containing Ag_2O at 550° C for 3 h. The residue thus obtained is dissolved in 1 M HOAc, diluted with deionized water, filtered, and analysed for SO_4^{2-} by a Dionex Model 2002i (Dionex Corp. Sunnyvale, Calif.) ion chromatograph fitted with a separator column (AS-2, 3 × 250 mm), packed with a low-capacity anion-exchange resin in the HCO_3^- form and a suppressor column (9 × 100 mm) packed with a high-capacity cation-exchange resin in the H^+ form. A precolumn (AG-3, 3 × 50 mm), containing the same resin as the separator column, was used to protect the separator column by removing potentially poisonous substances from the eluent stream (for components and operation of IC, see Section A.2.2). Results obtained by this IC method for seven Iowa and three Chilean soils agreed closely (Table A.1) with those obtained by reduction to H_2S and determination as methylene blue after alkaline oxidation with NaOBr (Tabatabai, Basta and Pirela, 1988).

Table A.1. Comparison of total S values of soils obtained by the methylene blue (MB) and ion chromatographic (IC) methods (Tabatabai, Basta and Pirela, 1988)

Soil	Total S by method indicated	
	MB[a] (mg kg^{-1})	IC (mg kg^{-1})
Iowa soils		
Canisteo	478	476
Downs	160	143
Fayette	241	240
Hayden	129	122
Muscatine	349	351
Nicollet	354	355
Tama	231	217
Chilean soils		
Agua del Gato	620	634
Constitucion	128	131
Osorno	914	926

[a]MB, methylene blue method after alkaline NaOBr oxidation (Tabatabai and Chae, 1982).

A.2.1.2 Automated Instrumental Methods

Several instrumental methods are available for determination of total S in soils. The LECO S analyser (Figure A.2) (LECO Corp., St Joseph, Mich.) was designed for determination of total S in steel, but because of its simplicity, speed, and convenience, this analyser has been used for S analysis of non-ferrous as well as ferrous materials. In total S analysis with this instrument, the sample is treated in a ceramic crucible with combustion accelerators (usually Fe plus Sn, Cu, or Sn-coated Cu), and the crucible is covered with a porous cover and heated to high temperature (*c.* 1600° C) in a stream of purified O_2, the high temperature being generated by using an induction furnace to induce an electrical field in the accelerator-treated sample. The SO_2 thus liberated is collected in dilute HCl containing KI, starch, and a trace amount of KIO_3 and is determined by titration with standard KIO_3 solution. The reactions involved in this titration procedure are as follows:

$$KIO_3 + 5KI + 6HCl = 3I_2 + 6KCl + 3H_2O \quad (A.3)$$

$$SO_2 + I_2 + 2H_2O = H_2SO_4 + 2HI \quad (A.4)$$

Before use, the burette of the automatic SO_2 titrator is calibrated by analysing S standards. The LECO Corp. supplies standards for this purpose. Evaluation of this analyser for rapid and precise determination of total S in soils has shown that its results are usually higher, with poor precision, than those obtained by wet or dry oxidation methods, followed by reduction and determination as methylene blue (Tabatabai and Bremner, 1970b).

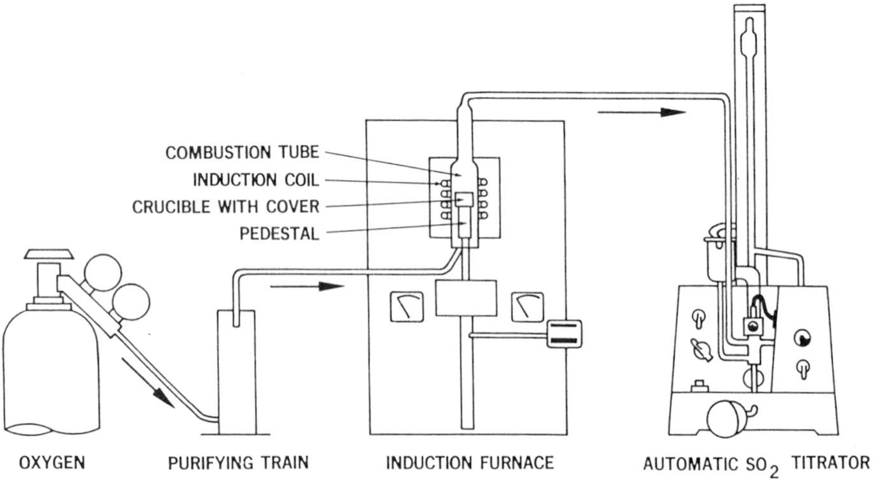

Figure A.2. Components of the LECO sulphur analyser

The LECO Model SC-132 automated total S analyser was recently evaluated for determination of total S in soils and sediments. In using this analyser, the sample is mixed with combustion accelerators in a ceramic boat and combusted in a resistance furnace at 1371° C in an O_2 atmosphere. The SO_2 thus produced is passed through an infrared (IR) cell. The IR cell is used as both a reference and a measure chamber. It detects total S, as SO_2, continuously. It consists of an IR source, a chopper motor, a precise wavelength filter, a condensing cone, and IR energy detector, and a cell body. The SO_2 collected absorbs the IR energy at a specific wavelength of the IR spectrum while in the cell. David *et al.* (1989) used this analyser for determination of total S in soils and sediments and showed that the results obtained by this instrument agreed with those by current methods used for determination of total S in such materials. Sample analysis takes < 3 min and the instrument provides a direct readout of total S values.

A.2.2 INORGANIC SULPHUR

The inorganic S fraction in soils may occur as sulphate and compounds of lower oxidation state such as sulphide, polysulphide, sulphite, thiosulphate, and elemental S. In well-drained, well-aerated soils, most of the inorganic S normally occurs as sulphate, and the amounts of reduced S compounds are generally $< 1\%$ (Freney, 1961). Under anaerobic conditions, particularly in tidal swamps and poorly drained or waterlogged soils, the main form of inorganic S in soils is sulphide and often elemental S (Brummer, Grunwaldt and Schroeder, 1971a, 1971b; Harmsen, 1954; Hart, 1959). Accurate determination of the reduced forms of inorganic S in soils is difficult, partly because of the ease with which they can be oxidized on exposure to air, but mainly because of the limitations of current analytical methods. No procedure has been entirely satisfactory for determination of sulphide in soils. Gilboa-Garber (1971) proposed a method for direct spectrophotometric determination of inorganic sulphide in biological materials. This method involves colorimetric determination of sulphide as methylene blue after extraction with Zn acetate. A modification of this method was successfully used by Howarth *et al.* (1983) for determination of sulphide in porewaters of core samples of a salt marsh. Smittenberg *et al.* (1951) proposed a method for determination of sulphide, sulphite, thiosulphate, polysulphide, elemental S, and organically bound S in soils. It involves digestion of a soil sample with HCl for determination of the monosulphidic S compounds and digestion with Sn and HCl for determination of the total oxidizable S. In both methods the H_2S released is determined colorimetrically as methylene blue. None of these methods, however, gives accurate results; digestion of soil with HCl does not release the S from acid-insoluble metal sulphides, causing underestimation of the sulphide present, and digestion with Sn and HCl releases S from organic

S, causing overestimation of the total oxidizable S (Melville, Freney and Williams, 1971). The nature of this S fraction, however, has not been identified. Much of the reducible S fraction in soils is extractable with 0.5 M NaOH and is distributed between both the fulvic acid and humic acid fractions. Therefore, digestion of soil with Sn and HCl may prove adequate where the amounts of oxidizable S are relatively large, but this method is unsatisfactory for use with soils containing small amounts of reduced inorganic S compounds, especially with surface soils containing appreciable amounts of organic S.

Studies of pyrite formation and the measurement of sulphate reduction in salt marsh sediment by Howarth and Merkel (1984) showed that the reduction with Cr(II) of pyrite and elemental S to H_2S is more specific and sensitive than oxidation of these forms of S to sulphate. The Cr(II) reduction method has been used in studies of inorganic S compounds in modern sediments and shales and of formation of ^{35}S-labelled elemental S and pyrite in coastal marine sediments during short-term $^{35}SO_4^{2-}$ reduction measurements (Howarth and Jørgensen, 1984; Canfield *et al.*, 1986).

The forms and concentrations of sulphur compounds in marine sediments are markedly different from those in soils. Zhabina and Volkov (1977) have developed a scheme for determination of various forms of sulphur compounds in sea sediments (Table A.2). Wieder, Lang and Granus (1985) studied the specificity and efficiency of the procedures for fractionation of total S into inorganic and organic constituents by analysing a series of known standards. They reported that acid volatilization was specific for FeS. Chromium reduction recovered > 90% of the S from FeS, S°, and FeS_2. Acetone extraction followed by Cr reduction of the filtrate was specific for S°. Hydriodic acid reduction recovered > 90% of the S from FeS, SO_4^{2-}, and *p*-nitrophenyl sulphate. The Zn–HCl procedure partially recovered S from SO_4^{2-}, S°, and FeS_2. None of the above procedures reduced L-methionine. Analysis of both moist and oven-dried peat by Wieder, Lang and Granus (1985) showed that oven-drying of peat samples increased the ester-sulphate S and SO_4^{2-} fractions and decreased the estimated C-bonded S, which was calculated from the difference between the total S and HI-reducible S.

Although well-drained, well-aerated soils do not contain reduced inorganic S compounds, thiosulphate and tetrathionate are produced in soils when treated with elemental S (Nor and Tabatabai, 1976, 1977). Also, such reduced S compounds are produced in tailing pond effluents from the mining industry, and it is believed that they may lead to acid pollution of rivers surrounding these industries. Concentrations of thiosulphate in tailing pond effluents have been estimated to be in the range of 500 mg l^{-1} (Wolkoff and Larose, 1975).

Several methods have been proposed for determination of thiosulphate and tetrathionate in milligram quantities. These include iodimetric titration,

Table A.2. Scheme of the determination of various forms of sulphur compounds in marine sediments (Zhabina and Volkov, 1977)

Wet sample			
↓			
HCl heating	$\longrightarrow$	$H_2S \longrightarrow$	Sulphide-S
↓			
Filtration	$\longrightarrow$ Solution $\longrightarrow$ $Sn(II) + H_3PO_4 \longrightarrow$	$H_2S \longrightarrow$	Sulphate-S
↓			
Precipitation			
↓			
Acetone extraction	$\longrightarrow$ extract $\nearrow$ $CrCl_2 \longrightarrow$	$H_2S \longrightarrow$	S^0
	extract $\searrow$ $HNO_3 + Br_2 \xrightarrow{SO_4^{2-}} Sn(II) + H_3PO_4 \longrightarrow$	$H_2S \longrightarrow$	Org. S
↓			
Precipitation			
↓			
$CrCl_2$ heating	$\longrightarrow$	$H_2S \longrightarrow$	Pyrite-S
↓			
Filtration	$\longrightarrow$ Precipitate $\longrightarrow$ $NHO_3 + KMnO_4 \longrightarrow HNO_3 + Br_2 \xrightarrow{SO_4^{2-}}$	$H_2S \longrightarrow$	Org. S

reacting thiosulphate with *o*-tolidine and Cu(II) ions (Haff, 1970), reduction with Zn and HCl and subsequent determination of the H_2S produced as methylene blue (Aspiras, Keeney and Chesters, 1972), and cyanolysis under various conditions (Sorbo, 1957; Kelley, Chambers and Trudinger, 1969; Nor and Tabatabai, 1975). Similar methods have been proposed for determination of tetrathionate (Kelley, Chambers and Trudinger, 1969; Nor and Tabatabai, 1975). Although some of these methods give quantitative results when used for determination of a specific reduced S compound, quantitative determination of thiosulphate and tetrathionate individually in the presence of a relatively large amount of sulphate is difficult. Some of these methods are not specific, and others require various cooling and heating conditions that make their application to soils or soil extracts very difficult (Aspiras, Keeney and Chesters, 1972; Kelley, Chambers and Trudinger, 1969). To overcome some of these problems, Nor and Tabatabai (1976) developed colorimetric methods for determination of thiosulphate and tetrathionate in soils. The methods involve extraction of these S forms from soils with 0.1 M LiCl or 0.15% $CaCl_2$ solution and alkaline cyanolysis of thiosulphate-S in the presence of Cu^{2+} and alkaline cyanolysis of tetrathionate-S in the absence of Cu^{2+}:

$$S_2O_3^{2-} + CN^- \xrightarrow{Cu^{2+}} SO_3^{2-} + SCN^- \qquad (A.5)$$

$$S_4O_6^{2-} + 3CN^- + H_2O \longrightarrow S_2O_3^{2-} + SO_4^{2-} + 2HCN + SCN^- \qquad (A.6)$$

and determination of the thiocyanate produced as ferric thiocyanate complex:

$$SCN^- + Fe^{3+} \longrightarrow \text{Fe–SNC complex} \qquad (A.7)$$

The absorbance of the reddish brown colour of the Fe–SCN complex formed is measured using a spectrophotometer adjusted to a wavelength of 460 nm. This method is rapid, sensitive, and accurate, and it permits determination of microgram quantities of thiosulphate, tetrathionate, or both in the presence of large amounts of sulphate.

Theoretically, for each mole of tetrathionate, two moles of thiocyanate are produced by alkaline cyanolysis in the presence of Cu^{2+} and one mole of thiocyanate is produced in the absence of Cu^{2+}. Therefore, a factor of 2 has been used to calculate the quantities of tetrathionate present in samples containing thiosulphate (Kelley, Chambers and Trudinger, 1969). Work in my laboratory, however, indicated that the theoretical factor 2 is never achieved when tetrathionate is analysed by this method and that the practical factor for calculating the quantities of tetrathionate is 1.75 (Nor and Tabatabai, 1975). Methods based on the principle described above have been successfully used for determination of thiosulphate and polythionate in porewaters of a New England Salt Marsh (Howarth *et al.*, 1983) and in soil extractants obtained in assay of rhodanese activity (Tabatabai and Singh, 1976) and in studies of the pathway of oxidation of elemental S in soils (Nor and Tabatabai, 1977).

A polarographic method for determination of the reduced S species in marine porewaters has been proposed by Luther, Giblin and Varsolona (1985). They have demonstrated that, with polarographic techniques, it is possible to measure thiosulphate, sulphide, bisulphide, and polysulphide ions with a mercury electrode. They considered the polysulphide ions, S_x^{2-}, to be composed of one S atom in the −2 oxidation state, S(2−), and the remaining $(x-1)$ S atoms in the zero-valent oxidation state, S(0). The number of S atoms in each is measurable by this technique. Tetrathionate and other polythionates can be measured by this technique, but Luther, Giblin and Varsolona (1985) could not detect these reduced S species in the porewaters that they tested. By using this differential pulse polarographic technique, Luther *et al.* showed that salt marsh and subtidal porewater profiles contain significant concentrations of thiosulphate, bisulphide, and polysulphide. This technique has been successfully used in demonstrating the seasonal cycling of S and Fe in porewaters of a Delaware salt marsh (Luther and Church, 1988).

Inorganic sulphate may occur in soils as water-soluble salts, may be absorbed by soil colloids, or may occur as insoluble forms. The fractions of inorganic sulphate present in each of these states are dependent on several soil properties (Tabatabai, 1987). Surface horizons of most well-drained temperate soils contain only small amounts of water-soluble sulphate. Large quantities of water-soluble sulphate, however, accumulate in soils of arid regions. Because of the effect of seasonal conditions on mineralization of organic S, leaching of soluble sulphate, and sulphate uptake by plants, it is very common to observe a considerable seasonable fluctuation in the amounts of water-soluble sulphate in surface soils. Other factors that may affect the concentration of water-soluble sulphate in soils are application of fertilizer materials and the sulphate content of rain and irrigation waters (Olson and Rehm, 1986; Stroehlein and Pennington, 1986).

Several factors affect inorganic sulphate adsorption by soils. Among these clay content, nature of the clay minerals, pH, and exchangeable cations are the most important. Adsorption of sulphate by soils is negligible above pH 6.5 and increases with decreasing pH below this value (Kamprath, Nelson and Fitts, 1956; Williams and Steinbergs, 1962). Sulphate adsorption is greater in soils containing large amounts of Al and Fe oxides (Berg and Thomas, 1959; Ensminger, 1954; Chao, Harward and Fang, 1962a, 1962b; Tabatabai, 1987); the former generally being more important. Adsorption of sulphate by soil constituents occurs at a number of energetically different reaction sites. This was demonstrated by the work reported by Alymore, Karim and Quirk (1967) showing that sulphate adsorbed on kaolinite clay is weakly held relative to that adsorbed on iron and aluminum oxides.

In addition to the water-soluble and absorbed sulphates fractions, several forms of insoluble sulphates are likely to occur in soils. The insoluble forms may include barium and strontium sulphates, sulphate co-precipitated with

$CaCO_3$, and basic iron and aluminum sulphates under certain conditions. Basic iron sulphate, jarosite [$KFe_3(OH)_6(SO_4)_2$], and coquinbite [$Fe_2(SO_4)_3 \cdot 5H_2O$], have been identified in separate tests taken from tidal marsh soils (Clark, Gobin and Sprout, 1961; Fleming and Alexander, 1961). Although no report is available on the detection of basalunite [$Al_4(OH)_{10} \cdot 5H_2O$] in soils, this compound has been prepared from $Al(OH)_3$ and sulphate ions in the presence of clay (Singh, 1967; Singh and Brydon, 1967, 1969).

The soluble sulphate fraction in soils is extractable with water and with salt solutions containing NaCl, LiCl, or $CaCl_2$. A salt solution, however, is frequently preferred because of dispersion difficulties associated with aqueous extractions. Arkley (1961) introduced the use of 0.1 M LiCl, and Williams and Steinbergs (1959) suggested the use of 0.15% $CaCl_2$ solution. The concentrations of electrolytes in these solutions are sufficient to keep the soil flocculated, but neither the hydrated Li nor the Cl ions would influence greatly the chemistry of the soil surface dominated with much more strongly adsorbed ions. The use of LiCl salt solution has the additional advantage that microbial activity is slowed or stopped during extraction of sulphate, or storage of the extract, because it has been shown that the Li ion is a metabolic inhibitor (King, 1953; Laties, 1959).

Two approaches have been used for estimation of the amount of adsorbed sulphate in soils. The first approach makes the use of the finding by Chao, Harward and Fang (1962a) that adsorbed sulphate is in kinetic equilibrium with solution sulphate and hence can be measured by isotope exchange. This approach was used by Hasan, Fox and Boyd (1970) to study sulphate desorption from soils. This method, however, overestimates the amount of adsorbed sulphate present in soils, because the radioactive sulphate may also exchange with S in soil organic matter (Freney, Melville and Williams, 1971). The second approach involves displacement of the adsorbed sulphate in soils by shaking with an appropriate solution. Several reagents have been proposed for extraction of soluble sulphate plus a fraction of the adsorbed sulphates. These include $NaHCO_3$ (Kilmer and Nearpass, 1960), $CaCO_3$ suspensions (Williams and Steinbergs, 1962), neutral 1 mol l^{-1} NH_4OAc (McClung, De Freitas and Lott, 1959), acid NH_4OAc (Bardsley and Lancaster, 1960; Stanford and Lancaster, 1962), and Morgan's solution (Chesnin and Yien, 1950). The adsorbed sulphate may also be displaced by solutions containing anions of greater coordinating ability according to the series (Bingham, Sims and Page, 1965; Chang and Thomas, 1963; Chao, Harward and Fang, 1964):

$$\text{Hydroxyl} > \text{Phosphate} > \text{Sulphate} = \text{Acetate} > \text{Nitrate} = \text{Chloride}$$

Alkaline solutions, such as the $NaHCO_3$ solution of Kilmer and Nearpass (1960) and the $CaCO_3$ suspension of Williams and Steinbergs (1962), are effective, but results obtained by these extractants are difficult to interpret because these reagents also extract part of the organic S. Calcium carbonate suspension, however, extracts less organic S than $NaHCO_3$. A solution of KH_2PO_4 containing 500 mg l^{-1} of P was recommended by Ensminger (1954) for extraction of soluble and adsorbed sulphate in soils. This solution extracts adsorbed sulphate effectively from most soils, but it has the disadvantage that the extracts are difficult to filter. Also, it may extract some S from the organic S fraction. These disadvantages are overcome by the 0.01 M Ca $(H_2PO_4)_2$ introduced by Fox, Olson and Rhoades (1964). This solution contains sufficient phosphate to displace most of the adsorbed sulphate of many soils. It is necessary to use a wide solution/soil ratio or repeated extractions to displace the adsorbed sulphate in soils having strong adsorption capacities. The amounts of adsorbed sulphate extracted by this reagent can vary from < 1 mg kg^{-1} to several thousand parts per million. Some strongly sulphate-adsorbing surface soils of Hawaii contain $>$ 1000 mg kg^{-1} and the subsurface soils $>$ 7000 mg kg^{-1}, of adsorbed SO_4^{2-}-S (Hasan, Fox and Boyd, 1970).

The sulphate thus extracted can be determined by one of a number of methods. Gravimetry, turbidimetry, titrimetry, and colorimetry are the most commonly used methods for estimation of sulphate in soil extracts. Selection of a procedure for sulphate analysis of aqueous solutions is dependent on the accuracy and precision desired. The advantages and limitations of these and other methods used for determination of sulphate have been reviewed by Beaton, Burns and Platou (1968) and are not covered here. Among these methods, however, the methylene blue colorimetric method developed by Johnson and Nishita (1952) is the most sensitive and accurate.

Analytical instruments such as ion chromatographs are commercially available. Since the introduction of ion chromatograph (IC) by Small, Stevens and Bauman (1975), research in the area of IC has made significant advances in separation and determination of ionic species. IC has become a rapid and sensitive technique for analysing complex mixtures of ions. Now, ion chromatographs are available that feature high-speed separation and continuous monitoring by dectector–analyser systems, which yield instantaneous readout of analytical data. The commercially available IC systems can be divided into two major groups: those that operate on the principle of eluent suppressors (dual-column system) and those with no suppressor column (single-column system). Because the electrical conductivity of a solution varies with the concentration of ions present, this principle is the basis of a universal IC detector system used for the determination of all ionic species in solution. The eluent suppressed IC instrument (Figure A.3) (Dionex Corp., Sunnyvale, Calif.) has been evaluated for determination of sulphate in soils, and it seems suitable for analysis of sulphate in a variety of soil extracts (Dick and Tabatabai, 1979; Tabatabai and Dick, 1979). These instruments are basically low-pressure liquid chromatographs consisting of a separator (or analytical)

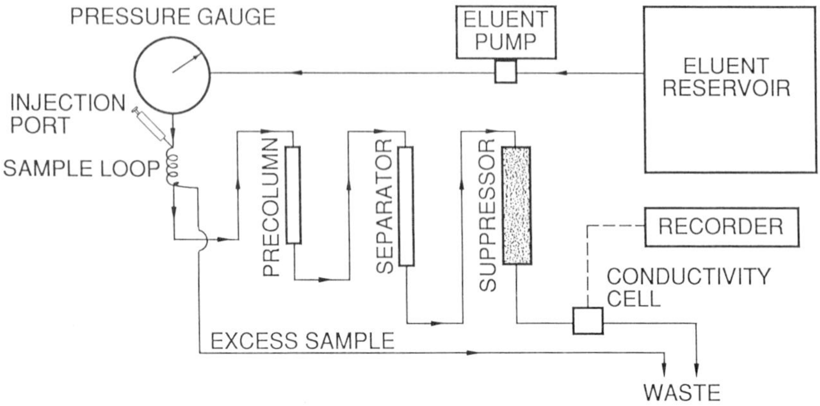

Figure A.3. Simplified schematic diagram of suppressed-type IC system (Dick and Tabatabai, 1979)

column packed with Dionex low-capacity anion exchange agglomerated resin, suppressor packed with Dionex high-capacity cation exchange resin, eluent and suppressor-column regenerating reagent (1 mol l^{-1} H_2SO_4) reservoirs, injection valve, adjustable sample loop, two pumps, conductivity cell, and valving system to direct the flow through various parts of the instrument. All valves are air-actuated Teflon slider valves. For simplicity, these components are not shown in Figure A.3. When combined with an externally attached strip chart recorder or a peak area integration system, the ion chromatographs are complete systems for performing ion chromatography.

For separation of anions by the dual-column IC instrument (Figure A.3.), the separator column contains a strong base anion exchange resin in the HCO_3^- form, and the suppressor column contains a strong acid cation exchange resin in the H^+ form. The eluent is 3.0 mM $NaHCO_3$ + 1.8 mM Na_2CO_3 at a flow rate of 3 ml min^{-1} and pump pressure of 31.6 kg cm^{-2} (450 psi). The eluent time is affected by the concentrations of $NaHCO_3$ and Na_2CO_3 and by the pump pressure used. When anions are separated by the use of a solution containing $NaHCO_3$ and Na_2CO_3 as the eluent, the eluent Na ions and sample cations exchange for H ions on the suppressor resin, leaving H_2CO_3 as the effluent. The suppressor column also converts the sample anions to acids before they enter the detector. The suppressor reactions are as follow:

Separator column:
Eluent:

$$NaHCO_3 + \text{Resin-}HCO_3 \rightleftharpoons \text{Resin-}HCO_3 + NaHCO_3 \qquad (A.8)$$

Sample:

$$MA + Resin\text{-}HCO_3 \rightleftharpoons Resin\text{-}A + MHCO_3 \quad (A.9)$$
$$NaHCO_3 + Resin\text{-}A \rightleftharpoons Resin\text{-}HCO_3 + NaA \quad (A.10)$$

Suppressor column:
Eluent:

$$NaHCO_3 + Resin\text{-}H \rightleftharpoons Resin\text{-}Na + H_2CO_3 \quad (A.11)$$

Sample:

$$MHCO_3 + Resin\text{-}H \rightleftharpoons Resin\text{-}M + H_2CO_3 \quad (A.12)$$
$$NaA + Resin\text{-}H \rightleftharpoons Resin\text{-}Na + HA$$
$$(H_2CO_3 \text{ has very low conductivity}) \quad (A.13)$$

where M and A represent cations and anions, respectively.

From calibration graphs obtained for the linear relationship between peak height and concentrations (parts per million) of SO_4^{2-}-S in standard solutions, SO_4^{2-}-S concentration of samples can be calculated. Although the sample loop of the injection valve contains a volume of 100 μl (the sample loop can be adjusted to give the desired volume), a 2 ml sample of soil extract is convenient to ensure proper flushing of the injection valve loop and lines. The retention times for selected anions are shown in Table A.3, and the chromatograms obtained for NO_3^- and SO_4^{2-} in soil extracts are shown in Figure A.4. The development of a chemical suppression device, the Micromembrane Suppressor (MMS, Dionex Corp.) has eliminated the regeneration of the suppressor column and extends the ability of the technique to determine inorganic anion and cation (Stillian, 1985). This device combines the high ion-exchange capacity of packed-bed suppressors with the constant regeneration feature of fibre suppressors.

The basic components of a non-suppressed-type (single-column system) ion chromatograph (SCIC) are shown in Figure A.5. The technique employs the following components: eluent pump and eluent reservoir, sample injection valve (a sample loop of $\geqslant$ 500 μl is normally used), and ion-exchange separator column and conductivity detector coupled to an output device. In this system, a low-capacity exchange column and low-conductivity eluent are used without the need for a suppressor column. Organic acids (phthalate, benzoate, or citrate) are often used in the mobile phase of SCIC, phthalic acid being the most common because of its wide range of retention control (via pH adjustment) and equivalent conductance. Anions of interest elute in the hydrogen form (e.g. H_2SO_4) against a background of ionized phthalate. Eliminating the suppressor column reduces the post-column dead volume, resulting in faster analysis, but the SCIC system is about 20 times less sensitive than the eluent-suppressed system. Description of the IC instruments and methods developed for the determination of anions, including

Table A.3. Retention times (minutes) for selected anions (Krupa and Tabatabai, 1986)

Anion or compounds	Ion-chromatographic conditions[a] A	B	C
$S_2O_3^{2-}$	1.2	ND	2.1(7.4)
S^{2-}	1.3	ND	ND
F^-	1.4	2.3	ND
Cl^-	1.9	3.7	ND
NO_2^-	2.3	4.5	ND
PO_4^{3-}	3.4	6.5	ND
NO_3^-	4.8	10.9	ND
SO_3^{2-}	5.2	ND	5.0(7.4)
$S_2O_5^{2-}$	5.2	ND	5.0(7.4)
$S_2O_6^{2-}$	R	R	R
SO_4^{2-}	8.2	15.7	7.4
$S_4O_6^{2-}$	R	R	R
SCN^-	R	R	R
Thiourea	R	R	R
Glucose-6-sulphate	R	R	R
p-Nitrophenyl sulphate	R	R	R

[a](A) IC conditions: 125 mm precolumn connected to 250 mm separator column; eluent = 3.0 mM $NaHCO_3$ + mM Na_2CO_3; flow rate = 182 ml h^{-1}; recorder = 3 μS full scale.

(B) IC conditions: 123 mm precolumn connected to 500 mm separator column; eluent = 3.0 mM $NaHCO_3$ + 2.4 Na_2CO_3; flow rate = 136 ml h^{-1}; recorder = 3 μS full scale.

(C) IC conditions: as in B, but without precolumn, figures in parentheses are retention times of SO_4^{2-} detected with the reduced form of S tested. R = retained by column; ND = not determined.

sulphate, in soils and soil solutions have been recently reviewed by Tabatabai and Basta (1991).

Evaluation of the eluent suppressed IC instrument showed that its results for extractable sulphate in a variety of soils agreed closely (Table A.4) with those obtained by the methylene blue method (Dick and Tabatabai, 1979). The ion chromatographic method has a high sensitivity, precision, and accuracy for analysis of SO_4^{2-}-S. It can detect as little as 0.2 ppm of SO_4^{2-}-S against a background of 10 mM chloride, phosphate, or acetate. Among the advantages of the IC method is that, in addition to sulphate, several other anions in soil extracts can be determined simultaneously in about 10 min.

A.2.3 ORGANIC SULPHUR

Although the chemical nature of organic S in soils has been under investigation for the past four decades, little is known about the organic S compounds

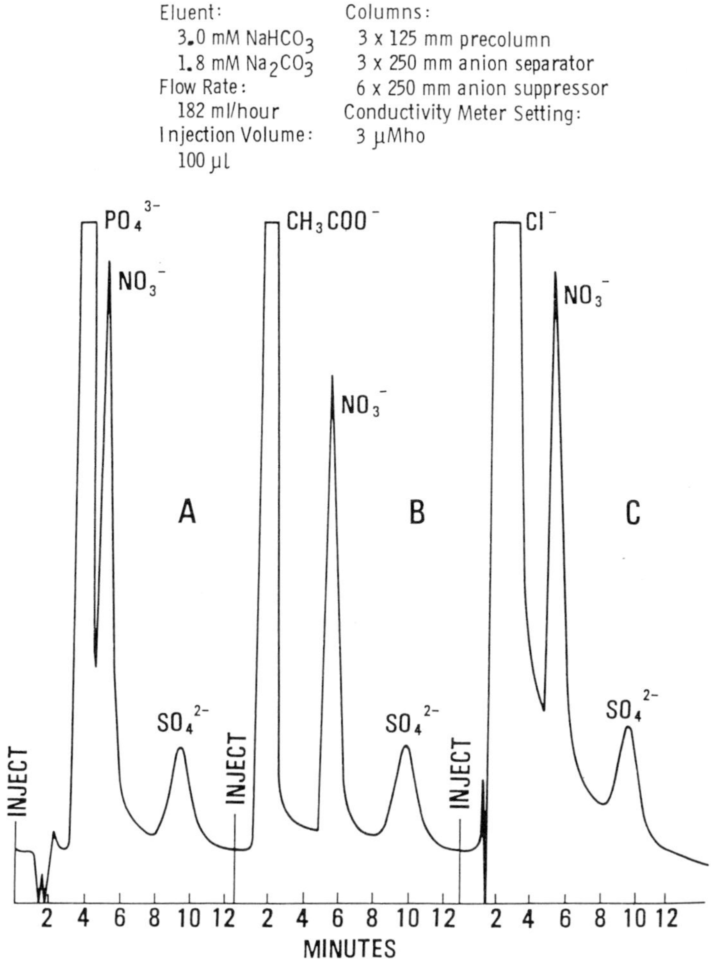

Figure A.4. Typical anion chromatograms of extractants containing 4 mg (NO_3^--N) l^{-1} and 1.5 mg (SO_4^{2-}-S) l^{-1}. The extractants were: (A) 100 mg (P) l^{-1} as $Ca(H_2PO_4)_2$, (B) 10 mM $Ca(C_2H_3O_2)_2$, (C) 0.15% $CaCl_2$ (Dick and Tabatabai, 1979)

present in soils. Early work showed the presence in soils of organic S compounds such as trithiobenzaldehyde (Shorey, 1913) and free cystine (Putnam and Schmidt, 1959). Other reports (Stevenson, 1956; Bremner, 1967; Kowalenko, 1978) show that cystine and methionine are present in small amounts in soil hydrolysates, and the reports indicate that they may occur as constituents of polypeptides in soils.

Although the chemical nature of organic S in soils cannot be defined as well as that of inorganic S, three broad groups of S compounds have been

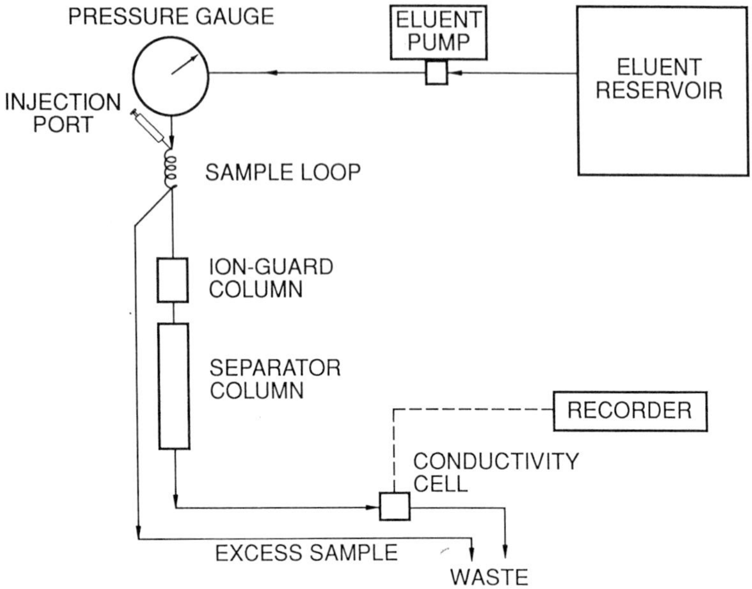

Figure A.5. Simplified schematic diagram of non-suppressed-type IC system (Tabatabai and Basta, 1991)

recognized:

(1) Organic S that is reduced to H_2S by HI. This S is not bonded directly to C and is believed to be largely in the form of ester sulphates (e.g. phenolic sulphates).
(2) Organic S that is reduced to inorganic sulphide by Raney Ni and that seems to consist almost entirely of S in the form of amino acid (e.g. cysteine and methionine).
(3) Organic S that is not reduced by either HI or Raney Ni. This fraction is assumed to consist of S bonded directly to C but not recoverable by current methods used for estimation of C-bonded S (fraction 2).

Unfortunately, no satisfactory direct method is available for determination of the total organic S or the organic S fraction reducible by HI. These fractions, however, can be determined indirectly. The total organic S can be determined in a soil residue after extraction of inorganic S or by difference involving subtraction of the inorganic S fraction from total S. The organic S reducible by HI or Raney Ni can be estimated by subtracting the extractable inorganic S fraction from the fraction of S reducible by HI and by subtracting inorganic reduced S compounds from the fraction of S reducible by Raney Ni (Tabatabai and Bremner, 1972; Neptune, Tabatabai and Hanway, 1975).

Table A.4. Comparison of sulphate-S in soils obtained by ion chromatographic and methylene blue methods (Dick and Tabatabai, 1979)

Soil	Method[a]	Sulphate-S extracted by reagent specified[b]						
		A	B	C	D	E	F	G
		($mg\ kg^{-1}$ soil)						
Ackmore	IC	5.6	5.5	5.7	5.6	5.5	5.6	5.6
	MB	5.5	5.7	5.5	5.5	5.7	5.7	5.5
Downs	IC	18	18	17	18	17	18	24
	MB	17	18	18	17	18	17	23
Webster	IC	5.2	5.1	5.0	5.2	5.2	5.1	6.8
	MB	5.1	5.2	5.2	5.0	4.9	5.0	6.7
Canisteo	IC	8.6	8.8	8.6	8.7	8.8	8.6	8.8
	MB	8.7	8.7	8.8	8.6	8.7	8.8	8.6
Okoboji	IC	8.9	9.0	9.0	8.9	9.0	8.9	9.0
	MB	8.8	8.9	9.0	8.8	8.9	9.0	8.9
Avg	IC	9.3	9.3	9.1	9.3	9.1	9.2	10.8
	MB	9.0	9.3	9.3	9.0	9.2	9.1	10.5

[a]MB = melthylene blue.
[b]A, water; B, 10 mM LiCl; C, 10 mM KCl; D, 0.15% $CaCl_2$; E, 10 mM $Ca(C_2H_3O_2)_2$; F, 100 ppm P as $Ca(H_2PO_4)_2$; G, 3.0 mM $NaHCO_3$ + 1.8 mM Na_2CO_3.

A.2.3.1 Hydriodic Acid-reducible Sulphur

The basic principle of the Johnson and Nishita (1952) method is the reduction of all inorganic S to H_2S by a strong acid reducing mixture containing HI, HCOOH, and H_3PO_2. The S evolved as H_2S is determined colorimetrically as methylene blue. This method does not recover the S in organic compounds in which S is directly bonded to C, such as cysteine, cystine, methionine, and taurine (Johnson and Nishita, 1952; Freney, 1958). The acid mixture, however, reduces a variety of organic S compounds in which S is present as ester sulphates (C—O—S linkage), such as diethyl sulphate, *p*-nitrophenyl sulphate, and chondroitin sulphate. Application of the Johnson and Nishita method directly to analysis of soils has shown that, on average, about 50% of the total S in soils of humid and temperate regions can be converted to H_2S, although analyses covering wide ranges of soils have indicated that the percentage may range from 30 to 80% (Freney, 1961; Tabatabai and Bremner, 1972; Neptune, Tabatabai and Hanway, 1975; Williams, 1967; Biederbeck, 1978). Because of the reduction of S in several model compounds and because of presence of some of these compounds in soils, it is believed that the major portion of HI-reducible S in soils is present as sulphated polysaccharides, phenolic sulphates, choline sulphate, or sulphated lipids (Freney, 1967). The fraction of S in soils reduced by HI includes inorganic S (mainly sulphate). Therefore, it is essential to correct the HI-reducible fraction for inorganic S when estimation of the ester sulphate fraction is desired. This can be accomplished by subtracting the value of inorganic S obtained by a separate analysis from the HI-reducible S value. The fraction of ester sulphate in soils is sometimes described as organic sulphate. This may be a more accurate description because the procedure used to determine HI-reducible S recovers S in organic sulphates containing C—N—S linkages (Freney, 1961), and there is no evidence that such sulphates do not occur in soils.

Because the S in this fraction can be readily hydrolysed to inorganic sulphate by acid or alkali, HI-reducible S is considered to be the most labile fraction of soil organic S (Spencer and Freney, 1960; Lowe, 1965; Freney, Melville and Williams, 1971; Cooper, 1972). This fraction is thought to be largely associated with certain side chain components of fulvic and humic materials (Bettany, Stewart and Halstead, 1973). Extracts obtained from soils with $NaHCO_3$ at pH 10 and with chelating resin (Freney, Melville and Williams, 1969) indicate that much of the HI-reducible S occurs in the high molecular weight (humic acid) fraction of the soil organic matter.

A.2.3.2 Carbon-bonded Sulphur

Application of Raney Ni to analysis of S directly bonded to C in soils was first proposed by DeLong and Lowe (1961), and since then it has been used by many workers for estimation of C-bonded S in soils. The method involves desulphurization by Raney Ni in the presence of alkali in the Johnson and Nishita apparatus, liberation of H_2S by acidification, and estimation of the sulphide released colorimetrically as methylene blue. Application of the method to a variety of organic S compounds showed that the method recovers all forms of organic S except S in inorganic sulphate and alkyl sulphone (Lowe and DeLong, 1963; Freney, Melville and Williams, 1970). It also recovers elemental S and reduced inorganic S (e.g. $S_2O_3^{2-}$, $S_2O_4^{2-}$, $S_4O_6^{2-}$, and SO_3^{2-}). But well-drained soils usually do not contain measurable amounts of these inorganic forms of S (Tabatabai and Bremner, 1972; Nor and Tabatabai, 1976, 1977). Lowe and DeLong (1963) found that $> 95\%$ of the total S in four of five Quebec soils could be accounted for as C-bonded S and HI-reducible S, but only 57% of the total S in one of these soils could be accounted for in these two forms, and they postulated the existence in this soil of an 'inert' form of S that is not recovered by the procedures used to determine HI-reducible and C-bonded S. Freney, Melville and Williams

(1970), however, showed that Fe and Mn can interfere with determination of C-bonded S in soils by the method of Lowe and DeLong (1963) and that a substantial fraction (avg 23%) of the organic S in 15 Australian surface soils could not be accounted for as (HI-reducible + C-bonded) S even when the method used to determine C-bonded S was modified to reduce interference by Fe and Mn. Other factors that may affect the results obtained for C-bonded S by the method of Lowe and DeLong (1963) are the amount of Raney Ni and concentration and amount of NaOH used (Freney, Melville and Williams, 1970).

It has been suggested that C-bonded S should be estimated as being the difference between total and HI-reducible S (Freney, 1967), and some workers have adopted this simple procedure. The assumption that all organic S in soils that is not HI reducible consists of C-bonded S, however, is probably an oversimplification and is not justified in view of the existence in soils of sizeable amounts of unidentified S with characteristics differing from those of HI-reducible and C-bonded S (Lowe, 1965; Freney, Melville and Williams, 1970; Tabatabai and Bremner, 1972; Neptune, Tabatabai and Hanway, 1975). The presence of this unidentified or residual fraction of soil organic S was first suggested by Lowe (1964) and later supported by results from other studies showing that this fraction accounted for between 3 and 59% of the total organic S in mineral soils (Biederbeck, 1978).

Based on analyses of soil hydrolysates, it has been estimated that about half of the C-bonded S occurs as amino acids (Freney, Stevenson and Beavers, 1972). The nature of the remainder (unidentified) of this fraction is unknown. A study by Biederbeck (1978) showed that C-bonded S accounts for between 5 and 54% of the total S in some mineral soils of Canada, Australia, United States, and Brazil. Lowe and DeLong (1963) reported concentrations of 47 to 58% C-bonded S in several organic soils.

A.3 SULPHUR IN PLANTS

Sulphur as an essential plant nutrient occurs in many different forms and in many plant compounds. Review of the chemistry of S in plants is beyond the scope of this chapter. Sulphur metabolism in plants and the roles and requirements of S in plant nutrition have been reviewed by Thompson, Smith and Madison (1986) and Duke and Reisenauer (1986) and will not be discussed here. Although S in plants may occur in a variety of organic and inorganic compounds, in most plant materials methionine and cysteine account for > 90% of the total S (Allaway and Thompson, 1966). Total S in oven-dry plant materials generally ranges from 0.10 to 1.5%. For herbage, the extreme values of total S content vary from 0.02 to 2.11%, but under normal agricultural conditions it is usually within the range of 0.20 to 0.45% (Beaton, Burns and Platou, 1968).

A.3.1 TOTAL SULPHUR

To estimate the total S in plant materials, it is necessary to convert all the S to one form that can be determined accurately. This can be accomplished by one of several dry ashing or wet digestion procedures. Determination of S in the ashed or digested material is then performed by a suitable analytical technique. Many of these procedures and associated analytical techniques have been summarized by Beaton, Burns and Platou (1968) and the advantages and disadvantages of each procedure are discussed. Most workers have concluded that conversion of total S in plant materials can be accomplished by oxidizing the organic S to SO_4^{2-} by dry ashing or alkaline digestion with NaOBr. The SO_4^{2-} thus formed is determined gravimetrically, turbidimetrically, or colorimetrically. Most of these methods, however, are time consuming, difficult to perform, susceptible to loss of S during oxidation and/or require a high degree of analytical skill.

Because digestion of plant materials with $HClO_4$ is unsafe and gives inaccurate results, Tabatabai and Chae (1982) developed a method for determination of total S in plant samples by the methylene blue method after alkaline oxidation with NaOBr. The results for total S by this method were

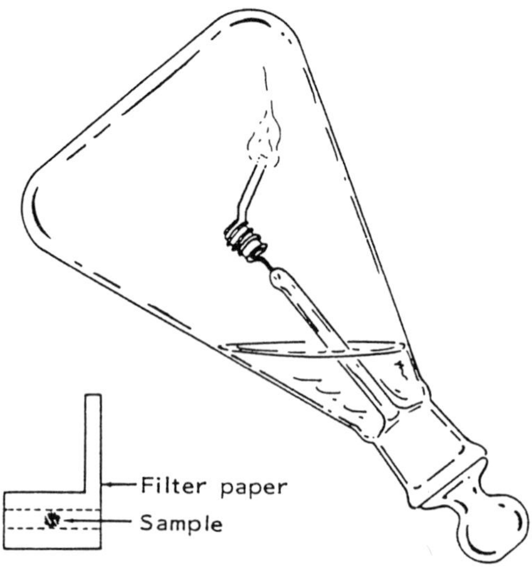

Figure A.6. Diagram of oxygen flask and the filter paper used for enclosing plant sample (Busman, Dick and Tabatabai, 1983)

generally higher than those obtained by the $HClO_4$-digestion method, suggesting possible loss of S as gases during digestion with $HClO_4$. In addition to being safe, digestion of plant samples with NaOBr is more rapid than with $HClO_4$.

Because of the advancement made on determination of SO_4^{2-} by ion chromatography (IC), Busman, Dick and Tabatabai (1983) developed an ion chromatographic method for simultaneous determination of total S and Cl in plant materials. This method involves combustion of a plant sample in a Schoniger-type oxygen flask containing deionized water (Figure A.6). The SO_4^{2-} and Cl^{1-} formed in the aqueous solution (the S and Cl gases produced are absorbed by water) were determined by using a Dionex Model 10 IC. Figure A.7 shows a typical chromatogram obtained by analysing the absorbing solution after igniting of a plant sample with and without additions of H_2O_2 to oxidize NO_2^- to NO_3^-. Results obtained for total S in a variety of plant materials by the IC method agreed closely with those obtained by the methylene blue method after digestion with NaOBr. The average total S values of 15 plant samples by the two methods were 0.255 and 0.259%, respectively (Table A.5).

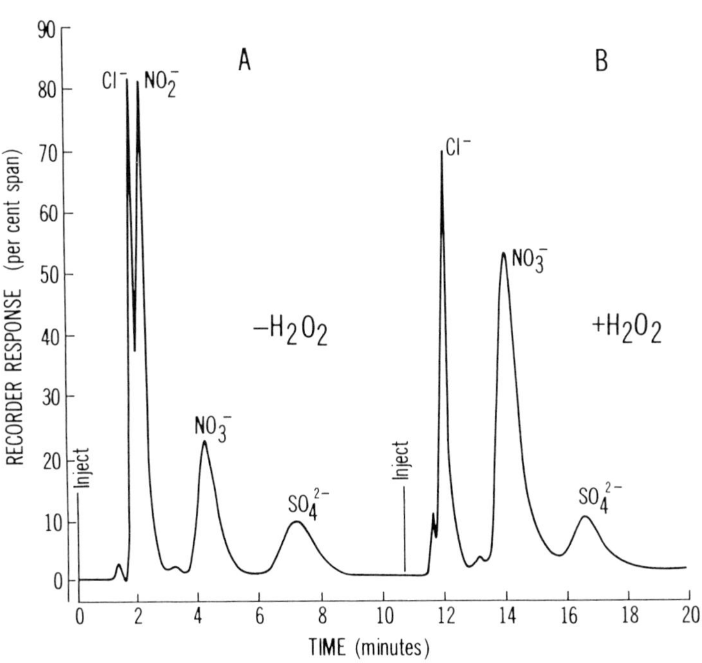

Figure A.7. Typical anion chromatograms resulting from injection of absorbing solution from oxygen flask (Figure A.6) before (A) and after (B) adding H_2O_2 to oxidize the nitrite to nitrate (Busman, Dick and Tabatabai, 1983)

Table A.5. Comparison of total S values obtained by colorimetric methods and by IC method (Busman, Dick and Tabatabai, 1983)

Sample	Total S: Methylene blue[a] (%)	Total S: Ion chromatography (%)
Leaves		
Ryegrass	0.397	0.370
Sorghum	0.148	0.144
Soybean	0.153	0.148
Corn	0.158	0.149
Crown vetch	0.264	0.268
Spinach	0.358	0.359
Lettuce	0.435	0.432
Alfalfa	0.512	0.496
Stalks or stems		
Soybean	0.079	0.080
Corn	0.077	0.065
Sorghum	0.113	0.110
Tomato	0.739	0.749
Grains		
Sorghum	0.082	0.095
Corn	0.141	0.143
Soybean	0.234	0.219
Average	0.259	0.255

[a]Tabatabai and Chae (1982).

Guthrie and Lowe (1984) evaluated the Fisher Model 475 sulphur analyser for determination of total S in plant materials. In this method, a sample of approximately 0.100 g was weighed into a ceramic combustion boat and covered with 100–200 mg of V_2O_5 powder, which acts as a flux to moderate sample combustion. The sample and boat were placed into a high temperature (1350° C) resistance type tube furnace, where all forms of S were converted to SO_2 in an oxygen atmosphere at reduced pressure. The SO_2 is absorbed into solution in a titration vessel and titrated using an automatic burette. The time required for each analysis depends on the sample size and S concentration, but was generally between 3 and 4 min.

Guthrie and Lowe (1984) compared three methods for determination of total S in tree foliage; ashing with $Mg(NO_3)_2$ followed by turbidimetry, NaOBr oxidation followed by colorimetry as BiS at 400 nm, and an instrumental method using the Fisher Model 475 S analyser. They reported that the

Table A.6. Comparison of total S values of plant materials obtained by methylene blue and IC methods (Tabatabai, Basta and Pirela, 1988)

Plant material	Total S by method indicated	
	MB[a] (%)	IC (%)
Alfalfa	0.277	0.246
Bromegrass	0.240	0.222
Corn leaves	0.405	0.406
Green bean leaves	0.296	0.302
Leaf lettuce	0.173	0.183
Red clover	0.192	0.189
Reed canarygrass	0.252	0.261
Ryegrass	0.375	0.386
Soybean leaves	0.226	0.224
Sudangrass	0.174	0.167

[a]MB, methylene blue method after alkaline NaOBr oxidation (Tabatabai and Chae, 1982).

Fisher S analyser was better than the other two methods with respect to accuracy and speed of analysis. Part of the problem in using the NaOBr oxidation method by Guthrie and Lowe (1984) could have been due to the use of BiS colorimetric method. It is well known that the BiS colorimetric method is less accurate and sensitive as compared with the methylene blue colorimetric method. They reported that the Fisher S analyser had the lowest operating cost per sample, largely owing to lower labour costs.

Recently, Tabatabai, Basta and Pirela (1988) developed an accurate and precise IC method for determination of total S in plant materials. It involves ignition of a mixture of plant material and $NaHCO_3$ containing Ag_2O at 550° C for 3 h. The residue was dissolved in 1 M HOAc, diluted with deionized water, filtered, and analysed for SO_4^{2-} by a Dionex 2202i ion chromatograph. Results by the IC method for 10 plant materials agreed closely with those obtained by the methylene blue colorimetric method after alkaline oxidation with NaOBr (Table A.6).

A.3.2 INORGANIC SULPHUR

The various extractants used for obtaining sulphate and extractable S from plant tissue are summarized by Beaton, Burns and Platou (1968). Among the extractants most frequently used are dilute HCl, direct treatment with the Johnson and Nishita (1952) reducing mixture, water, ethanol, and trichloroacetic acid. A number of other extractants including acetic acid, ammonium acetate plus acetic acid, sodium acetate plus acetic acid, sodium hydroxide, water plus magnesium carbonate, refluxing with water plus magnesium oxide and cadmium chloride or cadmium hydroxide, acetone and ninhydrin have also been used to remove sulphate and extractable S from plant tissue. Sulphate was also extracted with electrodialysis.

In using the Johnson and Nishita (1952) reduction apparatus and determination of the H_2S released as methylene blue, it is possible to determine inorganic sulphate directly in dried plant tissue samples without prior ashing. Organic S is then calculated from the difference between total S determined by the methylene blue after digestion of plant material and the amount of inorganic S measured by treating the dry plant sample directly with the reducing reagent used in the methylene blue colorimetric method (Johnson and Ulrich, 1959; Ulrich *et al.*, 1967).

Analytical techniques for determination of the extracted sulphate included chromatographic, colorimetric, gravimetric, microbiological assay, radioactivity assay, spectrographic, titrimetric, and turbidimetric procedures. The titrimetric procedures included acidometry, barium chromate, barium percholate with thorin indicator, and barium chloride with nitrochromazo indica-

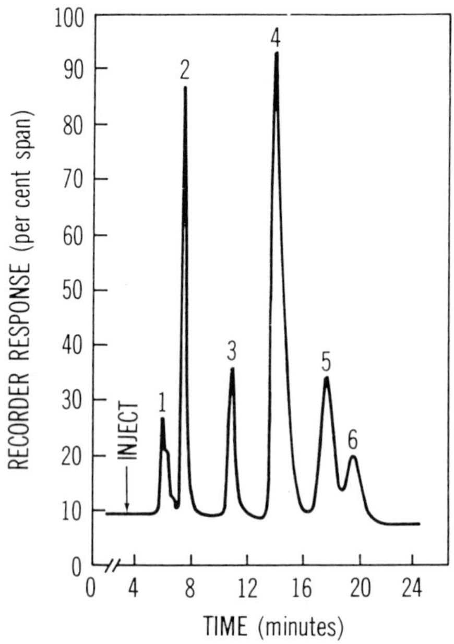

Figure A.8. Typical anion chromatograms of plant extract (ryegrass). The peaks are as follows: 1, F^-; 2, Cl^-; 3, PO_4^{3-}; 4, NO_3^-; 5, malic acid; 6, SO_4^{2-} (Kalbasi and Tabatabai, 1985)

tor. Methylene blue and barium chloranilate were among the colorimetric methods used (Beaton, Burns and Platou, 1968).

A method based on simultaneous determination of anions (NO_3^-, Cl^-, SO_4^{2-}, and PO_4^{3-}) in plant materials by IC is also available (Kalbasi and Tabatabai, 1985). It involves extraction of these anions with deionized water and their simultaneous determination by IC. Figure A.8 shows a typical chromatogram of plant extract (ryegrass) obtained after appropriate dilution. The average of 10 plant samples by the IC method agreed closely with those obtained by steam-distillation method for nitrate (0.158 *vs* 0.164%), titrimetric method using $AgNO_3$ and K_2CrO_4 for Cl^- (1.22 *vs* 1.19%), and colorimetric heteropoly blue method for PO_4^{3-} (0.335 *vs* 0.330%). The average SO_4^{2-}-S value of 10 plant samples by the IC method (0.067%) was somewhat lower than the corresponding value (0.075%) obtained by reduction and colorimetric determination as methylene blue.

A.4 SULPHUR IN WATERS

Sulphur enters natural waters from various sources. The weathering of rocks and soil directly exposed to surface waters is usually the largest natural source. Atmospheric fallout of particulate matter from natural sources as well as from anthropogenic inputs, which include the combustion of fossil fuels and mineral processing, can introduce relatively large quantities of S compounds into the natural-water systems. Other anthropogenic inputs, including the discharge of various treated and untreated waters to the receiving waters or the contamination of the aquatic systems through mining operations, can introduce large quantities of S compounds into natural waters. Degradation products of plant and animal tissues also contribute organic and inorganic S compounds to receiving waters. Other sources of S in natural waters are the S-containing fertilizers added to agricultural lands. The organic and inorganic S in natural waters are subject to complex abiotic and biological interactions (Strumm and Morgan, 1970).

A.4.1 FORMS OF SULPHUR IN WATERS

Most of the S in natural waters occurs as SO_4^{2-}. Reduced inorganic S species are present in natural waters under certain conditions. Sulphide is present in water generally as the result of industrial waste contamination or reduction of SO_4^{2-} by anaerobic bacterial action. The occurrence of SO_3^{2-} in natural water is rare because SO_3^{2-} is so readily oxidized by air to SO_4^{2-}. Although drainage water from agricultural land does not contain reduced inorganic S compounds, $S_2O_3^{2-}$ and $S_4O_6^{2-}$ are produced in soils when treated with S^0 (Nor and Tabatabai, 1976, 1977). Also, such reduced S compounds are produced in tailing pond effluents from the mining industry, and it is believed that they may lead to acid pollution of rivers surrounding these industries. Concentrations of $S_2O_3^{2-}$ in tailing pond effluents have been estimated to be in the range of 500 mg l^{-1} (Wolkoff and Larose, 1975). In addition to the inorganic S forms mentioned, depending on the source, some natural waters may contain small amounts of organic S. The suspended materials in natural waters may contain organic S such as those described by Tabatabai (1982) for soils.

A.4.2 ANALYTICAL METHODS

Sulphate, being a stable highly oxidized and soluble form of S, is the form in which the element is generally present in natural surface and groundwaters. Sulphates in water are generally bound to alkali and alkaline earth metals and are readily soluble. Its concentration in natural waters, precipitation, and wastewaters will vary depending on the source. The amount present in natural waters will depend largely on the type of terrain to which it has been exposed. High concentrations may be found in some well waters and surface waters in arid regions where sulphate minerals abound.

Methods in common use for the determination of SO_4^{2-} concentration in water include measurements by colorimetric, gravimetric, turbidimetric, titrimetric, conductimetric, potentiometric titration, and ion chromatographic techniques. The gravimetric, turbidimetric, and titrimetric (volumetric) methods are generally considered to be standard methods for SO_4^{2-} determination (American Public Health Association, 1975; American Society for Testing and Materials, 1977) but as Philbert (1982) indicated, they each have certain features which make them somewhat unattractive from a practical standpoint.

Of the various colorimetric methods recommended for determination of SO_4^{2-}, the procedure developed by Johnson and Nishita (1952) is the most sensitive and accurate method available. The procedure, reagents, and reactions involved were discussed in Section A.2.1.1. Unlike other methods, however, this method requires analytical skill in handling the apparatus used for reduction of SO_4^{2-}. Also, in addition to SO_4^{2-}, it recovers a variety of reduced S compounds and organic SO_4^{2-} (Tabatabai, 1982). But reduced inorganic S species are not present in most natural waters. Other colorimetric and titrimetric methods are described by Philbert (1982).

The gravimetric method is based on the quantitative precipitation of SO_4^{2-} by $BaCl_2$ in a HCl medium, after removal of Si and suspended solids (e.g. $CaSO_4 + BaCl_2 = BaSO_4 + CaCl_2$). The method is subject to interferences from SO_3^{2-}, S^{2-}, Si, NO_3^-, certain heavy metals such as Cr and Fe, and suspended matter. According to American Public Health (1975), the gravimetric method is the most accurate method for determining SO_4^{2-} concentrations $> 10\,mg\ SO_4^{2-}\ l^{-1}$. However, the method is time consuming and normally requires relatively large volumes of samples. Moreover, the quality of results depends to some large measure on the skill and experience of the analyst. The relatively high detection limit of the method precludes its application to waters containing $< 10\,mg\ SO_4^{2-}\ l^{-1}$. American Public Health Association (1975) reports a between-laboratory relative SD and CV of 4.7% and 1.9%, respectively, for 32 laboratories analysing an unknown synthetic sample with $259\,mg\ SO_4^{2-}\ l^{-1}$.

The turbidimetric method is based on the same principle as the gravimetric method, except the precipitated $BaSO_4$ is measured by a nephelometer or transmission photometer. The resulting turbidity in the solution is proportional to the SO_4^{2-} concentration, which is determined by converting the measured turbidity through a calibration curve prepared by following the same procedure with a series of SO_4^{2-} standards. The method is much more rapid than the gravimetric procedure, but is usually less accurate than the latter at concentrations $> 10\,mg\ SO_4^{2-}\ l^{-1}$. The precision and accuracy with which this could be accomplished is highly dependent on the skill and experience of the analyst. In addition to the above interferences, natural sample colour may also interfere.

A simple and accurate method for determination of SO_4^{2-} in natural waters, including rainfall and snowmelt, was developed by Tabatabai (1974). The method is based on measuring the turbidity formed when $BaCl_2$ and gelatine are added to an acidified water sample. The results obtained by this method are comparable to those obtained by the methylene blue method of Johnson and Nishita (1952).

American Public Health Association (1975) reports a between-laboratory relative SD and CV of 9.1% and 1.2%, respectively, for 19 laboratories analysing an unknown synthetic sample with $259\,mg\ SO_4^{2-}\ l^{-1}$. The US Environmental Protection Agency (1973) reports a between-laboratory relative SD of 26.7% and 5.9% for 16 laboratories analysing a synthetic sample containing 8.6 and $199\,mg\ SO_4^{2-}\ l^{-1}$, respectively. The biases observed at those levels were −3.7% and −1.7%, respectively.

The conductimetric method entails the titration of SO_4^{2-} present in the sample with a standard $BaCl_2$ solution in the presence of $BaSO_4$ adsorption (precipitation) support. The latter is prepared by mixing a solution of H_2SO_4 and $BaCl_2$ and allowing the $BaSO_4$ precipitate to stand for about 24 h, after which the supernatant liquid is decanted. The precipitate is then washed several times to get rid of all SO_4^{2-}. The SO_4^{2-} concentration in the sample is determined by measuring the solution conductivity with successive additions of $BaCl_2$ titrant and plotting a conductivity curve, which normally consists of two straight lines intersecting at the equivalence point. Knowing the titer of the $BaCl_2$ solution from titration of a standard SO_4^{2-} solution, the concentration of the sample may be calculated.

A potentiometric method is available (Vogel, 1961) for titration of SO_4^{2-} in solution with a standard $Pb(NO_3)_2$ solution after addition of potassium ferrocyanide and potassium ferricyanide. Titration is done potentionmetrically in a 1 : 1 water–ethanol solution with a Pt wire electrode and a Ag reference electrode. The endpoint is directed by a sharp change in the electrode potential due to the precipitation of lead ferrocyanide. Cortellesa and Napoli (1968) have used an adaptation of this procedure for the determination of SO_4^{2-} at considerably low concentrations in solution.

Potentiometric methods have been proposed for determination of S^{2-} and SO_4^{2-} in water samples. Although SO_4^{2-} sensitive electrodes have been developed (Rechnitz, Lin and Zamochnick, 1967; Mohan and Rechnitz, 1973), they have relatively poor selectivity and working ranges and they are not available commercially. Silver ion has been used to titrate S^{2-} by using the sulphide electrode, but the titration curve is asymmetrical and the equivalence point does not coincide with the point of inflection. Titration with Cd^{2+} or Pb^{2+} gives a symmetrical titration curve that is much less affected by the presence of other anions. Ehman (1976) and Green and Schnitker (1974) preferred Cd^{2+} for the titration of very dilute sulphide solutions. Sulphide electrodes have been used for determination of S^{2-} in soil solution and seawater (Allam, Pitts and Hollis, 1972; Mor *et al.*, 1975).

The most commonly used method for total and dissolved sulphides is a colorimetric procedure based on the fact that *p*-aminodimethylaniline, Fe^{3+},

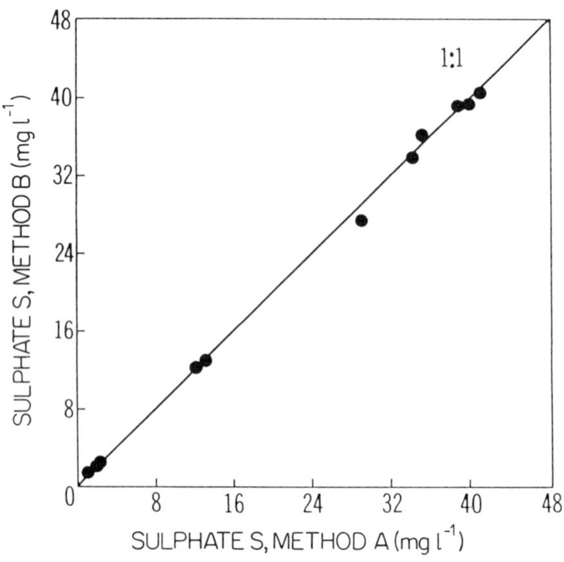

Figure A.9. Comparison of SO_4^{2-}-S values obtained for water samples by IC method (A) and by methylene blue method (B) (Tabatabai and Dick, 1983)

and the S^{2-} ion react in acid solution to produce methylene blue. The concentration range of the method is 0.1 to 20 mg S^{2-} l^{-1}, and the accuracy ± 0.1 mg l^{-1} for clear water (Gilboa-Garber, 1971; American Society for Testing and Materials, 1969).

Tabatabai and Dick (1983) compared the results obtained by the IC method for SO_4^{2-} in natural waters with those by the methylene blue method. Figure A.9 shows the results obtained by the IC method agreed closely with those obtained by the methylene blue method. Tests indicated that the detection limit of the IC method for determination of SO_4^{2-} in natural waters is 0.1 mg S l^{-1}. The method is highly sensitive and accurate. The concentrations of Cl^-, NO_3^-, SO_4^{2-}-S, and PO_4^{3-}-P in the rainwater sample shown in Figure A.10 were 1.33, 0.09, 2.37, and 0.01 mg l^{-1}, respectively.

Single column ion chromatographs (without suppressor columns) are also available (Fritz, Gjerde and Pohlandt, 1982), but in general, these systems have lower detection limits for most inorganic anions than those of the dual column IC system. Recently Nieto and Frankenberger (1985) showed that

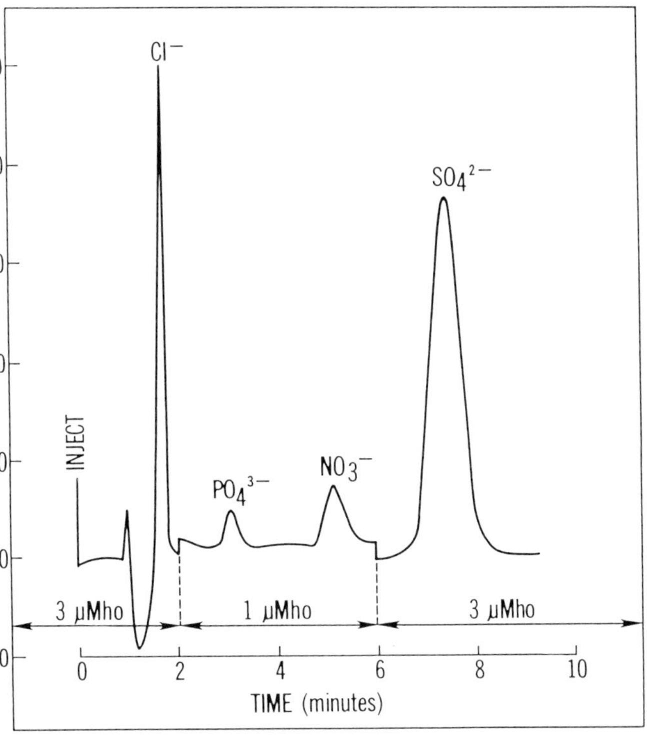

Figure A.10. Typical anion chromatogram of water sample (rainfall) (Tabatabai and Dick, 1983)

high performance liquid chromatographs can be used as a single column system (Figure A.5) for determination of a variety of inorganic anions, including SO_3^{2-} and SO_4^{2-} in soil extracts. Therefore, this system should be useful for water analysis.

REFERENCES

Allam, A., Pitts, I. G. and Hollis, J. P. (1972). Sulfide determination in submerged soils with an ion selective electrode. *Soil Sci.*, **114**, 456–67.

Allaway, W. H. and Thompson, J. F. (1966). Sulfur in the nutrition of plants and animals. *Soil Sci.*, **101**, 240–7.

Alymore, L. A. G., Karim, M. and Quirk, J. P. (1967). Adsorption and desorption of sulfate ions by soil constituents. *Soil Sci.*, **103**, 10–15.

American Public Health Association (1975). *Standard Methods for the Examination of Water and Wastewater*, 14 edn. APHA-AWWA-WPCF. American Public Health Assoc., Washington DC.

Arkley, T. H. (1961). Sulfur compounds of soil systems. *Ph.D. Thesis*, University of California, Berkeley.

Aspiras, R. B., Keeney, D. R. and Chesters, G. (1972). Determination of reduced inorganic sulfur forms as sulfide by zinc-hydrochloric acid distillation. *Anal. Lett.*, **5**, 425–32.

Association of Official Agricultural Chemists (1955). *Official Methods of Analysis*, 8th edn. AOAC, Washington DC.

Bardsley, C. E. and Lancaster, J. D. (1960). Determination of reserve sulfur and soluble sulfates in soils. *Soil. Sci. Soc. Am. Proc.*, **24**, 265–8.

Beaton, J. D., Burns, G. R. and Platou, Jr (1968). Determination of sulphur in soils and plant material. *Tech. Bull.*, **14**, The Sulphur Inst., Washington DC.

Bentley, C. F., Hoff, D. J. and Scott, D. B. (1955). Fertilizer studies with radioactive sulphur. II. *Can. J. Agric. Sci.*, **35**, 264–81.

Berg, W. A. and Thomas, G. W. (1959). Anion elution patterns from soils and soil clays. *Soil Sci. Soc. Am. Proc.*, **23**, 348–50.

Bertrand, G. and Silberstein, L. (1927). Sur la dosage due soufre contenu dans la terre arable. *Ann. Sci. Agron. Fr. Estrang.*, **43**, 71–7.

Bettany, J. R., Stewart, J. W. B. and Halstead, E. H. (1973). Sulfur fractions and carbon, nitrogen, and sulfur relationships in grassland, forest, and associated transitional soils. *Soil Sci. Soc. Am. Proc.*, **37**, 915–18.

Biederbeck, V. O. (1978). Soil organic sulfur and fertility. In: M. Schnitzer and S. M. Khan (Eds). *Soil Organic Matter*. Elsevier North-Holland, New York, pp. 273–310.

Bingham, F. T., Sims, R. J. and Page, A. L. (1965). Retention of acetate by montmorillonite. *Soil Sci. Soc. Am. Proc.*, **29**, 670–2.

Blanchar, R. W. (1986). Measurement of sulfur in soils and plants. In: Tabatabai, M. A. (Ed.). *Sulfur in Agriculture, Monograph No. 27*, Am. Soc. Agron., Madison, Wis., pp. 455–90.

Bloomfield, C. (1962). A colorimetric method for determining total sulfur in soils. *Analyst (London)*, **87**, 586–9.

Bremner, J. M. (1967). Nitrogenous compounds. In: A. D. McLaren and G. H. Peterson (Ed.). *Soil Biochemistry*, Vol. 1. Marcel Dekker, New York, pp. 19–66.

Brümmer, G., Grunwaldt, H. S. and Schroeder, D. (1971a). Contributions to the

genesis and classification of Marsh soils. II. On the sulphur metabolism of muds and salt marshes. *Z. Pflanz. Bodenkd.*, **128**, 208–220.
Brümmer, G., Grunwaldt, H. S. and Schroeder, D. (1971b). Contributions to the genesis and classification of Marsh soils. III. Contents, oxidation status, and mechanisms of bounding of sulphur in polder soils. *Z. Pflanz. Bodenkd.*, **129**, 92–108.
Busman, L. M., Dick, R. P. and Tabatabai, M. A. (1983). Determination of total sulfur and chlorine in plant materials by ion chromatography. *Soil Sci. Soc. Am. J.*, **47**, 1167–70.
Canfield, D. E., Raiswell, R., Westrich, J. T., Reaves, C. M. and Berner, R. A. (1986). The use of chromium reduction in the analysis of reduced inorganic sulfur in sediments and shales. *Chem. Geol.*, **54**, 149–55.
Chang, M. L. and Thomas, G. W. (1963). A suggested mechanism for sulfate adsorption by soils. *Soil Sci. Soc. Am. Proc.*, **27**, 281–3.
Chao, T. T., Harward, M. E. and Fang, S. C. (1962a). Movement of S^{35} tagged sulfate through soil columns. *Soil Sci. Soc. Am. Proc.*, **26**, 27–32.
Chao, T. T., Harward, M. E. and Fang, S. C. (1962b). Soil constituents and properties in the adsorption of sulfate ions. *Soil Sci.*, **94**, 276–83.
Chao, T. T., Harward, M. E. and Fang, S. C. (1964). Iron and aluminum coatings in relation to sulfate adsorption. *Soil Sci. Soc. Am. Proc.*, **28**, 632–5.
Chesnin, L. and Yien, C. H. (1950). Turbidimetric determination of available sulfates. *Soil Sci. Soc. Am. Proc.*, **15**, 149–51.
Clark, J. S., Gobin, C. A. and Sprout, D. N. (1961). Yellow mottles in some poorly drained soils of the Lower Fraser Valley, British Columbia. *Can. J. Soil Sci.*, **41**, 218–27.
Cooper, P. J. M. (1972). Aryl sulphatase activity in northern Nigerian soils. *Soil Biol. Biochem.*, **4**, 333–7.
Cortellesa, G. C. and Napali, C. A. (1968). Potentiometric micro determination of the sulphate ion. *Analyst (London)*, **93**, 546–9.
David, M. B., Mitchell, M. J., Aldcorn, D. and Harrison, R. B. (1989). Analysis of sulfur in soil, plant and sediment materials: Sample handling and use of an automated analyzer. *Soil Biol. Biochem.*, **21**, 119–23.
DeLong, W. A. and Lowe, L. E. (1961). Note on carbon-bonded sulphur in soil. *Can. J. Soil Sci.*, **42**, 223.
Dick, W. A. and Tabatabai, M. A. (1979). Ion chromatographic determination of sulfate and nitrate in soils. *Soil Sci. Soc. Am. J.*, **43**, 899–904.
Duke, S. H. and Reisenauer, H. M. (1986). Role and requirements of sulfur in plant nutrition. In: Tabatabai, M. A. (Ed.). *Sulfur in Agriculture. Monograph No. 27.* Am. Soc. Agron., Madison, Wis., pp. 123–68.
Ensminger, L. E. (1954). Some factors affecting the adsorption of sulfate by Alabama soils. *Soil Sci. Soc. Am. Proc.*, **18**, 259–64.
Evans, C. A. and Rost, C. O. (1945). Total organic sulfur and humus in Minnesota soils. *Soil Sci.*, **59**, 125–7.
Fleming, J. F. and Alexander, L. T. (1961). Sulfur activity in South Carolina tidal marsh soils. *Soil Sci. Soc. Am. Proc.*, **25**, 94–8.
Fox, R. L., Olson, R. A. and Rhoades, H. F. (1964). Evaluating the sulfur status of soils by plants and soil tests. *Soil Sci. Soc. Am. Proc.*, **28**, 243–6.
Freney, J. R. (1958). Determination of water-soluble sulphate in soils. *Soil Sci.*, **86**, 241–4.
Freney, J. R. (1961). Some observations on the nature of organic sulphur compounds in soils. *Aust. J. Agric. Res.*, **21**, 424–32.

Freney, J. R. (1967). Sulfur-containing organics. In: A. D. McLaren and G. H. Peterson (Ed.). *Soil Biochemistry*. Marcel Dekker, New York, pp. 220–59.

Freney, J. R., Melville, G. E. and Williams, C. H. (1969). Extraction, chemical nature, and properties of soil organic sulphur. *J. Sci. Food Agric.*, **20**, 440–5.

Freney, J. R., Melville, G. E. and Williams C. H. (1970). The determination of carbon bonded sulfur in soil. *Soil Sci.*, **109**, 310–18.

Freney, J. R., Melville, G. E. and Williams, C. H. (1971). Organic sulphur fractions labeled by addition of ^{35}S-sulphate to soils. *Soil Biol. Biochem.*, **3**, 133–41.

Freney, J. R., Stevenson, F. J. and Beavers, A. H. (1972). Sulfur-containing amino acid in soil hydrolysates. *Soil Sci.*, **114**, 468–76.

Fritz, J. S., Gjerde, D. T. and Pohlandt, C. (1982). *Ion Chromatography*. Huthig, New York.

Gilboa-Garbar, N. (1971). Direct spectrophotometric determination of inorganic sulfide in biological materials and in other complex mixtures. *Anal. Biochem.*, **43**, 129–33.

Green, E. J. and Schnitker, D. (1974). The direct titration of water-soluble sulfide in estuarine muds of Montsweag Bay, Maine. *Mar. Chem.*, **2**, 111–24.

Gustafsson, L. (1960). Determination of ultramicro amounts of sulphate as methylene blue. I. The colour reaction. *Talanta*, **4**, 227–35.

Guthrie, T. F. and Lowe, L. E. (1983). A comparison of methods for total sulphur analysis of tree foliage. *Can. J. For. Res.*, **14**, 470–3.

Haff, L. V. (1970). Oxygen-containing inorganic compounds. In: Karchmer, J. H. (Ed.). *The Analytical Chemistry of Sulphur and its Compounds*. Wiley-Interscience, New York, pp. 183–282.

Harmsen, G. W. (1954). Observations on the formation and oxidation of pyrite in the soil. *Plant Soil*, **5**, 324–48.

Hart, E. B. and Peterson, W. A. (1911). Sulfur requirements of farm crops in relation to the soil and air supply. *Wisconsin Agric. Exp. Stn. Res. Bull.*, 14.

Hart, M. G. R. (1959). Sulphur oxidation in tidal mangrove soils of Sierra Leone. *Plant Soil*, **11**, 215–36.

Hasan, S. M., Fox, R. L. and Boyd, C. C. (1970). Solubility and availability of sorbed sulfate in Hawaiian soils. *Soil Sci. Soc. Am. Proc.*, **34**, 897–901.

Howarth, R. W. and Jørgensen, B. B. (1985). Formation of ^{35}S-labelled elemental sulfur and pyrite in coastal marine sediments (Limfjorden and Kysing Fjord, Denmark), during short-term $^{35}SO_4^{2-}$ reduction measurements. *Geochim. Cosmochim. Acta*, **48**, 1807–18.

Howarth, R. W. and Merkel, S. (1984). Pyrite formation and the measurement of sulfate reduction in salt marsh sediments. *Limnol. Oceanogr.*, **29**, 598–608.

Howarth, R. W., Giblin, A., Gale, J., Peterson, B. J. and Luther III, G. W. (1983). Reduced sulfur compounds in the pore waters of a New England salt marsh. *Environ. Biogeochem.*, **35**, 135–52.

Jenkinson, D. S. (1968). A titrimetric method for determining total sulphur in mineral soils. *Analyst (London)*, **93**, 535–39.

Johnson, C. M. and Nishita, H. (1952). Microestimation of sulfur in plant materials, soils, and irrigation waters. *Anal. Chem.*, **24**, 736–42.

Johnson, C. M. and Ulrich, A. (1959). Analytical methods for use in plant analysis. *Calif. Agric. Exp. Stn. Bull.*, 766.

Kalbasi, M. and Tabatabai, M. A. (1985). Simultaneous determination of nitrate, chloride, sulfate, and phosphate in plant materials by ion chromatography. *Commun. Soil Sci. Plant Anal.*, **16**, 787–800.

Kamprath, E. L., Nelson, W. L. and Fitts, J. W. (1956). The effect of pH, sulfate and phosphate concentrations on the adsorption of sulfate by soils. *Soil Sci. Soc. Am. Proc.*, **20**, 463–6.

Kelley, D. P., Chambers, L. A. and Trudinger, P. A. (1969). Cyanaolysis and spectrophotometric estimation of trithionate in mixture with thiosulfate and tetrathionate. *Anal. Chem.*, **41**, 898–901.

Kilmer, V. J. and Nearpass, D. C. (1960). The determination of available sulfur in soils. *Soil Sci. Soc. Am. Proc.*, **24**, 337–40.

King, R. C. (1953). Effects of alkali metal ions on development of drosophilla, with special reference to lithium induced abnormalities. *Proc. Natl. Acad. Sci. USA*, **39**, 403–6.

Kowalenko, C. G. (1978). Organic nitrogen, phosphorus, and sulfur in soils. In: M. Schnitzer and S. M. Khan (Eds). *Soil Organic Matter*. Elsevier North-Holland, New York, pp. 95–126.

Krupa, S. V. and Tabatabai, M. A. (1986). Measurement of sulfur in the atmosphere and natural waters. In: Tabatabai, M. A. (Ed.). *Sulfur in Agriculture. Monograph No. 27.* Am. Soc. Agron., Madison, Wis., pp. 491–548.

Laties, G. G. (1959). The development and control of coexisting respiratory systems in slices of chicory root. *Arch. Biochem. Biophys.*, **79**, 378–91.

Little, R. C. (1957). Sulphur in soils. II. Determination of the total sulphur content of soil. *J. Sci. Food Agric.*, **8**, 271–9.

Lowe, L. E. (1964). An approach to the study of the sulphur status of soils and its application to selected Quebec soils. *Can. J. Soil. Sci.*, **44**, 176–9.

Lowe, L. E. (1965). Sulphur fractions of selected Alberta solid profiles of the chernozemic and podzolic orders. *Can. J. Soil. Sci.*, **45**, 297–303.

Lowe, L. E. and Bustin, R. M. (1985). Distribution of sulphur in six facies of peats of the Fraser River Delta. *Can. J. Soil. Sci.*, **65**, 531–41.

Lowe, L. E. and Bustin, R. M. (1989). Forms and hydrolytic behavior of sulphur in humic acid and residue fractions of four peats from the Fraser Loweland. *Can. J. Soil Sci.*, **69**, 287–93.

Lowe, L. E. and DeLong, W. A. (1963). Carbon bonded sulfur in selected Quebec soils. *Can. J. Soil Sci.*, **43**, 151–5.

Luther, III, G. W. and Church, T. M. (1988). Seasonal cycling of sulfur and iron in porewaters of a Delaware salt marsh. *Mar. Chem.*, **23**, 295–309.

Luther, III, G. W., Giblin, E. A. and Varsolona, R. (1985). Polarographic analysis of sulfur species in marine porewaters. *Limnol. Oceanogr.*, **30**, 727–36.

McClung, A. C., DeFreitas, L. M. and Lott, W. L. (1959). Analyses of several Brazilian soils in relation to plant responses to sulfur. *Soil Sci. Soc. Am. Proc.*, **23**, 221–4.

Melville, G. E., Freney, J. R. and Williams, C. H. (1971). Reaction of organic sulfur compounds in soil with tin and hydrochloride acid. *Soil Sci.*, **112**, 245–8.

Mohan, M. S. and Rechnitz, G. A. (1973). Preparation and properties of the sulfate ion-selective membrane electrode. *Anal. Chem.*, **45**, 1323–6.

Mor, E., Scotto, V., Marcenaro, G. and Alabiso, G. (1975). The use of membrane electrodes in the determination of sulfides in sea waters. *Anal. Chim. Acta*, **75**, 159–67.

Neptune, A. M. L., Tabatabai, M. A. and Hanaway, J. J. (1975). Sulfur fractions and carbon–nitrogen–phosphorus–sulfur relationships in some Brazilian and Iowa soils. *Soil Sci. Soc. Am. Proc.*, **39**, 51–5.

Nieto, K. F. and Frankenberger, Jr, W. T. (1985). Single column ion chromatography. I. Analysis of inorganic anions in soils. *Soil Sci. Soc. Am. J.*, **49**, 587–92.

Nor, Y. M. and Tabatabai, M. A. (1975). Colorimetric determination of microgram quantities of thiosulfate and tetrathionate. *Anal. Lett.*, **8**, 537–47.

Nor, Y. M. and Tabatabai, M. A. (1976). Extraction and colorimetric determination of thiosulfate and tetrathionate in soils. *Soil Sci.*, **122**, 171–8.

Nor, Y. M. and Tabatabai, M. A. (1977). Oxidation of elemental sulfur in soils. *Soil Sci. Soc. Am. J.*, **41**, 736–41.

Olson, R. A. and Rhem, G. W. (1986). Sulfur in precipitation and irrigation waters and its effects on soils and plants. In: Tabatabai, M. A. (Ed.). *Sulfur in Agriculture. Monograph No. 27.* Am. Soc. Agron., Madison, Wis., pp. 279–93.

Philbert, F. J. (1982). Major ions. In: Van Loon, J. C. (Ed.). *Chemical Analysis of Inorganic Constituents of Water*. CRC Press, Boca Raton, FL, pp. 67–129.

Putnam, H. D. and Schmidt, E. L. (1959). Studies of the free amino acid fraction of soils. *Soil Sci.*, **87**, 22–7.

Rechnitz, G. A., Lin, Z. F. and Zamochnick, S. B. (1967). Potentiometric measurements with SO_4^{2-} and PO_4^{3-} sensitive membrane electrode. *Anal. Lett.*, **1**, 24–33.

Robinson, W. O. (1930). Method and procedure of soil analysis used in the division of soil chemistry and physics. US Dept of Agriculture Circ. 139.

St Lorant, I. (1929). Über eine neue colorimetrische mikromethode zur bestimmung des schwefels in sulfiden, sulfaten usw. *Z. Physiol. Chem.*, **185**, 245–66.

Shorey, E. C. (1913). Some organic soil constituents. US Div. of Soils Bull. No. 88.

Singh, S. S. (1967). Sulfate ions and ion activity product $(Al)(OH)^3$ in Wyoming bentonite suspensions. *Soil Sci.*, **104**, 433–8.

Singh, S. S. and Brydon, J. E. (1967). Precipitation of aluminum by calcium hydroxide in the presence of Wyoming bentonite and sulfate ions. *Soil Sci.*, **103**, 162–7.

Skerrett, E. J. and Dickes, G. J. (1961). The determination of microgram amounts of sulphur-containing compounds. *Analyst (London)*, **86**, 69–71.

Small, H., Stevens, T. S. and Bauman, W. C. (1975). Novel ion exchange chromatographic method using conductimetric detection. *Anal. Chem.*, **47**, 1801–9.

Smittenberg, J., Harmsen, G. W., Quispel, A. and Otzen, D. (1951). Rapid methods for determining different types of sulphur compounds in soil. *Plant Soil*, **3**, 353–60.

Sörbo, B. (1957). A colorimetric method for the determination of thiosulfate. *Biochim. Biophys. Acta*, **23**, 412–16.

Spencer, K. and Freney, J. R. (1960). A comparison of several procedures for estimating the sulphur status of soils. *Aust. J. Agric. Res.*, **11**, 948–59.

Stanford, J. O. and Lancaster, J. D. (1962). Biological and chemical evaluation of the readily available sulfur status of Mississippi soils. *Soil Sci. Soc. Am. Proc.*, **26**, 63–5.

Steinbergs, A. (1955). A method for the determination of total sulphur in soils. *Analyst (London)*, **80**, 457–61.

Steinbergs, A., Iismaa, O., Freney, J. R. and Barrow, N. J. (1962). Determination of total sulphur in soil and plant material. *Anal. Chim. Acta*, **27**, 158–64.

Stevenson, F. J. (1956). Isolation and identification of some amino compounds in soils. *Soil Sci. Soc. Am. Proc.*, **20**, 210–4.

Stillian, J. (1985). An improved suppressor for ion chromatography. *LC*, September, 1985.

Strochlein, J. L. and Pennington, D. A. (1986). Use of sulfur compounds for soil and irrigation water treatments. In: Tabatabai, M. A. (Ed.). *Sulfur in Agriculture. Monograph No. 27.* Am. Soc. Agron., Madison, Wis., pp. 435–54.

Stumm, W. and Morgan, J. J. (1970). *Aquatic Chemistry*. Wiley-Interscience, New York.

Tabatabai, M. A. (1974). Determination of sulphate in water samples. *Sulphur Inst. J.*, **10**, 11–13.

Tabatabai, M. A. (1982). Sulfur. In: Page, A. L., Miller, R. H. and Keeney. D. R. (Eds.). *Methods of Soil Analysis*, Part 2, 2nd edn. *Agronomy*, **9**, 501–38.

Tabatabai, M. A. (1987). Physicochemical fate of sulfate in sois. *JAPCA*, **37**, 34–8.
Tabatabai, M. A. and Basta, N. T. (1991). Ion chromatography. In: Smith, K. A. (Ed.). *Soil Analysis: Modern instrumental techniques*, 2nd edn. Dekker, New York, pp. 229–59.
Tabatabai, M. A., Basta, N. T. and Pirela, H. J. (1988). Determination of total sulfur in soils and plant materials by ion chromatography. *Commun. Soil Sci. Plant Anal.*, **19**, 1701–14.
Tabatabai, M. A. and Bremner, J. M. (1970a). An alkaline oxidation method for determination of total sulfur in soils. *Soil Sci. Soc. Am. Proc.*, **34**, 62–5.
Tabatabai, M. A. and Bremner, J. M. (1970b). Comparison of some methods for determination of total sulfur in soils. *Soil Sci. Soc. Am. Proc.*, **34**, 417–20.
Tabatabai, M. A. and Bremner, J. M. (1972). Forms of sulfur, and carbon, nitrogen, and sulfur relationships in Iowa soils. *Soil Sci.*, **114**, 380–6.
Tabatabai, M. A. and Bremner, J. M. (1991). Automated instruments for determination of total carbon, nitrogen, and sulfur in soils by combustion techniques. In: Smith, K. A. (Ed.). *Soil Analysis: Modern instrumental techniques*, 2nd edn. Dekker, New York, pp. 261–86.
Tabatabai, M. A. and Chae, Y. M. (1982). Alkaline oxidation method for determination of total sulfur in plant materials. *Agron. J.*, **74**, 404–6.
Tabatabai, M. A. and Dick, W. A. (1979). Ion chromatographic analysis of sulfate and nitrate in soils. In: J. D. Mulik and E. Sawicki (Eds). *Ion Chromatographic Analysis of Environmental Pollutants*, Vol. 2. Ann Arbor Science Publishers, Ann Arbor, Mich. pp. 361–70.
Tabatabai, M. A. and Dick, W. A. (1983). Simultaneous determination of nitrate, chloride, sulfate, and phosphate in natural waters by ion chromatography. *J. Environ. Equal.*, **12**, 209–13.
Tabatabai, M. A. and Singh, B. B. (1976). Rhodanese activity in soils. *Soil Sci. Soc. Am. J.*, **40**, 381–5.
Thompson, J. F., Smith, I. K. and Madison, J. T. (1986). Sulfur metabolism in plants. In: Tabatabai, M. A. (Ed.). *Sulfur in Agriculture. Monograph No. 27.* Am. Soc. Agron., Madison, Wis., pp. 57–121.
Ulrich, A., Tabatabai, M. A., Ohki, K. and Johnson, C. M. (1967). Sulfur content in alfalfa in relation to growth in filtered and unfiltered air. *Plant Soil*, **26**, 235–52.
US Environmental Protection Agency (1973). *Handbook for Monitoring Industrial Waste Water*. US Environmental Protection Agency, Office of Technology Transfer, Washington DC.
Vogel, A. I. (1961). *A Textbook of Quantitative Inorganic Analysis Including Elementary Instrumental Analysis*, 3rd edn. Longmans, London.
Wieder, R. K. and Lang, G. E. (1988). Cycling of inorganic and organic sulfur in peat from Big Run Bog, West Virginia. *Biogeochemistry*, **5**, 221–42.
Wieder, R. K., Lang, G. E. and Granus, V. A. (1985). An evaluation of wet chemical methods for quantifying sulfur fractions in freshwater wetland peat. *Limnol. Oceanogr.*, **30**, 1109–15.
Williams, C. H. (1967). Some factors affecting the mineralization of organic sulphur in soils. *Plant Soil*, **26**, 205–23.
Williams, C. H. and Steinbergs, H. (1959). Soil sulphur fractions as chemical indices of available sulphur in some Australian soils. *Aust. J. Agric. Res.*, **10**, 340–52.
Williams, C. H. and Steinbergs, A. (1962). The valuation of plant-available sulphur in soils. I: The chemical nature of sulphate in some Australian soils. *Plant Soil*, **17**, 279–94.
Wolkoff, A. W. and Larose, R. H. (1975). Separation and detection of low concentra-

tions of polythionates by high speed anion exchange liquid chromatography. *Anal. Chem.*, **47**, 1003–8.

Zhabina, N. N. and Volkov, I. I. (1978). A method for determination of various sulfur compounds in sea sediments and rocks. In: Krumbein, W. E. (Ed.). *Environmental Biogeochemistry and Geomicrobiology*, Vol. 3. Ann Arbor Sci., Mich. pp. 735–46.

Index